Industry Experience

Power Generation

- *Turbine lube oil systems utilizing leading edge fluid conditioning*
- *Gate control systems for hydro-electric as well as fuel handling and control systems for alternative energy co-generation facilities*
- *Sophisticated turbine control systems and blade control systems for wind turbines*
- *Reconditioning of transformer fluids*

Oil From Crude To The Finished Product

- *Explosion proof and intrinsically safe systems at the drill rig*
- *Down hole hydraulic systems utilizing "mud" as the power transfer medium or filtration systems for fuel at the dispensing pump and most applications in between*
- *Hydraulic power packages to sophisticated pneumatic logic controls*

Wood Products

- *45 years of experience in design of innovative solutions for lumber industries problems*
- *Our experience spans harvesting equipment through sophisticated motion control systems for producing dimensioned lumber*
- *Wood yard handling, reduction of wood fiber, and production of finished paper*

Industry Experience

Plastics Glass and Fiberglass Molding

- *State-of-the-art controls and components for plastic injection molding operations*
- *Systems for plants manufacturing floating sheet glass as well as molded glass and fiberglass products*

Metal Forming & Machining

- *Aluminum stretching presses of aircraft skin, powdered metal turbine blade forming presses, steel forging presses, aluminum die casting machines for custom automotive rims, can forming machines, tube bending machines or aluminum drawing machines*
- *Experience in retrofitting control systems for presses and metal forming equipment*

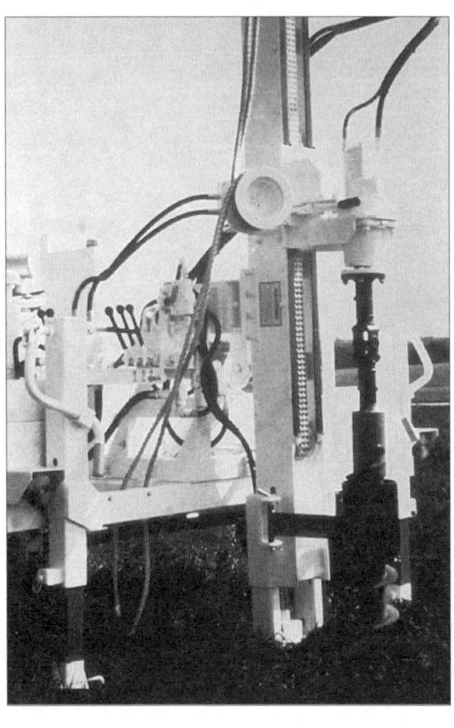

Mining & Construction Equipment

- *Large canal construction equipment, horizontal drilling equipment and paving equipment*
- *Integrated rotary, linear and remotely controlled motion control systems*

Industry Experience

Food Processing

- *Design and component manufacturing for specialty harvesting equipment*
- *Integrated fluid control manifolds*
- *Timing of operating equipment through sophisticated controls which can be changed instantaneously when basic products are changed*
- *From vine products, to tree fruits, to bread making, to tortilla presses*

Refuse and Waste Treatment

- *Refuse and waste treatment from the automation of curbside pick-up equipment to the handling of sludge in sewage treatment plants*
- *Integrated systems for the transfer and flow control of waste materials in waste treatment plants*
- *Pumping and handling of hazardous waste residuals*

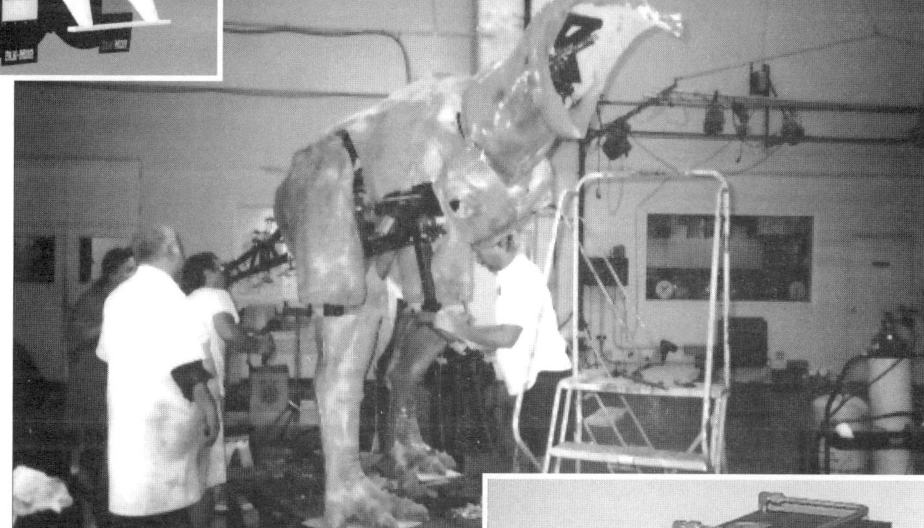

Animation and Special Effects

- *Specialists in the field of entertainment automation*
- *Movie special effects and animation*
- *New design or the redesign and rebuild of many existing theme park attractions*
- *Systems operating characters ranging from aliens, monsters, animals, fish, and even animatronic people for stunts*

Fluid Power
Design Handbook

First Edition

Publisher
B&T Hydraulics & Pneumatics, Inc.

Printed in the United States of America

Library of Congress Catalog Number: 96-85826

ISBN 0-9653599-0-5

Limits Of Liability
The Author and Publisher of this book have used their best efforts in preparing the book and information contained herein. Every effort has been made to test, research and verify this information. However, the publisher disclaims any warranty expressed or implied or any liability or responsibility for the accuracy or correctness of any calculation or information published in this document. Design calculations utilizing the information contained in this publication should be independently verified.

Fluid Power
Design Handbook

PUBLISHER
B&T Hydraulics & Pneumatics, Inc.

EDITORIAL STAFF
Editor-in-Chief
Barbara Munson, B&T Hydraulics & Pneumatics, Inc.

Managing Editor
Michael Jacobs, B&T Hydraulics & Pneumatics, Inc.

Contributing Editors
Douglas A. Ostlind, B&T Hydraulics & Pneumatics, Inc.
Scott B. Sauer P.E., HECO, Inc.

The publisher wishes to acknowledge with sincere thanks the contributions and cooperation of the following concerns and individuals who have aided in the preparation of this publication with information, graphics and editorial review.

Aeroquip Corporation Fluid Connector Section
American Welding Society General Engineering Section
Machine Tool Builders Association Noise Section
Pall Industrial Hydraulics Servo Valve and Filtration Sections
Ross Controls Pneumatics Section
Carolina Fluid Components, Inc. Hydraulic Safety Section

Chris Cesar Mechanical Engineering Consultant
K. Sean Cowan B&T Hydraulics & Pneumatics
David Fursh P.E. Westinghouse Electric Corp.
Curt Lomax P.E. N.A.S.A.
Jim Ott Pacific Fluid Systems
Ken Adamson Carolina Fluid Components, Inc.
Bernie Clarke Taylor-Newcomb Industries
Steve Jenkens Carolina Fluid Components, Inc.
Fred Lehrer Carolina Fluid Çomponents, Inc.
Jack Marks Carolina Fluid Components, Inc.
Todd Mathewson Equipment Consultant
Scott Ruszala B&T Hydraulics & Pneumatics
J.R. "Bob" Webber The Machine Design Barn
Craig Weitz B&T Hydraulics & Pneumatics
Tom Wells Pacific Fluid Systems
Patrick Whelan P.E. Schilling Robotic Systems
Mark McDonald Carolina Fluid Components, Inc.
Chuck Shelton Carolina Fluid Components, Inc.
Howard Taylor Taylor-Newcomb Industries
John Taylor Taylor-Newcomb Industries
Michael Wolfe Pacific Fluid Systems
Cheryl Stinson B&T Hydraulics & Pneumatics

Fluid Power
Design Handbook

Catalog Association Members

To obtain more information on the Fluid Power Design Handbook or to order additional copies, please contact the distributor in your region.

Atlantic Fluid Power
111 Bridge Road
Hauppauge, NY 11788
(516) 234-3131
FAX:(516) 234-3283

B&T Hydraulics & Pneumatics
2324 Del Monte Street
P.O. Box 835 (95691)
West Sacramento, CA 95691
(916) 372-0660
FAX:(916) 372-0492

Carolina Fluid Components
9309 Stockport Place
Charlotte, NC 28273
(704) 588-6101
FAX:(704) 588-6115

Ralph W. Earl Co., Inc.
5930 E. Molloy Road
P.O. Box 2369 (13220)
Syracuse, NY 13211
(315) 454-4431
FAX:(315) 454-0977

Fluid Power Inc.
534 Township Line Road
Blue Bell, PA 19422
(215) 643-0350
FAX:(215) 643-4017

Gulf Controls Corporation
5201 Tampa West Blvd.
P.O. Box 15100 (33684)
Tampa, FL 33634
(800) 282-9125
FAX:(800) 282-9120

Hydraquip Corporation
4723 Pinemont Drive
Houston, TX 77092
(713) 680-1951
FAX:(713) 680-9799

Midwest Fluid Power
5702 Opportunity Drive
Toldeo, OH 43612
(419) 478-9086
FAX:(419) 478-4839

Pacific Fluid Systems
1925 N.W. Quimby Street
Portland, OR 97209
(505) 222-3295
FAX:(505) 228-6036

Pearse-Pearson Company, Inc.
22 Tobey Road
Bloomfield, CT 06002
(860) 242-7777
FAX:(860) 242-2673

Price Engineering
22577 Johnson Drive
Waukesha, WI 53186
(414) 547-2700
FAX:(414) 547-0416

RHM Fluid Power
375 Manufacturers Drive
Westland, MI 48185
(313) 326-5400
FAX:(313) 326-0339

Taylor-Newcomb Industries
5501 N. Oxford
P.O. Box 20223 (46220)
Indianapolis, IN 46220
(317) 257-1441
FAX:(317) 252-4329

Williams Equipment Co., Inc.
14808 W. 117th Street
P.O. Box 3237 (66063)
Olathe, KS 66063
(913) 764-9326
FAX:(913) 764-8372

Fluid Power
Design Handbook

Table of Contents

Quick Reference Formulas

CYLINDER			
Pressure (lb/in²)	$P = \frac{F}{A}$	Circular Area (in²)	$A = \pi \times r^2$
Force (lb)	$F = P \times A$	Power (hp)	$hp = \frac{P \times Q}{1714}$
Velocity (ft/sec)	$v = \frac{231 \times Q}{12 \times 60 \times A}$	V = Volume (gallons)	$V = \frac{A \times L}{231}$
Flow Rate (gal/min)	$Q = \frac{12 \times 60 \times v \times A}{231}$		

MOTOR			
Power (hp)	$hp = \frac{T \times n}{63025}$	Speed[1] (rpm)	$n = \frac{231 \times Q}{d}$
Torque[1] (lb - in)	$T = \frac{P \times d}{2 \times \pi}$	Torque (lb - in)	$T = \frac{hp \times 63025}{n}$
Flow rate[1] (gal/min)	$Q = \frac{n \times d}{231}$	Bearing life (hrs)	$B_{10} = L_{MFG} \times \frac{n_R}{n} \times \left(\frac{f_r}{f}\right)^{10/3}$
Displacement	$d = \frac{T \times 2 \times \pi}{P}$		

PUMP			
Power (hp)	$hp = \frac{T \times n}{63025}$	Speed[1] (rpm)	$n = \frac{231 \times Q}{d}$
Torque[1] (lb - in)	$T = \frac{P \times d}{2 \times \pi}$	Torque (lb - in)	$T = \frac{hp \times 63025}{n}$
Flow rate[1] (gal/min)	$Q = \frac{n \times d}{231}$	Bearing life (hrs)	$B_{10} = L_{MFG} \times \frac{n_R}{n} \times \left(\frac{f_r}{f}\right)^{10/3}$

MISCELLANEOUS	
Heat Capacity of a Steel Reservoir $hp = 0.001 \times A \times \Delta t$	**Heating Hydraulic Fluid** 1 watt will raise 1 gallon 1°F per hour
Estimating Pump Drive Horsepower 1 hp for each gpm at 1500 psi	**Boyles Law** $P_1V_1 = P_2V_2$ Subscript_1 = Initial conditions Subscript_2 = Final conditions
Pressure Drop Through an Orifice $Q = 26.3 \times A \times \sqrt{\text{Pressure Drop, psi}}$ Q = Flow in gpm A = Area of orifice in sq. inch (Specific gravity of fluid = .88) (C = .65)	

[1] The formulas provided assume 100% efficiency. Contact manufacturer for actual performance data.

- A = Area (in²)
- I = amperes
- D = Diameter (in)
- d = Displacement (in³/rev)
- Δ = Delta
- E = Volts
- eff = efficiency
- f_r = Force on bearing, at mfg rating
- F = Force (lb)
- f = Force on bearing, actual
- hp = Horsepower
- Kva = Kilovolt-amperes
- KW = Kilowatts
- L = Length (in)
- L_{MFG} = L_{10} life per mfg catalog
- n = speed (rpm)
- n = rpm of bearing, actual
- n_R = rpm of bearing at Mfg rating
- P = Pressure (lb/in²)
- π = 3.1415926
- pf = power factor
- Q = Flow rate (gal/min)
- r = radius (in)
- T = Torque (lb -in)
- t = Temperature (°F)
- V = Volume (in³)
- v = Velocity (ft/sec)

Quick Reference Formulas

ELECTRICAL			
Single-Phase		**Three-Phase**	
Current (Amperes) (hp known)	$I = \frac{hp \times 746}{E \times eff \times pf}$	Current (Amperes) (hp known)	$I = \frac{hp \times 746}{1.73 \times E \times eff \times pf}$
Current (Amperes) (KW known)	$I = \frac{KW \times 1000}{E \times pf}$	Current (Amperes) (KW known)	$I = \frac{KW \times 1000}{1.73 \times E \times pf}$
Current (Amperes) (Kva known)	$I = \frac{Kva \times 1000}{E}$	Current (Amperes) (Kva known)	$I = \frac{Kva \times 1000}{1.73 \times E}$
KW	$KW = \frac{I \times E \times pf}{1000}$	KW	$KW = \frac{1.73 \times I \times E \times pf}{1000}$
Kva	$Kva = \frac{I \times E}{1000}$	Kva	$Kva = \frac{1.73 \times I \times E}{1000}$
hp	$hp = \frac{I \times E \times eff \times pf}{746}$	hp	$hp = \frac{1.73 \times I \times E \times eff \times pf}{746}$

Electrical Rules of Thumb

Synch Speed rpm	Approx. Torque lb-ft per hp
3600	1.4
1800	3
1200	4.5
900	5.8

Rated Voltage	Approximate Amps/hp	
	Single-Phase	Three-Phase
115	10	
230	5	2.5
460		1.25
575		1.00

VISCOSITY CONVERSION		
32 to 100 SUS	100 to 240 SUS	240 SUS+
$Cst = .2253 \times SUS - \frac{194.4}{SUS}$	$Cst = .2193 \times SUS - \frac{134.6}{SUS}$	$Cst = \frac{SUS}{4.635}$

MISCELLANEOUS FLUID DATA

Weight of 1 cu ft of water = 62.4 (lb/ft^3)

Bulk Modulus of Elasticity
Hydraulic Oil = 280,000 psi (range of 150,000 - 350,000 psi)
Water = 320,000 psi

Specific Gravity
Hydraulic Oil = .88 (.87 to .90)

$$\text{Specific Gravity} = \frac{141.5}{131.5 + {}^0\text{A.P.I.}}$$

A = Area (in^2)
I = amperes
D = Diameter (in)
d = Displacement (in^3/rev)
Δ = Delta
E = Volts
eff = efficiency
f_r = Force on bearing, at mfg rating
F = Force (lb)
f = Force on bearing, actual
hp = Horsepower
Kva = Kilovolt-amperes
KW = Kilowatts
L = Length (in)
L_{MFG} = L_{10} life per mfg catalog
n = speed (rpm)
n = rpm of bearing, actual
n_r = rpm of bearing at mfg rating
P = Pressure (lb/in^2)
π = 3.1415926
pf = power factor
Q = Flow rate (gal/min)
r = radius (in)
T = Torque (lb-in)
t = Temperature (^{0}F)
V = Volume (in^3)
v = Velocity (ft/sec)

Formulas For Hydraulic Cylinders

CYLINDER

PRESSURE (lb/in²) $P = \frac{F}{A}$

P = Pressure (lb/in²)
F = Force (lb)
A = Area (in²)

CIRCULAR AREA (in²) $A = \pi \times r^2$

A = Area (in²)
π = 3.14159 (no units)
r = Radius (in)

FORCE (lb) $F = P \times A$

F = Force (lb)
P = Pressure (lb/in²)
A = Area (in²)

POWER (hp) $hp = \frac{P \times Q}{1714}$

hp = horsepower
Q = Flow rate (gal/min)
P = Pressure (lb/in²)
1714 = Conversion {(in/ft) x ((lb-ft/min)/(hp))}/(in³/gal) (12 x 33,000)/(231) Where 12 is used to convert from in to ft, 33,000 is to convert from lb-ft/min to hp, and 231 is used to convert from gal to in³.

VELOCITY (ft/sec) $v = \frac{231 \times Q}{12 \times 60 \times A} = \frac{Q}{3.117 \times A}$

v = Velocity, cylinder piston (ft/sec)
231 = Conversion factor, gal to in³ (in³/gal)
Q = Flow rate (gal/min)
12 = Conversion factor, ft to inches (in/ft)
60 = Conversion factor, min to sec (sec/min)
A = Area (in²)

VOLUME $V = \frac{A \times L}{231}$

V = Volume, cylinder capacity (gallons)
A = Area (in²)
L = Stroke Length (in)
231 = Conversion factor, gal to in³ (in³/gal)

FLOW RATE (gal/min) $Q = \frac{12 \times 60 \times v \times A}{231}$

$= 3.117 \times v \times A$

Q = Flow rate (gal/min)
12 = Conversion, ft to inches (in/ft)
60 = Conversion, min to sec (sec/min)
v = Velocity, cylinder piston (ft/sec)
A = Area (in²)
231 = Conversion, gal to in³ (in³/gal)

Hydraulic Cylinders

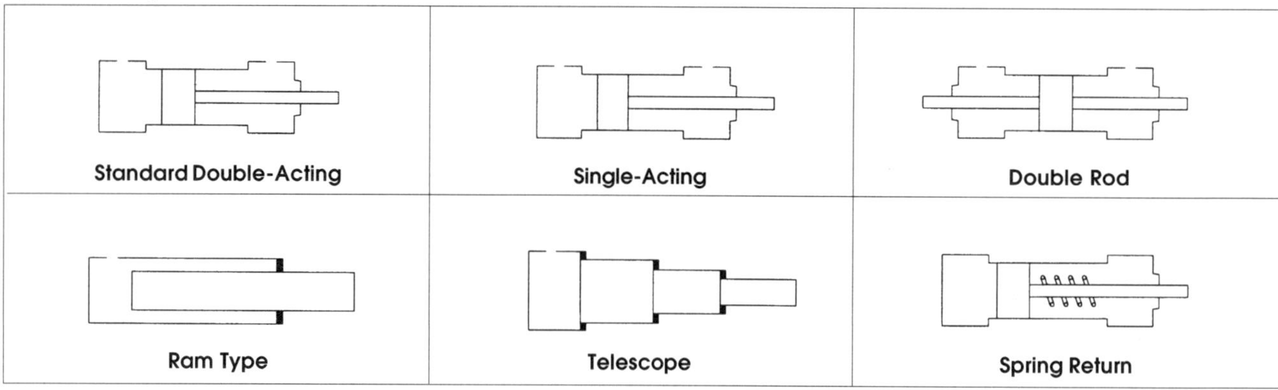

Fundamental Cylinders

STANDARD DOUBLE-ACTING CYLINDER

The power stroke is in both directions. There is a differential area for the extend and retract sides of the piston due to the area of the rod. The extend side provides the greater force but is the slower moving, and vice-versa for the retract side which is the lower force, but the faster moving. This is the typical hydraulic cylinder used in most applications.

SINGLE-ACTING CYLINDER

When a force is required in one direction only, a double-acting cylinder can be used in a single-acting mode. Because of the large number of double-acting cylinders manufactured, in some applications it is cost effective to use them in single acting mode over a ram or spring type single-acting cylinder. The unused side of the cylinder is vented to the atmosphere through a breather for a pneumatic application, or to a reservoir below the oil level in a hydraulic application.

DOUBLE-ROD CYLINDER

When equal displacement or forces are required for both extend and retract or a double push/pull application, a double-rod cylinder may be used. Other applications include operation of cams etc., during the load cycle.

RAM, SINGLE-ACTING CYLINDER

Ram cylinders have only one fluid chamber and rely on an external force (normally the weight of an object) to return the cylinder to the retracted position and are therefore normally mounted vertically. Typical applications are car lifts, etc.

TELESCOPING CYLINDER

As the name implies, the telescoping cylinder is made up of several interlocked pieces that telescope under pressure. The main advantage is the very short retracted length compared to the total cylinder stroke.

SPRING RETURN SINGLE-ACTING CYLINDER

Typically used in short stroke cylinder applications only, such as clamping and holding.

CUSHIONS

Cylinders can be designed with a cushion installed in either the extend or retract side of the cylinder (or both) to control the deceleration rate of the rod during the final part of its travel.

LINEAR MEASUREMENT

1 $^{3}/_{8}$ diameter rod is the smallest rod that can be gun drilled for standard linear transducers.

Determining Proper Cylinder Bore Size Estimation Table

(Push Force = Bore Area x Working Pressure)

Cyl. Bores dia. in.	Piston Area Sq. in.	Push Stroke Forces In Pounds — Pressures of Operating Medium (psi) 50	60	80	100	200	250	500	750	1000	1500	2000	3000	4000	5000	6000	Oil consumption per inch of stroke in one direction — Gals. Dis-placed
1 1/2	1.767	88	106	141	177	353	442	884	1,325	1,767	2,651	3,534	5,301	7,068	8,835	10,602	.00765
2	3.142	157	189	251	314	628	786	1,571	2,357	3,142	4,713	6,283	9,426	12,566	15,710	18,852	.01360
2 1/2	4.909	245	295	393	491	982	1,227	2,455	3,682	4,909	7,364	9,818	14,727	19,636	24,545	29,454	.0213
3 1/4	8.296	415	498	664	830	1,659	2,074	4,148	6,222	8,296	12,444	16,592	24,888	33,184	41,480	49,776	.0359
4	12.566	628	754	1,005	1,257	2,513	3,141	6,283	9,425	12,566	18,849	25,132	37,698	50,264	62,830	75,396	.0544
5	19.635	982	1,178	1,571	1,964	3,927	4,909	9,818	14,726	19,635	29,453	39,270	58,905	78,540	98,175	117,810	.0850
6	28.274	1,414	1,696	2,262	2,827	5,657	7,071	14,137	21,205	28,274	42,411	56,548	84,822	113,096	141,370	169,644	.1224
7	38.485	1,924	2,309	3,079	3,849	7,697	9,621	19,242	28,864	38,485	57,728	76,970	115,455	153,940	192,425	230,910	.1666
8	50.265	2,513	3,016	4,021	5,027	10,053	12,566	25,133	37,699	50,265	75,398	100,530	150,795	201,060	251,325	301,590	.2176
10	78.54	3,927	4,712	6,283	7,854	15,710	19,635	39,270	58,905	78,540	117,810	157,080	235,620	314,160	392,700	471,240	.3400
12	113.10	5,655	6,786	9,048	11,310	22,620	28,275	56,550	84,825	113,100	169,650	226,200	339,300	452,400	565,500	678,600	.4896
14	153.94	7,697	9,236	12,315	15,394	30,790	38,485	76,970	115,455	153,940	230,910	307,880	461,820	615,760	769,700	923,640	.6664
16	201.06	10,053	12,064	16,085	20,106	40,211	50,265	100,530	150,796	201,060	301,590	402,120	603,180	804,240	1,005 E^3	1,206,360	.8704
18	254.47	12,723	15,268	20,358	25,447	50,890	63,615	127,235	190,852	254,470	381,705	508,940	763,410	1,017 E^3	1,272 E^3	1,526,820	1.1016
20	314.16	15,708	18,850	25,133	31,416	62,830	78,540	157,080	235,620	314,160	471,240	628,320	942,480	1,256 E^3	1,570 E^3	1,884,960	1.3600

Thrusts for operating pressures not shown in the table may be calculated by multiplying the operating pressures by the piston areas.

Pull Stroke Cylinder Bores and Forces

(Pull Force = (Bore Area - Rod Area) x Working Pressure)

Pull Stroke Forces In Pounds — Deduct the following thrusts or consumptions corresponding to rod size from push stroke pressures or consumptions to determine pull stroke pressure or consumptions

Piston rod dia. in.	Piston Area Sq. in.	Pressures of Operating Medium (psi) 50	60	80	100	200	250	500	750	1000	1500	2000	3000	4000	5000	6000	Oil consumption per inch of stroke in one direction — Gals. Dis-placed
5/8	.307	15	18	25	31	61	77	154	230	307	461	614	921	1,228	1,535	1,842	.00133
1	.785	39	47	63	79	157	196	393	589	785	1,176	1,570	2,355	3,140	3,925	4,710	.0034
1 3/8	1.485	74	89	119	149	297	371	743	1,114	1,485	2,228	2,970	4,455	5,940	7,425	8,910	.00673
1 3/4	2.405	120	144	192	241	481	601	1,203	1,804	2,405	3,608	4,810	7,215	9,620	12,025	14,450	.01041
2	3.142	157	189	251	314	628	786	1,571	2,357	3,142	4,713	6,283	9,426	12,566	15,710	18,852	.01360
2 1/2	4.900	245	294	392	491	980	1,225	2,450	3,675	4,900	7,364	9,800	14,700	19,600	24,500	29,400	.0213
3	7.069	353	424	566	707	1,414	1,767	3,535	5,302	7,069	10,604	14,138	21,207	28,276	35,345	42,414	.0306
3 1/2	9.621	481	577	770	962	1,924	2,405	4,811	7,216	9,621	14,432	19,242	28,863	38,484	48,105	57,726	.0417
4	12.566	628	754	1,005	1,257	2,513	3,141	6,283	9,425	12,566	18,849	25,132	37,698	50,264	62,830	75,396	.0544
4 1/2	15.904	795	954	1,272	1,590	3,181	3,976	7,952	11,928	15,904	23,856	31,808	47,712	63,616	79,520	95,424	.0688
5	19.635	982	1,178	1,571	1,964	3,927	4,909	9,818	14,726	19,635	29,453	39,270	58,905	78,540	98,175	117,810	.0850
5 1/2	23.758	1,188	1,425	1,901	2,376	4,752	5,940	11,879	17,819	23,758	35,657	47,516	71,274	95,032	118,790	142,548	.1028
7	38.485	1,924	2,309	3,079	3,849	7,697	9,621	19,242	28,864	38,485	57,728	76,970	115,455	153,940	192,425	230,910	.1666
8	50.265	2,513	3,016	4,021	5,027	10,053	12,566	25,133	37,699	50,265	75,398	100,530	150,795	201,060	251,325	301,590	.2176
9	63.617	3,180	3,817	5,089	6,361	12,722	15,900	31,800	47,712	63,617	95,400	127,234	190,850	254,468	318,085	381,700	.2754
10	78.54	3,927	4,712	6,283	7,854	15,710	19,635	39,270	58,905	78,540	117,810	157,080	235,620	314,160	392,700	471,240	.3400

Determining Cylinder Thrust Required

Horizontal Motion

F = W x f

Vertical Motion

For deceleration downward or acceleration upward F = W x f +W
For deceleration upward or acceleration downward F = W x f - W

Vm — Maximum velocity
W — Weight of object (lbs)
f — Acceleration or deceleration force factor
d — Acceleration or deceleration distance (inches)

To determine force required for acceleration (to go from zero to desired speed) and deceleration (to reduce speed from maximum to zero) for a given mass, determine force factor "f" from the chart under the column headed by the distance in which acceleration or deceleration is desired. Use figure on same horizontal line as desired maximum velocity.

The formula neglects friction, therefore the "f" force determined for acceleration must be increased to overcome friction of the moving mass and mechanism and "f" force factor for deceleration may be reduced by the amount of the friction force of the moving mass and mechanism. If motion is horizontal, use formula F = W x f, to determine force in pounds to accelerate or decelerate. If motion is vertical, use formula F = W x f + W for acceleration upwards and deceleration downwards, and use formula F = W x f - W for deceleration upwards and acceleration downwards. (In some cases involving vertical motion a negative answer may be obtained, which indicates that gravity alone is sufficient to cause acceleration or deceleration).

If exact acceleration or deceleration distance desired is not shown in the chart, adjust any force factor proportionately (i.e., note force factor for 1" distance is twice the force factor for 2" distance. Similarly force factor for 1" distance is three times the force factor for a 3" distance).

Example: Motion horizontal: Formula F = W x f is used
Vm (desired maximum velocity) = 100 ft./min.
W (weight of mass in pounds) = 3000 pounds
d (distance) = 1/4"
F = 3000 pounds x 2.073 force factor = 6219 pounds

Determining Distances and Forces Required to Start Cylinder Piston Travel When Loads and Speeds are Known

Vm Max. Velocity		Force Factor "f" — d, Acceleration and Deceleration Distance in Inches								
Ft. Per Min.	Ft. Per Sec.	1/8	1/4	1/2	5/8	3/4	1	1 1/2	2	3
15	.250	.092	.046	.023	.018	.016	.012	.008	.006	.004
20	.333	.166	.083	.041	.033	.028	.021	.014	.010	.007
25	.416	.260	.130	.065	.052	.043	.032	.022	.016	.011
30	.500	.372	.186	.093	.074	.062	.047	.031	.023	.016
35	.583	.508	.254	.127	.102	.085	.063	.042	.032	.021
40	.666	.664	.332	.166	.133	.111	.083	.055	.041	.028
45	.750	.840	.420	.210	.168	.140	.105	.070	.052	.035
50	.833	1.036	.518	.259	.207	.173	.130	.086	.065	.043
55	.916	1.254	.627	.314	.251	.209	.157	.105	.078	.052
60	1.000	1.492	.746	.373	.298	.249	.187	.124	.093	.062
65	1.083	1.752	.876	.438	.350	.292	.219	.146	.109	.073
70	1.166	2.032	1.016	.508	.406	.339	.254	.169	.127	.085
75	1.250	2.332	1.166	.583	.466	.389	.292	.194	.146	.097
80	1.333	2.654	1.327	.663	.531	.442	.331	.221	.166	.111
85	1.416	2.996	1.498	.749	.599	.499	.374	.249	.187	.125
90	1.500	3.358	1.679	.840	.672	.560	.420	.280	.210	.140
95	1.583	3.742	1.871	.935	.748	.624	.468	.312	.234	.156
100	1.666	4.146	2.073	1.036	.829	.691	.518	.345	.259	.173
105	1.750	4.570	2.285	1.142	.914	.762	.571	.381	.286	.190
120	2.000	5.970	2.985	1.493	1.194	.995	.746	.498	.373	.249
150	2.500	9.326	4.663	2.332	1.865	1.554	1.166	.777	.583	.387
180	3.000	13.432	6.716	3.358	2.686	2.239	1.679	1.119	.840	.560
210	3.500	18.284	9.142	4.571	3.657	3.047	2.285	1.524	1.143	.762
240	4.000	23.880	11.940	5.970	4.776	3.980	2.985	1.990	1.493	.995
270	4.500	30.224	15.112	7.556	6.045	5.037	3.778	2.519	1.889	1.259
300	5.000	37.312	18.656	9.328	7.462	6.219	4.664	3.109	2.332	1.555
330	5.500	45.150	22.575	11.287	9.030	7.525	5.644	3.762	2.822	1.881
360	6.000	53.732	26.866	13.433	10.746	8.955	6.716	4.478	3.358	2.239
420	7.000	73.134	36.567	18.283	14.627	12.189	9.142	6.094	4.571	3.047
480	8.000	95.522	47.761	23.881	19.104	15.920	11.940	7.960	5.970	3.980
540	9.000	120.896	60.448	30.224	24.179	20.149	15.112	10.075	7.556	5.037
600	10.000	149.254	74.627	37.314	29.851	24.876	18.657	12.438	9.328	6.219
660	11.000	180.598	90.299	45.149	36.120	30.099	22.575	15.049	11.287	7.525
720	12.000	214.92	107.46	53.731	42.98	35.821	26.866	17.910	13.433	8.955
840	14.000	292.54	146.27	73.133	58.51	48.755	36.567	24.378	18.283	12.189
960	16.000	382.26	191.13	95.562	76.45	63.682	47.781	31.841	23.881	15.920
1080	18.000	483.58	241.79	120.9	96.72	80.597	60.448	40.229	30.224	20.149
1200	20.000	597.02	298.51	149.25	119.40	99.502	74.627	49.751	37.313	24.876

NOTE: Use 1/2" distance 'd' for cylinders having self-regulating cushions.

Determining Proper Port and Pipe Size

Having chosen your bore size and selected a particular cylinder, you are now ready to verify whether the standard port size will provide the cylinder speed you require. Follow these simple steps to find out:

1. Divide your cylinder stroke length (in inches) by the length of time (in seconds) permitted for the stroke to take place. Multiply the result by 60 to obtain the stroke speed in inches per minute.

Example: 20 inch stroke, 2 seconds
20 ÷ 2 = 10 inches/second
10 x 60 = 600 inches/minute

2. In the Bore Size Table on Page 5, right hand column of the cylinder chart, find the oil consumption per inch stroke for the cylinder bore you have chosen. Multiply the stroke speed calculated in Step 1 by the oil consumption per inch of stroke. This will give you the oil consumption per minute.

Example: 3 1/4 inch bore, push stroke cylinder
600 x 0.0359 = 21.54 gallons/minute

3. Find the standard port size for the cylinder you have chosen, in the manufacturers literature.

Example: 3 1/4 inch bore, cylinder has a standard port size of 1/2 inch
(and has 14 NPT threads per inch)

4. In the Pipe and Port Size Table below find the oil flow (in gallons per minute) which is in the column corresponding to your desired fluid velocity and the row corresponding to your standard port size.

Example: 1/2 inch
Table row "1/2S", column "15 ft. per sec."
shows 14.31 gallons per mintue

5. If the flow rate found in Step 4 is greater than or equal to that calculated in Step 2, then the standard size port is adequate, and you should proceed to the next step. If the Step 4 rate is less than the Step 2 rate, then return to the Pipe and Port Size Table and find the flow rate value in the column corresponding to your desired fluid velocity which is equal to (or larger than but closest to) the Step 2 flow rate. In the same row, move across the table to the first column and find the proper oversize port dimension.

NOTE: For an air/over oil system the oil velocity should be kept below 7 ft/sec.

Example: 25.17 is the number shown in the "15 ft. per second column" which is larger than but closest to 21.54. Looking across to the first column 3/4S, which means oversize ports of 3/4" should be ordered.

Pipe and Port Size Table

Butt Welded Steel Clean Pipe			Oil flow (gallons per minute) and friction pressure drop (pounds per square inch) per foot length pipe												Equivalent Length of Straight Pipe In Feet For Various Fittings		
			Velocity = 5 ft. per sec.		Velocity = 10 ft. per sec.		Velocity = 15 ft. per sec.		Velocity = 20 ft. per sec.		Velocity = 25 ft. per sec.		Velocity = 30 ft. per sec.		See page 42 for additional fittings		
Pipe Size	Internal Dia. In.	Internal Area Sq. In.	Gals. Per Min.	Press. Drop in psi	Gals. Per Min.	Press. Drop in psi	Gals. Per Min.	Press. Drop in psi	Gals. Per Min.	Press. Drop in psi	Gals. Per Min.	Press. Drop in psi	Gals. Per Min.	Press. Drop in psi	Pipe Size	Cyl. & 2&3 Way Valves	4 Way Valves
3/8S	.493	.191	2.99	.58	5.98	1.99	8.97	2.35	11.96	3.71	14.95	5.44	17.94	7.31	3/8	5-25	10-50
1/2S	.622	.304	4.74	.39	9.48	.82	14.31	1.65	18.96	2.75	23.70	4.00	28.44	5.36	1/2	6-30	12-60
3/4X	.742	.433	6.76	.27	13.52	.69	20.28	1.38	27.04	2.15	33.80	3.12	40.56	4.15			
3/4S	.824	.533	8.39	.22	16.78	.59	25.17	1.17	33.56	1.80	41.95	2.60	50.34	3.44	3/4	10-50	20-100
1 1/4XX	.896	.630	9.83	.18	19.66	.54	29.49	1.07	39.32	1.64	49.15	2.37	58.98	3.13			
1X	.957	.719	11.21	.16	22.42	.49	33.63	.97	44.84	1.54	56.05	2.22	67.26	2.93			
1S	1.049	.864	13.59	.14	27.18	.43	40.77	.85	54.36	1.40	67.95	1.94	81.54	2.67	1	13-65	25-125
1 1/2XX	1.100	.950	14.81	.13	29.62	.41	44.43	.81	59.24	1.34	74.05	1.86	88.86	2.44			
1 1/4X	1.278	1.283	20.15	.11	40.30	.33	60.45	.65	80.60	1.07	100.75	1.53	120.90	2.00			
1 1/4S	1.380	1.495	23.48	.10	46.96	.31	70.44	.60	93.92	.91	117.40	1.29	140.88	1.76	1 1/4	15-75	30-150
1 1/2X	1.500	1.767	27.49	.09	54.98	.28	82.47	.53	109.96	.84	137.45	1.19	164.94	1.62			
2XX	1.503	1.774	27.59	.09	55.18	.28	82.77	.53	110.36	.84	137.95	1.19	165.54	1.62			
1 1/2S	1.610	2.036	31.68	.08	63.36	.25	95.04	.48	126.72	.71	158.40	1.05	190.08	1.43	1 1/2	20-100	40-200
2 1/2XX	1.771	2.464	38.32	.07	76.64	.22	114.96	.40	153.28	.64	191.60	.94	229.92	1.30			
2X	1.939	2.953	45.98	.06	91.96	.19	137.94	.36	183.92	.59	229.90	.87	275.88	1.11			
2S	2.067	3.355	52.36	.05	104.72	.17	157.08	.32	209.44	.53	261.80	.77	314.16	1.05	2	25-125	50-250
3XX	2.300	4.155	64.87	.05	129.74	.15	194.61	.28	259.48	.47	324.35	.70	389.22	1.00			
2 1/2X	2.323	4.238	66.55	.05	133.10	.15	199.65	.28	266.20	.47	332.75	.69	399.30	.97			
2 1/2S	2.469	4.788	74.75	.04	149.50	.14	224.25	.26	299.00	.44	373.75	.65	448.50	.88	2 1/2	30-150	60-300
3X	2.900	6.605	103.34	.04	206.68	.11	310.02	.21	413.36	.35	516.70	.52	620.04	.72			
3S	3.068	7.393	114.89	.03	229.78	.10	344.67	.19	459.56	.33	574.45	.49	689.34	.66	3	35-175	70-350

S = Standard Weight Pipe **X = Extra Strong** **XX = Double Extra Strong**

The pressure drop shown in the above table is for butt welded clean steel pipe. Pressure drop is the same regardless of operating pressure. Avoid large pressure drops in LOW PRESSURE SYSTEMS. Note that oil flows at high velocity (up to 30 ft. per sec.) with small pressure drop loss through large pipes but results in prohibitive loss in small pipes at this speed. Pipe line velocities in excess of 15 ft. per sec. might result in excessive shock loading. Hydraulic oil shown is approximately 225 S.S.U. at 100ºF — .88 specific gravity at 60ºF. Approximate specific gravity at 100ºF — 865. For water, the above figures are conservative (contact your distributor for detailed calculations).

Stop Tubes for Long Push Stroke Cylinders

The use of a stop tube is a generally accepted and preferred method for reducing piston and bearing loads on long push stroke cylinders and, additionally, for preventing jack-knifing or buckling of horizontally mounted, long stroke cylinders on push stroke. Stop tubes are more effective, less costly, and lighter in weight than oversize piston rods.

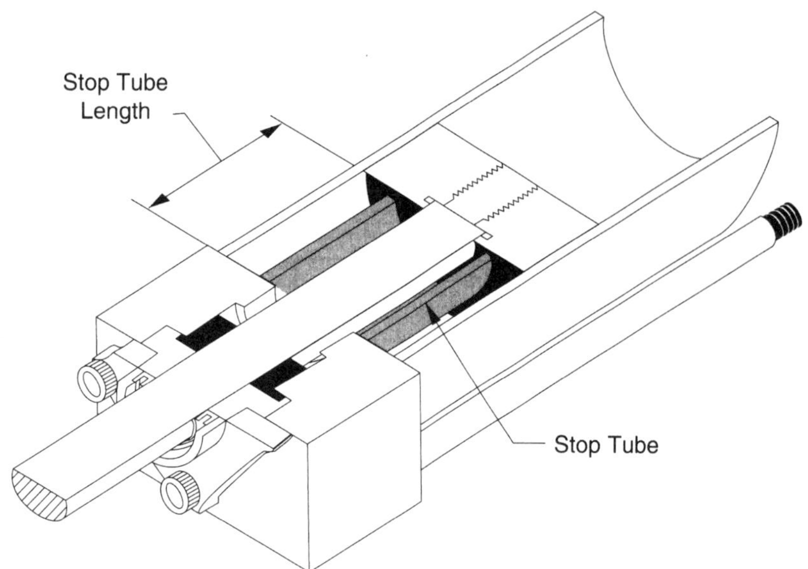

Stop Tube Table

"L" (inches)	Stop Tube Length (inches)	"L" (inches)	Stop Tube Length (inches)
0-40	0	171-180	14
41-50	1	181-190	15
51-60	2	191-200	16
61-70	3	201-210	17
71-80	4	211-220	18
81-90	5	221-230	19
91-100	6	231-240	20
101-110	7	241-250	21
111-120	8	251-260	22
121-130	9	261-270	23
131-140	10	271-280	24
141-150	11	281-290	25
151-160	12	291-300	26
161-170	13	301-310	27

NOTE: 'L' or 'D' are calculated from mounting point with rod extended (See page 9).

Positioned between the piston and cylinder head, a stop tube restricts the extended position of the rod so that the added distance between the piston and bushing results in less strain, wear, and bearing load.

Determining the Length of Stop Tube

1. Referring to the illustration which corresponds to your cylinder application on the following page, determine the value of "L". Be certain to include the thickness of the cylinder head, cap and piston assembly plus twice the length of the cylinder stroke. Then go down the first column of the Stop Tube Table and find the range which encompasses that value of "L". The number shown to the right in the second column is the length of stop tube your cylinder requires.

2. Add the stop tube length to your "L" dimension to obtain an "Adjusted L Dimension". This dimension will be used in the procedures on this page to determine whether your cylinder requires an oversize piston rod in addition to the stop tube.

Buckling

For short rods the limiting load is controlled by the yield point of the rod material and an appropriate factor of safety. For long rods the limiting load is controlled by the rods buckling capacity and a factor of safety. The method of support for both ends and the method of load application is very important in the buckling calculation. The generally accepted formula for calculation of the buckling capacity is the "Euler column formula":

$$P_{CR} = \frac{n\pi^2 EI}{L^2} \times \frac{1}{FS}$$

P_{CR} = Critical Load (lb)

n = End-condition constant (no units)

Typical Values:

Column End Condition	Conservative	Typical Value
fixed-free	1/4	1/4
rounded-rounded	1	1
fixed-rounded	1	1.2
fixed-fixed	1	1.2

π = 3.1415926 (no units)

E = Modulus of Elasticity (psi)

typical values: Steel — 30×10^6 psi

I = Area of inertia (in^4)

"I" for a circle = $\frac{\pi \times D^4}{64}$ D = Rod diameter (in)

L = Column length (in)

FS = Factor of Safety (no units)

Column Length "L" in Inches

The following chart is based on an "n" value of 1.0 and a Factor of Safety of 4. See page 9 for definition of L.

Thrust in pounds	Piston Rod Diameter							
	5/8	1	1 3/8	1 3/4	2	2 1/2	3	3 1/2
200	53	135						
300	43	110						
400	37	95						
500	33	85	161					
750	27	70	132					
1000	24	60	114	185	241	377		
2000	17	43	81	131	170	266	384	
3000	14	35	66	107	139	218	313	
4000	12	30	57	92	121	188	271	369
5000	11	27	51	83	108	169	242	330
7500		22	42	67	88	138	189	270
10,000		19	36	58	76	119	172	234
20,000			25	41	54	84	121	165
30,000				34	44	69	99	135
40,000				29	38	60	86	117
50,000					34	53	77	104
75,000						44	63	85
100,000							54	74
150,000								60

The chart is provided as a quick reference only. Each particular application should be reviewed considering actual end conditions, side loading and eccentric loading.

Fundamental Cylinders

Group A

With piston rod extended. To be checked for rod diameter only. Stop tube not required.

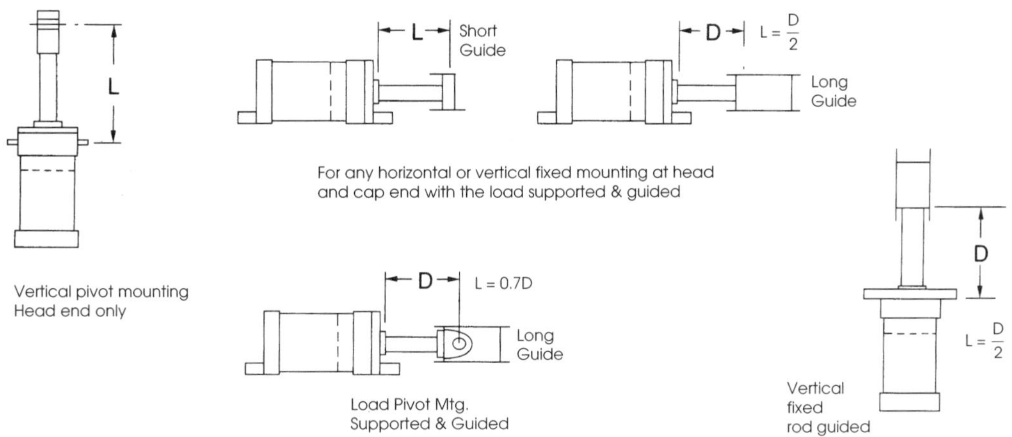

Group B

To avoid rod buckling or cylinder jackknifing, check for stop tube and rod diameter requirements with piston rod extended. Use cylinder dimensional charts. No stop tube required if cylinder operates on pull stroke only.

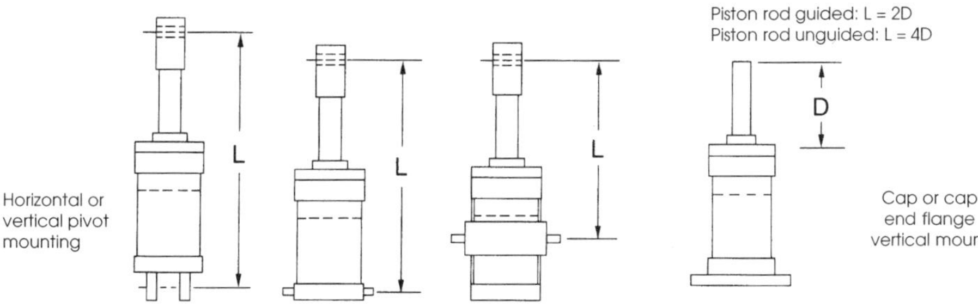

Group C

To be checked for stop tube length and piston rod diameter to eliminate buckling or jackknifing with piston rod extended.

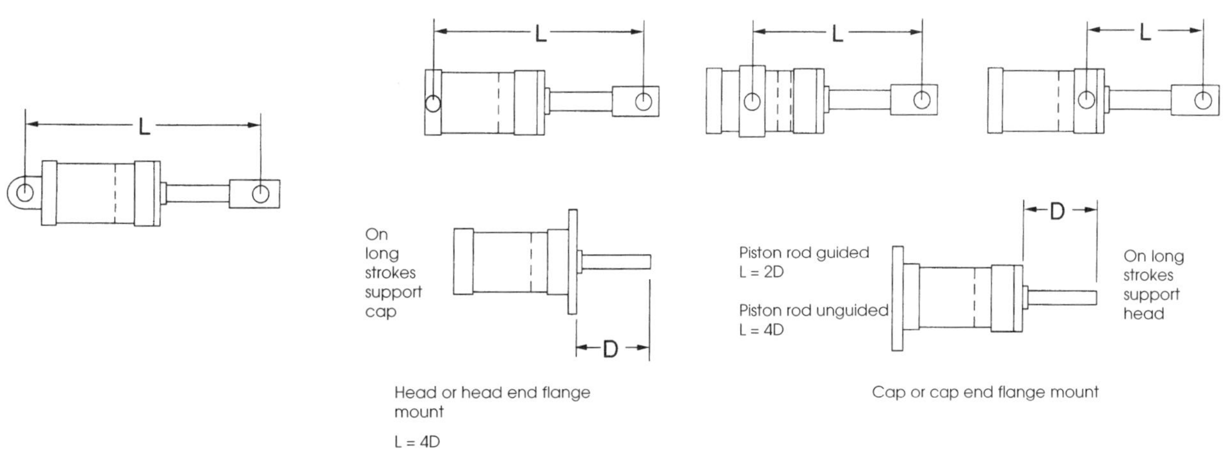

Formulas For Pumps

Pumps

POWER[1]
Required to Drive a Pump

$$hp = \frac{Q \times p}{1714}$$

hp = horsepower
P = Pressure (lb/in^2)
1714 = Conversion {(in/ft) x ((lb-ft/min)/(hp))}/(in^3/gal) (12 x 33,000)/(231) Where 12 is used to convert from in to ft, 33,000 is to convert from lb-ft/min to hp, and 231 is used to convert from gal to in^3.
Q = flow rate, pump output (gal/min)

SPEED[1]
of a Hydraulic Pump

$$n = \frac{231 \times Q}{d}$$

n = speed (rpm)
231 = Conversion, gal to in^3 (in^3/gal)
Q = flow rate, pump output (gal/min)
d = displacement, pump (in^3/rev

TORQUE[1]
of a Hydraulic Pump

$$T = \frac{P \times d}{2 \times \pi}$$

or

$$T = \frac{hp \times 63025}{n}$$

T = torque (lb - in)
P = pressure (lb/in^2)
d = displacement, pump (in^3/rev)
π = 3.14159 (no units)
hp = horsepower
63025 = Conversion ((lb-ft/min/hp) x (in/ft)/(no units)) ((33,000 x 12)/(2 x π)) where 12 is used to convert from inches to feet, (2 x π) is used to convert to angular velocity, and 33,000 is used to convert from (lb - ft)/min to hp
n = speed (rpm)

FLOW[1]
Output of Pump

$$Q = \frac{n \times d}{231}$$

Q = Flow rate, pump output (gal/min)
n = speed (rpm)
d = displacement, pump (in^3/rev)
231 = Conversion, gal to in^3 (in^3/gal)

LIFE
Bearing

$$L_{10} = L_{MFG} \times \frac{n_R}{n} \times \left(\frac{P_R}{P}\right)^{10/3}$$

L_{10} = Bearing life (rev or hrs)
The bearing life associated with 90% reliability
L_{MFG} = Manufacturers rated life (rev or hrs)
n_R = Speed, rated (rpm)
n = Speed, working (rpm)
P = Pressure, working (psi)
P_R = Pressure, rated (psi)

PUMP EFFICIENCY

MECHANICAL
Mechanical Efficiency

$$E_M = \frac{T_{ACTUAL}}{T_{THEO}}$$

E_M = Mechanical efficiency, pump (no units)
T_{THEO} = Torque theoretical (lb - in)
T_{ACTUAL} = Torque actual (lb - in)

VOLUMETRIC
Efficiency

$$E_v = \frac{Q_{ACT}}{Q_{THEO}}$$

E_V = Volumetric efficiency, pump (no units)
Q_{ACT} = Flow rate, actual output (gpm)
Q_{THEO} = Flow rate, theoretical output (gpm)

OVERALL
Efficiency

$$E_O = \frac{hp_{out}}{hp_{IN}}$$

E_O = Overall efficiency, pump (no units)
hp_{OUT} = Horsepower out of the pump (hp)
hp_{IN} = Horsepower into the pump (hp)

$$E_O = E_V \times E_M$$

E_O = Overall efficiency, pump (no units)
E_V = Volumetric efficiency, pump (no units)
E_M = Mechanical efficiency, pump (no units)

[1] The formulas provided assume 100% efficiency. Contact manufacturer for actual performance data.

Recommended Maximum Flows For Pressure Lines

Pump Suction
Size to maintain the velocity in the range of 2 to 4 feet per second.

Oil Return Lines
Size to maintain the velocity in the range of 10 to 15 feet per second.

Medium Pressure Lines
Size to maintain the velocity in the range of 15 to 20 feet per second.

High Pressure Lines
Size to maintain the velocity at a maximum of 30 feet per second.

Safe Pump Inlet Vacuum

	Gear Pumps	Vane Pumps	Piston Pumps
Max. Safe Inlet Vacuum, psi	3 to 5	2 to 3	2
Max. Safe Inlet Vacuum, "Hg	6 to 10	4 to 6	4

NOTE: The suction strainer should be cleaned or replaced when inlet vacuum on a hydraulic pump reaches these values. Sustained operation at these vacuums may damage the pump. When suction strainer is clean, inlet vacuum should not be more than 1/3 of these values.

Horsepower Required to Drive a Hydraulic Pump

100% Efficient Pump

Pump Flow	System Pressure (psi)							
gpm	100	250	500	1000	2000	3000	4000	5000
0.50	0.03	0.07	0.15	0.29	0.58	0.88	1.17	1.46
1.00	0.06	0.15	0.29	0.58	1.17	1.75	2.33	2.92
2.00	0.12	0.29	0.58	1.17	2.33	3.50	4.67	5.83
3.00	0.18	0.44	0.88	1.75	3.50	5.25	7.00	8.75
4.00	0.23	0.58	1.17	2.33	4.67	7.00	9.33	11.67
5.00	0.29	0.73	1.46	2.92	5.83	8.75	11.67	14.59
6.00	0.35	0.88	1.75	3.50	7.00	10.50	14.00	17.50
7.00	0.41	1.02	2.04	4.08	8.17	12.25	16.34	20.42
8.00	0.47	1.17	2.33	4.67	9.33	14.00	18.67	23.34
9.00	0.53	1.31	2.63	5.25	10.50	15.75	21.00	26.25
10.00	0.58	1.46	2.92	5.83	11.67	17.50	23.34	29.17
20.00	1.17	2.92	5.83	11.67	23.34	35.01	46.67	58.34
30.00	1.75	4.38	8.75	17.50	35.01	52.51	70.01	87.51
40.00	2.33	5.83	11.67	23.34	46.67	70.01	93.35	116.69
50.00	2.92	7.29	14.59	29.17	58.34	87.51	116.69	145.86
60.00	3.50	8.75	17.50	35.01	70.01	105.02	140.02	175.03
70.00	4.08	10.21	20.42	40.84	81.68	122.52	163.36	204.20
80.00	4.67	11.67	23.34	46.67	93.35	140.02	186.70	233.37
90.00	5.25	13.13	26.25	52.51	105.02	157.53	210.04	262.54
100.00	5.83	14.59	29.17	58.34	116.69	175.03	233.37	291.72

$$\text{Horsepower} = \frac{\text{gpm x psi}}{\text{1714 x pump efficiency}^{1}}$$

[1] Chart assumes 100% efficiency. Consult your pump manufacturer for actual pump efficiency in your application. For pump flows not listed combine horsepower of flows and pressures to find total horsepower requirement.

Suction Pressure

The identification of the pump suction requirement is significant in any pump application. Specifying a higher suction lift than actually exists results in selection of a pump at a lower speed than necessary. Not only does this mean a larger, more expensive pump but also a costlier driver. Should the suction lift requirement be higher than specified, the outcome could be a noisy installation due to pump cavitation.

There is a common misconception that pumps "pull" fluid into the inlet opening unassisted by any outside force. Actually, fluid flows into the pump due to a difference in pressure between pump inlet and the fluid source. A primary step in pump selection is the calculation of system Net Inlet Pressure, sometimes called Net Positive Inlet Pressure. This is the absolute pressure above fluid vapor pressure at pump inlet and is determined as follows:

- Atmospheric pressure (at jobsite altitude — psia)
  Plus: static head (minimum level of fluid above pump inlet — psi) or,
- Minus: static lift (maximum level of fluid below pump inlet — psi)
- Minus inlet line friction losses including entrance losses including entrance loss from reservoir to pipe, pressure drops through valves, fittings, strainers, etc. (at maximum viscosity — psi)
- Minus fluid vapor pressure (usually at maximum pumping temperature — psia)
- Equals system Net Inlet Pressure in psia

NOTE: When Inlet Pressure is expressed in feet of liquid, it is called Net Positive Suction Pressure

System Net Inlet Pressure available must always equal or exceed pump Net Inlet Pressure required.

Suction conditions of rotary pumps are normally rated as suction lift capability in inches of mercury vacuum (with air-free oil of negligible vapor pressure at sea level). This can be expressed as Net Inlet Pressure required by subtracting suction lift capability from 30" Hg and converting the remainder to psia. Pump Net Inlet Pressure required must always be less than or equal to system Net Inlet Pressure available.

Suction condition is the most frequently over-looked parameter in pump selection. Time spent determining it accurately can optimize pump selection and result in a quiet installation.

Pressure Equivalents

Gage	Absolute	Gage	Absolute	
0" hg	30" hg	0 psig	14.7 psia	1 Atmosphere at sea level
- 30" hg	0" hg	-14.7 psig	0 psia	

Properties of the Lower Atmosphere

Altitude Ft.	Pressure psia	Pressure in. Hg at 32°F	Altitude Ft.	Pressure psia	Pressure in. Hg at 32°F
0	14.696	29.921	2000	13.66	27.82
100	14.64	29.81	2200	13.56	27.62
200	14.59	29.71	2400	13.47	27.42
300	14.54	29.60	2600	13.37	27.21
400	14.48	29.49	2800	13.27	27.02
500	14.43	29.38	3000	13.17	26.82
600	14.38	29.28	3500	12.93	26.33
700	14.33	29.17	4000	12.69	25.84
800	14.28	29.07	4500	12.46	25.37
900	14.22	28.96	5000	12.23	24.90
1000	14.17	28.86	6000	11.78	23.98
1100	14.12	28.75	7000	11.34	23.09
1200	14.07	28.65	8000	10.92	22.22
1300	14.02	28.54	9000	10.50	21.39
1400	13.97	28.44	10000	10.11	20.58
1500	13.92	28.33	12000	9.346	19.03
1600	13.87	28.23	14000	8.633	17.58
1700	13.82	28.13	15000	8.293	16.89
1800	13.76	28.02	16000	7.965	16.21
1900	13.71	27.92	18000	7.339	14.94

Pump Comparison Chart

Pump Type	Cost	Efficiency	Contamination Tolerance	Pressure Range
Gear	Low	Low	Best	0 to 3,000
Vane	Med-High	Med	Med	0 to 2,500
Piston	Med-High	High	Worst	0 to 7,500

Pump/Motor Shafts & Flanges

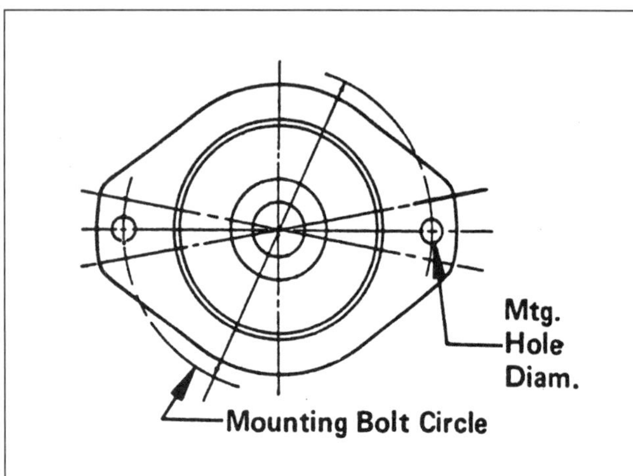

NFPA Standard
Two-Bolt Pump/Motor Mounting Flanges

Flange Code	SAE No.	SAE hp Rating[1]	Mtg. Bolt Circle	Mtg. Hole Dia.	Pilot Dia.	Pilot Ht.
50-2	N/A	N/A	3 1/4	13/32	2	1/4
82-2	A	10	4 3/16	7/16	3 1/4	1/4
101-2	B	25	5 3/4	9/16	4	3/8
127-2	C	50	7 1/8	11/16	5	1/2
152-2	D	100	9	13/16	6	1/2
165-2	E	200	12 1/2	1 1/16	6 1/2	5/8
177-2	F	300	13 25/32	1 1/16	7	5/8

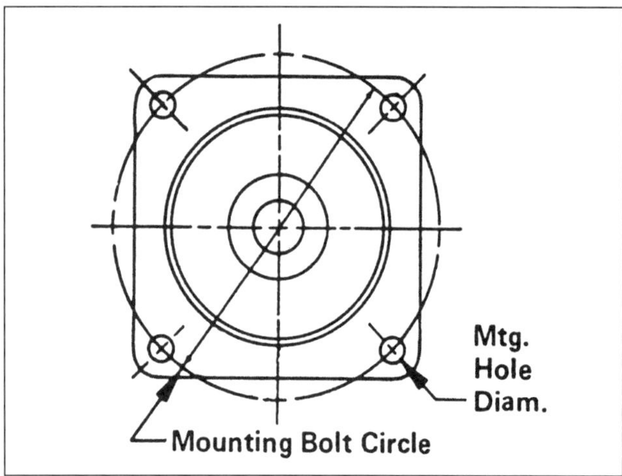

NFPA Standard
Four-Bolt Pump/Motor Mounting Flanges

Flange Code	SAE No.	SAE hp Rating[1]	Mtg. Bolt Circle	Mtg. Hole Dia.	Pilot Dia.	Pilot Ht.
101-4	B	25	5	9/16	4	3/8
127-4	C	50	6 3/8	9/16	5	1/2
152-4	D	100	9	13/16	6	1/2
165-4	E	200	12 1/2	13/16	6 1/2	5/8
177-4	F	300	13 25/32	1 1/16	7	5/8

[1] 1750 rpm electric motor

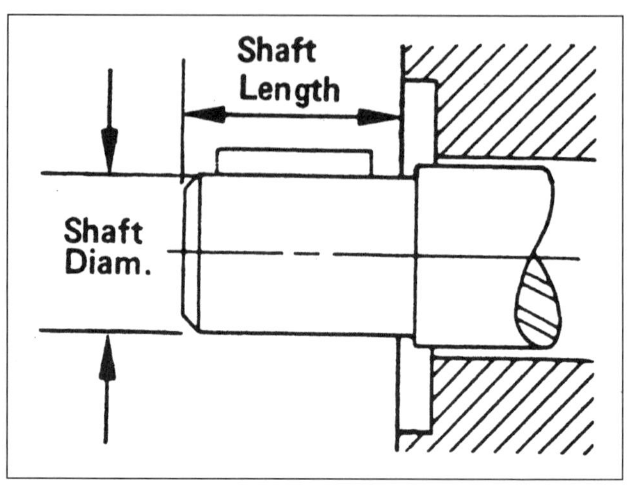

NFPA Standard
Straight Shafts Without Thread

Shaft Code	SAE Reference[2]	Shaft Dia.	Short Shaft Length	Long Shaft Length	Key Width Inches
13-1	N/A	1/2	3/4	N/A	1/8
16-1	A	5/8	15/16	2	5/32
22-1	B	7/8	1 5/16	2 1/2	1/4
25-1	N/A	1	1 1/2	2 3/4	1/4
32-1	C	1 1/4	1 7/8	3	5/16
38-1	N/A	1 1/2	2 1/8	3 1/4	3/8
44-1	D, E	1 3/4	2 5/8	3 5/8	7/16

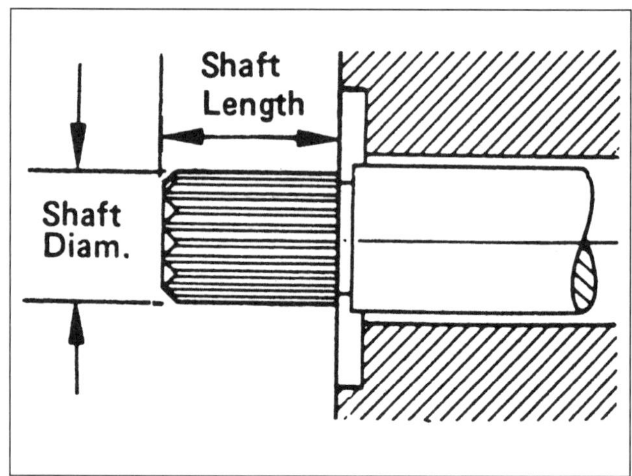

NFPA Standard
30-Degree Involute Spline Shafts

Shaft Code	SAE No.	Shaft Diam.	Shaft Length	Spline Specifications
13-4	N/A	1/2	3/4	9T, 20/40 DP
16-4	A	5/8	15/16	9T, 16/32 DP
22-4	B	7/8	1 5/16	13T, 16/32 DP
25-4	N/A	1	1 1/2	15T, 16/32 DP
32-4	C	1 1/4	1 7/8	14T, 12/24 DP
38-4	N/A	1 1/2	2 1/8	17T, 12/24 DP
44-4	D, E	1 3/4	2 5/8	13T, 8/16 DP
50-4	N/A	2	3 1/8	15T, 8/16, DP

[2] Indicates matching SAE front flange for each shaft diameter.

Formulas For Hydraulic Motors

Motors

POWER
Output of a Hydraulic Motor $hp = \frac{T \times n}{63025}$

hp = horsepower
T = torque (lb - in)
n = speed (rpm)
63025 = Conversion ((lb-ft/min/hp) x (in/ft)/(no units)) ((33,000 x 12)/(2 x π)) where 12 is used to convert from inch to feet, (2 x π) is used to convert to angular velocity, and 33,000 is used to convert from (lb - ft)/min to hp

SPEED[1]
of a Hydraulic Motor $n = \frac{231 \times Q}{d}$

n = speed (rpm)
231 = Conversion, gal to in^3 (in^3/gal)
Q = flow rate, (gal/min)
d = displacement, motor (in^3/rev)

TORQUE[1]
of a Hydraulic Motor $T = \frac{P \times d}{2 \times \pi}$

or

$T = \frac{hp \times 63025}{n}$

T = torque (lb - in)
P = pressure (lb/in^2)
d = displacement, motor (in^3/rev)
π = 3.14159 (no units)
hp = horsepower
63025 = Conversion ((lb-ft/min/hp) x (in/ft)/(no units)) ((33,000 x 12)/(2 x π)) where 12 is used to convert from inch to feet, (2 x π) is used to convert to velocity, and 33,000 is used to convert from angular (lb - ft)/min to hp
n = speed (rpm)

FLOW[1]
Input to Motor $Q = \frac{n \times d}{231}$

Q = Flow rate, motor input (gal/min)
n = speed (rpm)
d = displacement, motor (in^3/rev)
231 = Conversion, gal to in^3 (in^3/gal)

EFFICIENCY OF A HYDRAULIC MOTOR

MECHANICAL
Mechanical Efficiency $E_M = \frac{T_{ACTUAL}}{T_{THEO}}$

E_M = Mechanical efficiency, motor (no units)
T_{THEO} = Torque theoretical (lb - in)
T_{ACTUAL} = Torque actual (lb - in)

VOLUMETRIC
Efficiency $E_V = \frac{Q_{ACT}}{Q_{THEO}}$

E_V = Volumetric efficiency, motor (no units)
Q_{ACT} = Flow rate, actual input (gpm)
Q_{THEO} = Flow rate, theoretical input (gpm)

OVERALL
Efficiency $E_O = \frac{hp_{out}}{hp_{IN}}$

E_O = Overall efficiency, motor (no units)
hp_{OUT} = Horsepower out of the motor (hp)
hp_{IN} = Horsepower into the motor (hp)

$E_O = E_V \times E_M$

E_O = Overall efficiency, motor (no units)
E_V = Volumetric efficiency, motor (no units)
E_M = Mechanical efficiency, motor (no units)

LIFE
Bearing $L_{10} = L_{MFG} \times \frac{n_R}{n} \times \left(\frac{P_R}{P}\right)^{10/3}$

L_{10} = Bearing life (rev, or hrs)
The bearing life associated with 90% reliability
L_{MFG} = Manufacturers rated life (rev, or hrs)
n_R = Speed, rated (rpm)
P_R = Pressure, rated (psi)
n = Speed, working (rpm)
P = Pressure, working (psi)

DISPLACEMENT
of a Hydraulic Motor $d = \frac{T \times 2 \times \pi}{P}$

d = displacement, motor (in^3/rev)
T = torque (lb - in)
π = 3.14159 (no units)
P = Pressure, working (psi)

[1] The formulas provided assume 100% efficiency. Contact manufacturer for actual performance data.

Hydraulic Motor Torque in Pound-Inch (Theoretical)[1]

	Horsepower													
rpm	1	2	3	4	5	6	7	8	9	10	20	30	40	50
1	63025	126050	189075	252100	315125	378150	441175	504200	567225	630250	105042[2]	157563[2]	210083[2]	262604[2]
2	31513	63025	94538	126050	157563	189075	220588	252100	283613	315125	630250	945375	105042[2]	131302[2]
3	21008	42017	63025	84033	105042	126050	147058	168067	189075	210083	420167	630250	840333	87535[2]
4	15756	31513	47269	63025	78781	94538	110294	126050	141806	157563	315125	472688	630250	787813
5	12605	25210	37815	50420	63025	75630	88235	100840	113445	126050	252100	378150	504200	630250
6	10504	21008	31513	42017	52521	63025	73529	84033	94538	105042	210083	315125	420167	525208
7	9004	18007	27011	36014	45018	54021	63025	72029	81032	90036	180071	270107	360143	450179
8	7878	15756	23634	31513	39391	47269	55147	63025	70903	78781	157563	236344	315125	393906
9	7003	14006	21008	28011	35014	42017	49019	56022	63025	70028	140056	210083	280111	350139
10	6303	12605	18908	25210	31513	37815	44118	50420	56723	63025	126050	189075	252100	315125
15	4202	8403	12605	16807	21008	25210	29412	33613	37815	42017	84033	126050	168067	210083
20	3151	6303	9454	12605	15756	18908	22059	25210	28361	31513	63025	94538	126050	157563
25	2521	5042	7563	10084	12605	15126	17647	20168	22689	25210	50420	75630	100840	126050
30	2101	4202	6303	8403	10504	12605	14706	16807	18908	21008	42017	63025	84033	105042
40	1576	3151	4727	6303	7878	9454	11029	12605	14181	15756	31513	47269	63025	78781
50	1261	2521	3782	5042	6303	7563	8824	10084	11345	12605	25210	37815	50420	63025
100	630	1261	1891	2521	3151	3782	4412	5042	5672	6303	12605	18908	25210	31513
150	420	840	1261	1681	2101	2521	2941	3361	3782	4202	8403	12605	16807	21008
200	315	630	945	1261	1576	1891	2206	2521	2836	3151	6303	9454	12605	15756
300	210	420	630	840	1050	1261	1471	1681	1891	2101	4202	6303	8403	10504
400	158	315	473	630	788	945	1103	1261	1418	1576	3151	4727	6303	7878
500	126	252	378	504	630	756	882	1008	1134	1261	2521	3782	5042	6303

[1] This chart has not taken into account operating efficiencies of either the electric or hydraulic motor. Consult the manufacturer for these values and adjust the chart values for your application.
[2] Torque value is in foot-lbs.

Torque Conversion

lb-in	N-m	lb-in	N-m	lb-in	N-m	lb-in	N-m	lb-in	N-m
1	0.11	10	1.15	100	11.51	1000	115.12	10000	1151.27
2	0.23	20	2.30	200	23.02	2000	230.25	2000	2302.55
3	0.34	30	3.45	300	34.53	3000	345.38	30000	3453.83
4	0.46	40	4.60	400	60.51	4000	460.51	40000	4605.11
5	0.57	50	5.75	500	57.56	5000	575.63	50000	5756.39
6	0.69	60	6.90	600	69.07	6000	690.76	60000	6907.66
7	0.80	70	8.05	700	80.58	7000	805.89	70000	8058.94
8	0.92	80	9.21	800	92.10	8000	921.02	80000	9210.22
9	1.03	90	10.36	900	103.61	9000	1036.15	90000	10361.5

Pump/Motor Shafts & Flanges

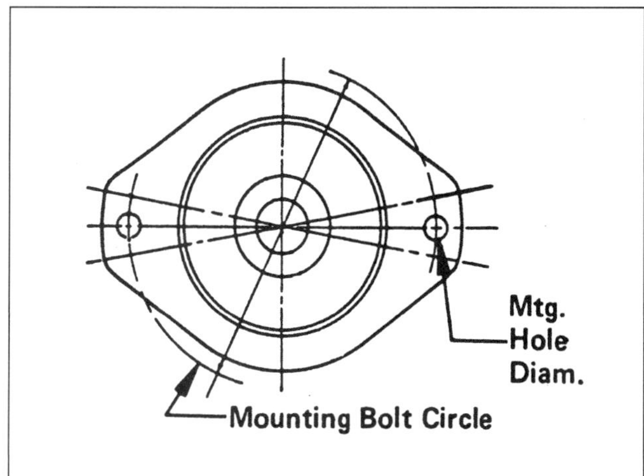

NFPA Standard
Two-Bolt Pump/Motor Mounting Flanges

Flange Code	SAE No.	SAE hp Rating[1]	Mtg. Bolt Circle	Mtg. Hole Dia.	Pilot Dia.	Pilot Ht.
50-2	N/A	N/A	$3\frac{1}{4}$	$\frac{13}{32}$	2	$\frac{1}{4}$
82-2	A	10	$4\frac{3}{16}$	$\frac{7}{16}$	$3\frac{1}{4}$	$\frac{1}{4}$
101-2	B	25	$5\frac{3}{4}$	$\frac{9}{16}$	4	$\frac{3}{8}$
127-2	C	50	$7\frac{1}{8}$	$\frac{11}{16}$	5	$\frac{1}{2}$
152-2	D	100	9	$\frac{13}{16}$	6	$\frac{1}{2}$
165-2	E	200	$12\frac{1}{2}$	$1\frac{1}{16}$	$6\frac{1}{2}$	$\frac{5}{8}$
177-2	F	300	$13\frac{25}{32}$	$1\frac{1}{16}$	7	$\frac{5}{8}$

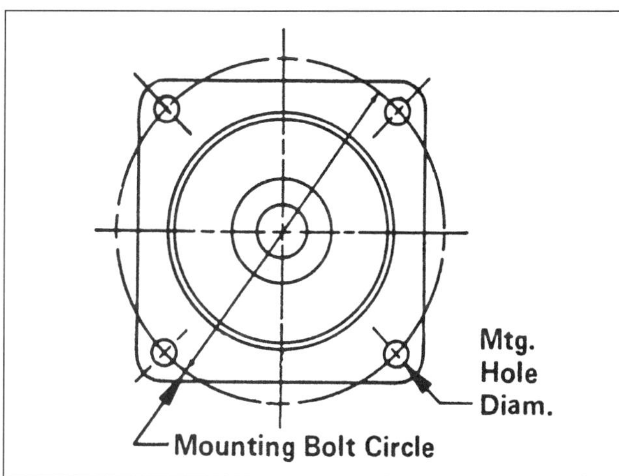

NFPA Standard
Four-Bolt Pump/Motor Mounting Flanges

Flange Code	SAE No.	SAE hp Rating[1]	Mtg. Bolt Circle	Mtg. Hole Dia.	Pilot Dia.	Pilot Ht.
101-4	B	25	5	$\frac{9}{16}$	4	$\frac{3}{8}$
127-4	C	50	$6\frac{3}{8}$	$\frac{9}{16}$	5	$\frac{1}{2}$
152-4	D	100	9	$\frac{13}{16}$	6	$\frac{1}{2}$
165-4	E	200	$12\frac{1}{2}$	$\frac{13}{16}$	$6\frac{1}{2}$	$\frac{5}{8}$
177-4	F	300	$13\frac{25}{32}$	$1\frac{1}{16}$	7	$\frac{5}{8}$

[1] 1750 rpm electric motor

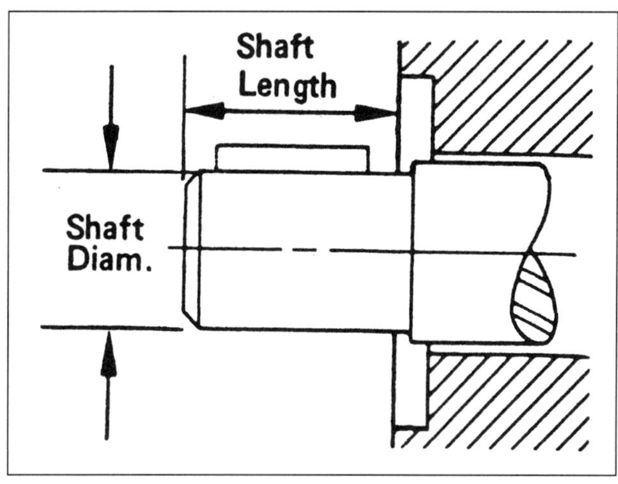

NFPA Standard
Straight Shafts Without Thread

Shaft Code	SAE Reference[2]	Shaft Dia.	Short Shaft Length	Long Shaft Length	Key Width Inches
13-1	N/A	$\frac{1}{2}$	$\frac{3}{4}$	N/A	$\frac{1}{8}$
16-1	A	$\frac{5}{8}$	$\frac{15}{16}$	2	$\frac{5}{32}$
22-1	B	$\frac{7}{8}$	$1\frac{5}{16}$	$2\frac{1}{2}$	$\frac{1}{4}$
25-1	N/A	1	$1\frac{1}{2}$	$2\frac{3}{4}$	$\frac{1}{4}$
32-1	C	$1\frac{1}{4}$	$1\frac{7}{8}$	3	$\frac{5}{16}$
38-1	N/A	$1\frac{1}{2}$	$2\frac{1}{8}$	$3\frac{1}{4}$	$\frac{3}{8}$
44-1	D, E	$1\frac{3}{4}$	$2\frac{5}{8}$	$3\frac{5}{8}$	$\frac{7}{16}$

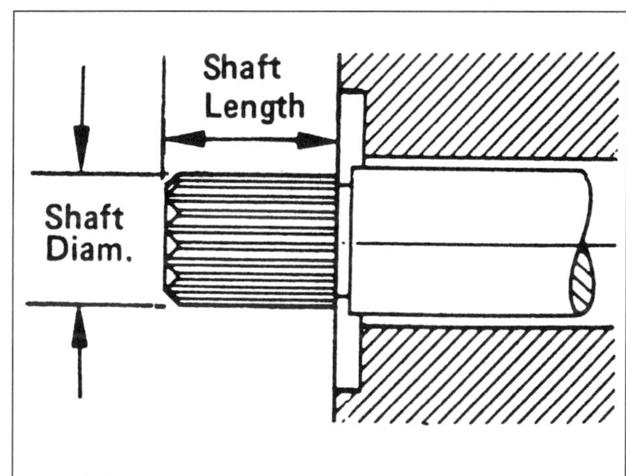

NFPA Standard
30-Degree Involute Spline Shafts

Shaft Code	SAE No.	Shaft Diam.	Shaft Length	Spline Specifications
13-4	N/A	$\frac{1}{2}$	$\frac{3}{4}$	9T, 20/40 DP
16-4	A	$\frac{5}{8}$	$\frac{15}{16}$	9T, 16/32 DP
22-4	B	$\frac{7}{8}$	$1\frac{5}{16}$	13T, 16/32 DP
25-4	N/A	1	$1\frac{1}{2}$	15T, 16/32 DP
32-4	C	$1\frac{1}{4}$	$1\frac{7}{8}$	14T, 12/24 DP
38-4	N/A	$1\frac{1}{2}$	$2\frac{1}{8}$	17T, 12/24 DP
44-4	D, E	$1\frac{3}{4}$	$2\frac{5}{8}$	13T, 8/16 DP
50-4	N/A	2	$3\frac{1}{8}$	15T, 8/16, DP

[2] Indicates matching SAE front flange for each shaft diameter.

Formulas For Vehicle Drives

VEHICLE DRIVE

To determine the size of the hydraulic motor for a Vehicle Drive System use the following steps:

(1) Determine the Total Tractive Effort
(2) Calculate the hydraulic motor speed
(3) Calculate the hydraulic motor torque
(4) Calculate the wheel slip and compare to the motor torque
(5) Calculate the radial load on the hydraulic motor
(6) Calculate the shock and brake loading on the hydraulic motor
(7) Select a hydraulic motor that meets the calculated requirements

TOTAL TRACTIVE EFFORT $TTE = RR + AF + GF + AR + DP$

TTE = Total tractive effort (lb)
RR = Rolling resistance[1] (lb)
AF = Acceleration force (lb)
GF = Grade resistance force (lb)
AR = Air resistance force (lb)
DP = Drawbar pull (lb)

[1] Force required to maintain steady vehicle motion on level ground

RR, ROLLING RESISTANCE $RR = \frac{GVW}{1{,}000} \times \text{Road Resistance Factor}$

RR = Rolling resistance (lb)
GVW = Gross vehicle weight (lb)
1,000 = Conversion from GVW to GVW/1,000

Road Resistance Factor

Concrete	10 to 20	Snow	25 to 37
Asphalt	12 to 22	Dirt	25 to 37
Mud	37 to 150	Sand	60 to 300

AF, ACCELERATION FORCE $AF = \frac{V_\Delta \times W}{G \times t}$

AF = Acceleration force (lb)
V_Δ = Change in velocity (ft/sec) (Final velocity - Initial velocity)
W = Weight (lb)
G = Acceleration due to gravity (ft/sec^2); 32.1725 used to convert weight (lbs) to mass (slugs)
t = Time (sec)

GF, GRADE RESISTANCE FORCE $GF = GR \times GVW$

GF = Grade resistance force (lb)
GR = Grade resistance (no units)
Grade resistance is grade expressed as a decimal value, i.e., 10 ft rise in 100 ft = 10% grade = .10 grade resistance
GVW = Gross vehicle weight (lb)

AR, AIR RESISTANCE FORCE $AR = FA \times .0025 \times VS^2$

AR = Air resistance force
FA = Vehicle front area (ft^2)
.0025 = Frictional factor (hr/{mile x ft^2})
VS = Vehicle speed (MPH)

Air resistance will be important only on fast moving vehicles (over 20-30 mph)

DRAWBAR PULL $DP = RR + AF + GF + AR$
or
DP = Designer requirement

DP = Drawbar pull
RR = Rolling resistance, trailer (lb)
AF = Acceleration force, trailer (lb)
GF = Grade resistance force, trailer (lb)
AR = Air resistance force, trailer (lb)

The drawbar pull is calculated with the same formulas as the Vehicle Total Tractive Effort except the Frontal Area used in the Air Resistance Force calculation for a trailer is the frontal area that is in excess of the vehicle frontal area only.

GRADE $\text{grade} = \frac{\text{rise}}{\text{run}}$

grade = Height increase over a distance expressed in %
rise = Height increase (or decrease)
run = Distance over which the height increases (decreases) (ft)

A 10% grade has a rise of 10 ft in a run distance of 100 ft $10\% = \frac{\text{rise}}{\text{run}}$

WHEEL SLIP
Or Loss of Traction $WS = \frac{W \times f \times R}{N \times M_G}$

WS = Wheel slip torque (lb-in)
W = Loaded vehicle weight over driven wheels (lb)
f = Coefficient of friction (no units)
f for rubber tire on:

Dirt	0.5
Hard surface	0.6-0.8
Cement	0.7

R = Wheel radius (in) See chart on page 17
N = Number of driving motors (no units)
M_G = Gear ratio (if required) (no units). Use 1 if no ratio

Should the torque required to slip the wheel (WS) be less than the torque ("T") calulated in the hydraulic motor, then the performance requirements cannot be met. However, it may be desirable that the wheels do slip when excessive loads are encountered. This may prevent hydraulic system overload if the vehicle becomes stalled. In this case "WS" should be slightly higher than "T".

POWER $HP = \frac{T \times n\ FS}{63025}$

HP = Horsepower (hp)
T = Torque (lb-in)
n = Speed (rpm)
FS = Factor of safety (no units)
In this case the factor of safety is to compensate for uncertainties in the system to ensure that the prime mover is capable of providing sufficient power to move the vehicle
63025 = Conversion (lb-ft/min/hp) x (in/ft)/(no units)) ((33,000 x 12)/(2 x π)) where 12 is used to convert from inches to feet, (2 x π) is used to convert to angular velocity and 33,000 is used to convert from (lb-ft)/min to HP

Formulas For Vehicle Drives

VEHICLE DRIVE

MOTOR SPEED

$$n = \frac{168 \times VS \times M_G}{R}$$

n = Hydraulic motor speed (rpm)
168 = Conversion factor (in/mile)/(min/hr)
$168 = 63360/(60 \times 2\pi)$ where 63360 converts from miles to inches, 60 converts from hours to min, and 2π converts from linear motion to angular rotation
VS = Vehicle speed (mph)
M_G = Gear ratio (if required) (no units) (use 1 if no ratio)
R = Rolling radius (Refer to chart below)

Rolling Radius

Tire Size	Ply Rating	Rolling Radius (in)	Tire Size	Ply Rating	Rolling Radius (in)
6.00 x 16	6	13.6	9.00 x 20	10 & 12	19.10
6.50 x 16	6	14.2	10.00 x 20	12 & 14	19.65
6.50 x 20	6 & 8	16.55	10.00 x 22	12	20.60
7.00 x 15	6 & 8	14.3	11.00 x 20	12 & 14	20.25
7.00 x 16	6 & 8	14.3	11.00 x 22	12 & 14	21.30
7.00 x 17	6 & 8	16.00	12.00 x 12	14	21.80
7.00 x 20	8	16.55	12.00 x 24	14	22.90
7.50 x 16	6	14.70	13.00 x 20	16	21.70
7.50 x 17	6 & 8	16.00	13.00 x 24	16	33.60
7.50 x 20	8 & 10	17.70	14.00 x 20	16	22.90
8.25 x 20	10 & 12	18.30	14.00 x 24	16	25.00

MOMENTUM

$$M = W \times VS \times \frac{1.467}{G}$$

M = Momentum (lb/sec)
W = Weight (lb)
VS = Vehicle speed (mph)
1.467 = Conversion (ft/sec)(mph)
1.467 = 5280/3600, where 5280 converts miles to feet and 1/3600 converts hours to seconds
G = Acceleration due to gravity (ft/sec_2)
32.1725 used to convert weight (lbs) to mass (slugs)

TORQUE

$$T = \frac{TTE \times R}{N \times M_G}$$

T = Torque at the axle (lb-in)
TTE = Total tractive effort (lb)
R = Wheel radius (in) See chart titled "Rolling Radius" in previous column
N = Number of driving motors (no units)
M_G = Gear ratio (if required) (no units) Use 1 if no ratio

PROPELLING FORCE

$$PF = \frac{T}{R}$$

PF = Propelling force (lb)
T = Torque (lb-in)
R = Wheel radius (in) See chart in previous column

RADIAL LOAD
On the Motor

$$RL = \sqrt{W^2 + (T/R)^2}$$

RL = Radial load on motor shaft (lb)
W = Loaded vehicle weight over driven wheels (lb)
T = Torque (in-lb)
R = Wheel radius (in) See chart above

The radial load calculated should not exceed the manufacturers rated capacity of the hydraulic motor

AXLES

$$A = 583.6 \times T \times \frac{L}{D^4 \times G}$$

A = Angle of deflection (deg)
T = Torque (lb-in)
L = Shaft length (in)
D = Shaft diameter (in)
G = Shear modulus of elasticity (psi)
(G = 12,000 psi for steel)

Axles and drive shafts must have a diameter large enough to transmit the torque without excessive deflection. The angle of deflection for a solid round shaft may be calculated with the axle formula. NOTE: Some authorities say that a steel shaft should be limited to an angular deflection of 0.08 degrees per foot of length to avoid failure.

TRAVEL SPEED

$$VS = WR \times DW/336$$

VS = Vehicle speed (mph)
WR = Wheel rpm (rpm)
DW = 2 x rolling radius (in)
336 = Conversion (min x mile)/(hr x inch)
$336 = 63360/(\pi \times 60)$ where π is 3.1415926, 60 is the conversion from min to hrs, and 1/63360 is the conversion from inches to miles

SHOCK

Shock loading should be considered when motors are used in direct wheel drive applications. A shock factor of 4 should be used for vehicles using pneumatic tires without a suspension. For applications using solid rubber tires without a suspension, a factor of 6 should be applied. The shock load should not be used in the radial calculation. The shock factor should be multiplied by the static weight on the wheel to determine a shock load on the wheel. The shock loading should not exceed twice the rated radial capacity of the motor.

BRAKING

If the hydraulic motor is to be used for braking the vehicle or if the motor can be back driven, the system pressure should be limited to 70% of the motor rating. A minimum pressure should be maintained at the inlet of the motor, or cavitation can occur during overrunning of the motor. Contact the motor manufacturer representative for detailed information.

SERVO VALVE, CUTAWAY

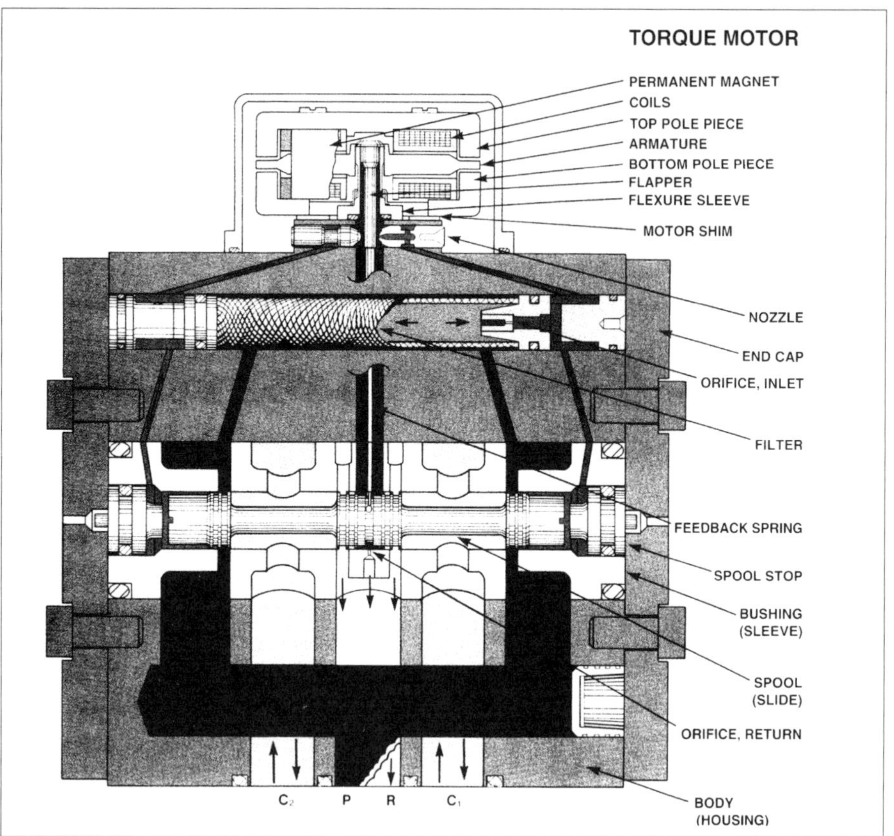

Figure 1A

PROPORTIONAL VALVE, CUTAWAY

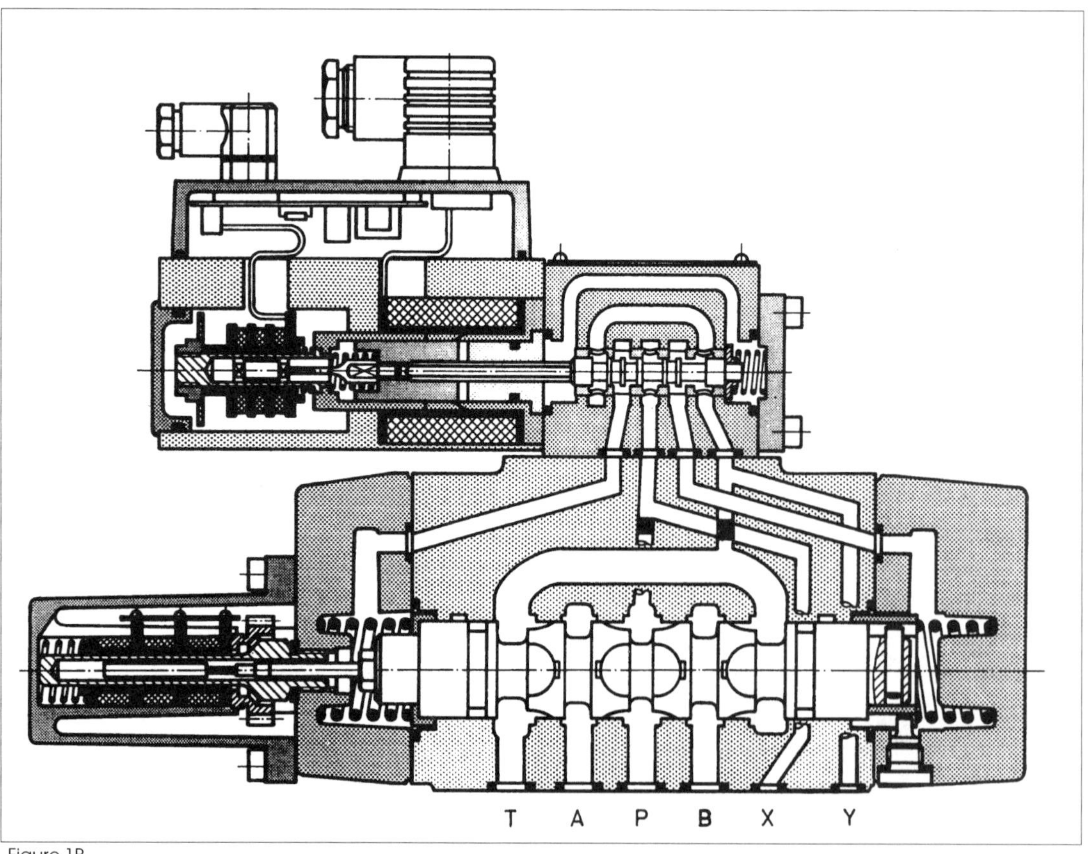

Figure 1B

SERVO VALVES VS. PROPORTIONAL VALVES
Description of and General Differences in Application of Both Type Valves

What is a Servo Valve?

An electrohydraulic servo valve produces a continuously controlled hydraulic output as a function of an electrical input signal (ref. Figure 1A). The controlled output may be flow rate, differential pressure, or a combination of the two. Servo valves are generally distinguished from other types of electrically controlled valves by these characteristics:

- Typically two or more stages
- Low-power electrical input signals
- Linear output versus input
- Negligible deadband
- High dynamic response with low phase shift
- Spool/sleeve typically zero lapped
- Sensitive to contamination

A two-stage servo valve has its output valving member positioned hydraulically by a pilot-stage valve which is connected directly to a proportional electromagnetic actuator. The pilot-stage valve acts as a hydraulic amplifier to increase substantially the actuation force of the output valving member over that obtainable directly from the electromagnetic actuator. Internal feedback between the main-stage valve and the pilot-stage valve produces an output valving member position and corresponding flow metering area proportional to the electrical input signal.

What is a Proportional Valve?

An electrohydraulic proportional valve produces a continuously controlled hydraulic output as a function of an electrical input signal (ref. Figure 1B). The controlled output may be flow rate, differential pressure, or a combination of both. Proportional valves generally are distinguishable from other types of electrically controlled valves by these characteristics:

- Typically one or two stages
- For single stage valves, the spool is directly connected to, and positioned by, a single proportional electrical solenoid
- The electrical control signal is amplified to drive the solenoid
- An LVDT[1] provides internal spool position feedback
- Deadband is common, except in high performance servo/proportional valves with negligible deadband
- Spool-sleeve overlap is common; however, recent developments of high performance servo/proportional valves have zero overlap commonly found in traditional servo valves
- Less sensitive to contamination than servo valves

Servo Valve Description

Although several different types of servo valves are used in industrial applications, the most common is the flapper/nozzle type illustrated in Figure 1A on the previous page.

This valve is a two stage servo valve in which a torque motor operates the first stage flapper. The flapper/nozzle assembly controls the pressure on either end of the main spool of the second stage causing it to move. Movement of the main spool creates a mechanical feedback movement of the flapper to the first stage (pilot stage).

Torque Motor

The torque motor consists of two permanent magnets, each surrounded by a frame (Figure 2). An armature and flapper assembly is supported between two magnetically charged frames by means of a flexure tube. The flexure tube allows the armature to move and also provides a fluid seal between the hydraulic and electrical areas of the valve. Surrounding the armature on either side of the flexure tube are two electrical coils. Passing an electrical current through the coils causes the armature to become magnetized. One end becomes a magnetic south pole; the other end becomes the north pole (Figure 3).

The armature now becomes attracted to one half of each frame and repelled by the other, creating movement (or torque) of the armature/flapper assembly. The magnitude of the torque is determined by the coil current; the current direction determines whether a clockwise or counterclockwise torque is created.

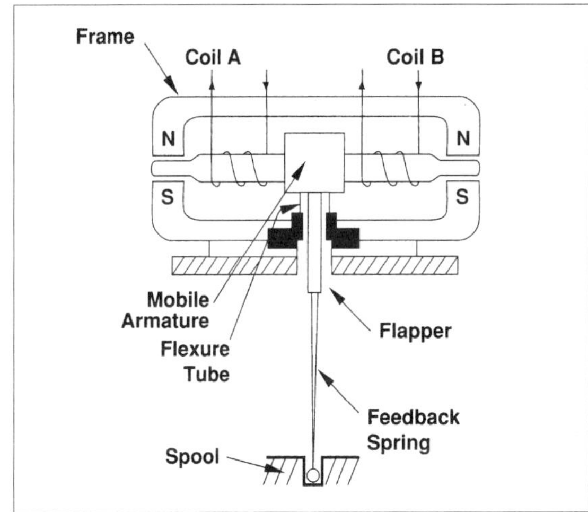

Figure 2 — Torque Motor

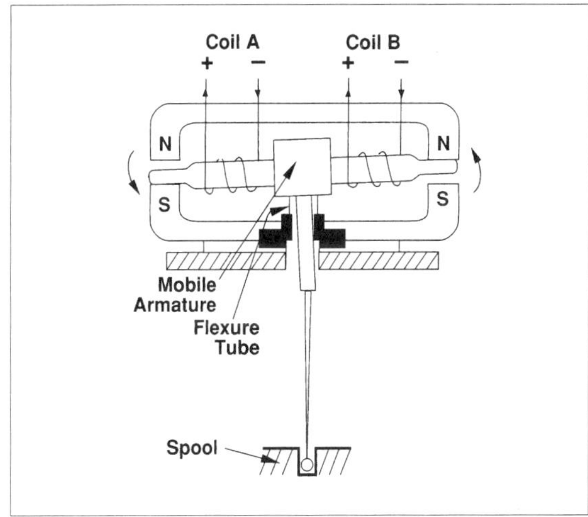

Figure 3 — Torque Motor Energized

Nozzle and Flapper Assembly

Attached to the center of the armature is a flapper and feedback spring assembly that extends down through the flexure tube. A ball on the end of the feedback spring is located in a recess in the main-stage spool. On each side of the flapper are two matched nozzles supplied with pressurized fluid (Figure 4 - next page).

The pressure feed for each of the two nozzles is taken from the supply pressure port of the valve through an integral Last Chance Filter (LCF) and fixed orifice. The fixed orifices limit the amount of flow to the nozzles. The LCF prevents large particles from entering this sensitive part of the valve. In some cases, the pilot feed pressure for the two nozzles can

[1] LVDT: Linear Variable Differential Transformer

be supplied by a separate pilot connection or fifth port. The fluid pressure in each end-chamber of the main stage spool is determined by the flow restriction of the nozzles. This, in turn, is controlled by the flapper position between two matched nozzles.

Spool and Sleeve

With a zero electrical signal to the torque motor, the flapper will be positioned midway between each nozzle. This creates equal nozzle restriction and equal pressure in each spool end-chamber. In this situation, the main spool will be centered, since any offset of the spool would move the flapper and unbalance the pressure in the spool end-chambers.

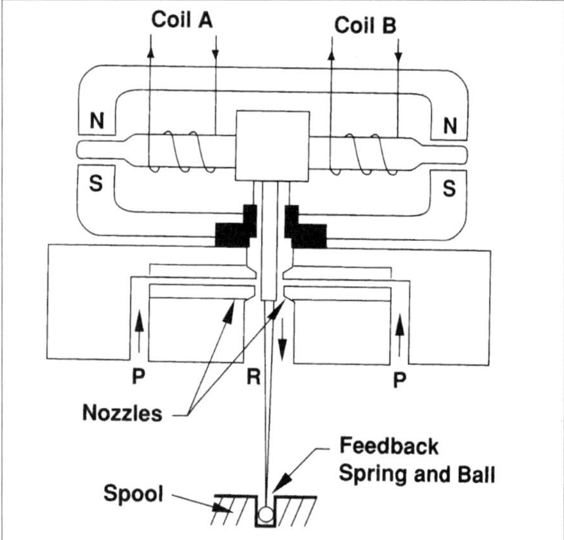

Figure 4 — Nozzle and Flapper Assembly

When an electrical signal is applied to the torque motor, the flapper moves, creating a greater restriction in one nozzle and a lesser restriction in the other. This unbalances the pressure in the spool end-chambers (Figure 5).

The unbalanced pressure now moves the main spool in the sleeve, creating a flow path through the valve. As the main spool moves across, the flapper is drawn back towards the midpoint of the two nozzles by the feedback spring. When the flapper is centered, the spool end-chamber pressures equalize (Figure 6).

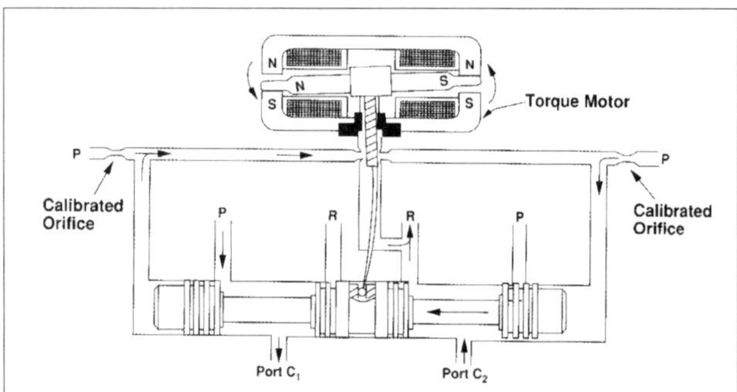

Figure 5 — Unbalanced pressure in spool end chambers

The main spool will move until the torque motor force and the mechanical spring force created on the flapper are balanced. Since the torque motor force is proportional to the coil current, the amount of spool movement will also be determined by the coil input current. Similarly, the direction of spool movement to either side of center will be determined by the direction of the coil current.

Jet Pipe Alternative to Nozzle and Flapper

The jet pipe (Figure 7) is an alternative to the nozzle and flapper first stage assembly. The torque motor moves a nozzle or a jet deflection device. Fluid flow from the nozzle impacts on a target bisected by a straight ridge. The flow on either side of the ridge controls the relative pressures in each of the two spool end-chambers. The jet pipe is at this time less commonly used than the nozzle and flapper first stage.

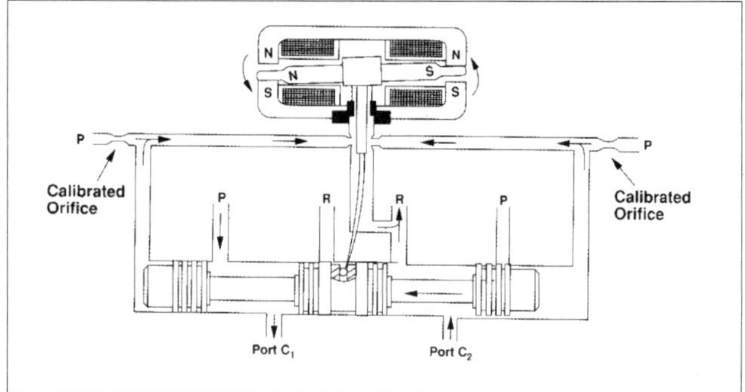

Figure 6 — Balanced pressure in spool end chambers

Null Adjustment of Spool/Sleeve Valve

In order to achieve accurate matching of the spool lands relative to the valve ports, the valve body is fitted with a sleeve incorporating precisely located flow ports. The spool slides inside this sleeve to provide flow openings which are precisely proportional to input current. By means of a null

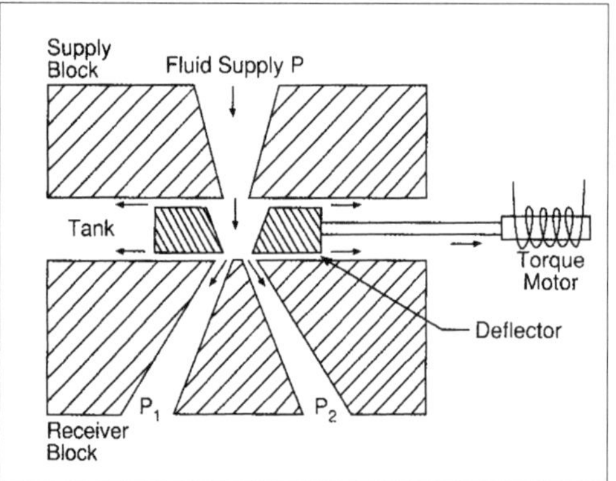

Figure 7 — Jet pipe Deflector Valve

adjustment assembly fitted in the valve body, the sleeve can be displaced slightly relative to the spool to ensure that, with zero electrical input signal applied to the valve, the spool is precisely centered in the sleeve. If proper null adjustment is impossible, this may indicate excessive spool/sleeve wear.

Proportional Valve Description

Internal Closed Loop Proportional Valves

For internal closed loop applications, single stage proportional valves use a standard directional valve body fitted with a spring-biased sliding spool and a position sensor (LVDT)(See figure 1B, page 18). To achieve accurate matching of the spool lands and valve ports, the body is fitted with a sleeve in which the ports are machined. The spool slides inside the sleeve and is positioned by means of a single proportional electrical solenoid. In most proportional valves, the spool is mechanically attached to the solenoid armature, and there is no torque motor. Current to the solenoid is provided by an electrical amplifier to position the valve spool. The spool returns to the null position under spring load with zero electrical input signal. The LVDT provides a spool position feedback signal to the valve control amplifier and allows the spool to be very accurately positioned via a closed loop. The spring-loaded spool provides a fail-safe condition with no power applied to the solenoid. The spring will push the spool fully over to provide a positive "power-off" condition (Figure 8).

Open Loop Proportional Valves

Open loop (non-feedback) proportional valves are less accurate, have a high hysteresis, and a lower response time than those with internal feedback.

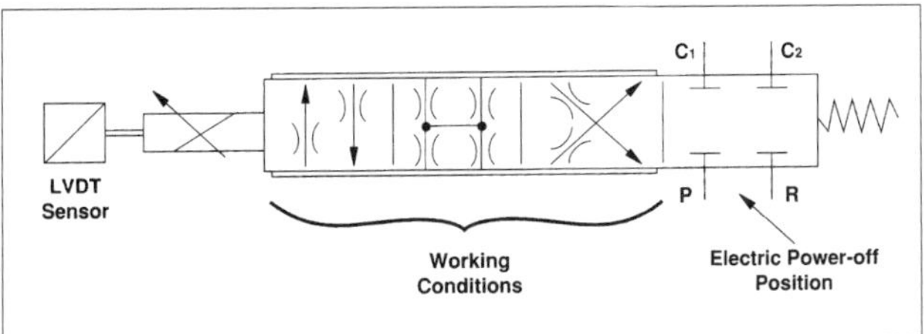

Figure 8 — Fail-safe mechanism of proportional valve

What Are the Significant Differences Between Servo and Proportional Valves?

Spool Lap Condition

Referring back to the servo valve in Figure 6, moving the main spool in the sleeve creates flow paths through the valve either P to C_1 and C_2 to R (spool movement towards the left as shown) or P to C_2 and C_1 to R (spool movement towards the right). Flow control is achieved by varying the amount of spool movement to uncover the port in the sleeve to a greater or lesser degree. An important characteristic of the valve is the relative positions of the edge of the spool land and the edge of the sleeve port in the null position. This is referred to as the spool lap condition.

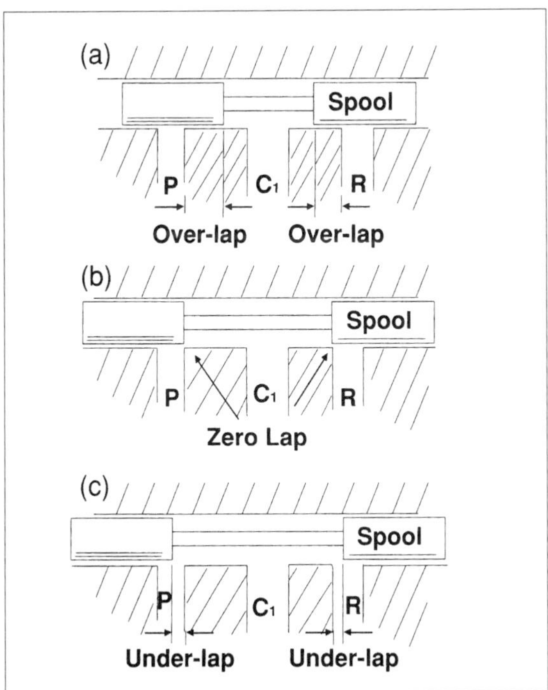

Figure 9 — Overlapped, (a), Zero lapped (b) and Underlapped valve spool (c).

Figure 9 (a) represents an overlapped spool where the spool land overlaps the edge of the port. Figure 9 (b) shows a zero-lap spool, where the edges of the spool lands align exactly with the edges of the ports. Figure 9 (c) is an underlapped spool where the edge of the spool land does not quite cover the port in the null position. The amount of overlap or underlap is normally expressed in terms of a percentage of the spool movement or as a percentage of the input signal. Typically, the proportional valves have spool overlap. An exception to this generalization is high performance proportional valves which have zero overlap. Servo valves typically have zero-lap spools.

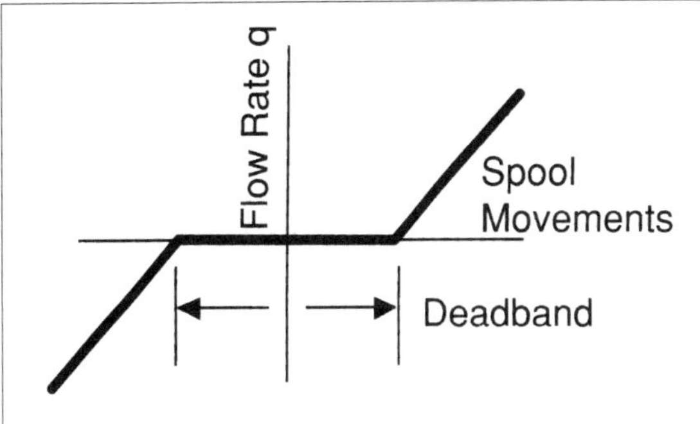

Figure 10 — Relation between flow and spool position for overlapped proportional valve

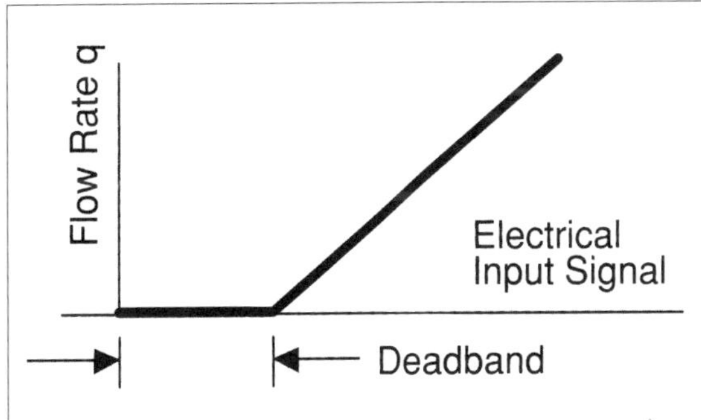

Figure 11 — Relation between flow and electrical input signal for overlapped proportional valve

Considering the characteristics of these arrangements and assuming spool movement towards the right for the overlapped spool in Figure 9 (a), there will obviously be a certain amount of deadband (zero flow) at the start of the spool movement. Flow through the valve will not occur until the edge of the spool land starts to uncover the port (Figure 10).

Once the spool has uncovered the port, the flow rate (q) will be proportional to the spool movement (s) for a given pressure differential. The relationship between spool movement and flow may not always be linear. The relationship will depend upon the shape of the ports in the valve sleeve and on whether notches are used in the spool lands. Generally, square ports and an un-notched spool will give an approximately straight line relationship between q and s.

A similar relationship will be obtained if the flow rate is related to valve input signal (Figure 11).

As a result of the spool overlap, there will be a minimum input signal (typically less than 4% of rated current) required to create flow through the valve. This means any electrical control signal less than this minimum level will not cause a flow through the valve, and the system will not be very sensitive to small input or error signals.

Lap condition is also important because the rate of internal leakage at the null position will also depend on overlap configuration. The general equation for internal leakage flow through the annular clearance as shown in Figure 12 is:

$$q = \frac{k \ \Delta P \ \varnothing \ r^3}{v \ L}$$

Where:
- q = null leakage flow through annular clearance
- k = constant
- ΔP = differential pressure
- $\varnothing$ = sleeve inside diameter
- r = radial clearance between spool & sleeve
- v = fluid viscosity
- L = overlap length

Figure 12 — Spool/sleeve clearances

If abrasive particles cause wear and increase the radial clearance, the rate of leakage will increase to the third power, rather than directly proportional to the change of clearance. For instance, if the radial clearance path is worn to twice the size, then the leakage flow through the annular clearance will increase eightfold.

From the equation listed above, it can be seen that increased overlap length reduces the amount of null leakage. This causes overlap type valves to be somewhat less sensitive to dirt. Overlap type valves are used mainly in manual controls where the zero position needs to be well defined. A good example is a variable displacement piston pump swash plate used in a vehicle hydrostatic transmission. The operator will move the joystick in order to have full forward, neutral or reverse movement of a vehicle. The deadband caused by valve overlap corresponds to the neutral transmission position.

In contrast, zero overlap valves are more sensitive to dirt because any abrasion or erosion of the spool metering edges will lead to a greater increase in valve null leakage. This will generate valve drifting and hunting, and the associated actuator will oscillate.

Spool Sleeve Configuration

A major difference between servo valves and proportional valves is the spool configuration of the two valve types. Proportional valves might be characterized as power delivery valves, whereas servo valves could be characterized as precision flow metering valves. Proportional valve spools are designed in such a manner that when fully open, the valve pressure drop is only 5 to 10% of system supply pressure. Output spools for servo valves are designed for a fully open pressure drop of 33% of system supply pressure. The higher pressure drop provides precision flow metering characteristics.

Another difference is that low performance proportional valves usually have a flow deadband of about 20 to 30% of spool stroke. Servo valves and high performance proportional valves have negligible flow deadband. While a wide flow deadband ensures a reasonable positive null position and is desirable for open loop/manual control, it is highly undesirable for closed loop control. Deadband has pure phase lag characteristics, and its presence in a control loop makes compensation difficult in situations requiring high gain with stability.

In all fairness to low performance proportional valves, it should be mentioned that they can be used in closed loop control systems if non-linear electronic controller techniques are also used. For the most part, however, closed loop control systems using proportional valves are limited to those requirements with low to modest dynamic response frequency band width and modest position accuracy requirements. More demanding requirements are satisfied by using a servo or a high response proportional valve.

Another distinct difference between the valves is that a proportional valve has very low null leakage as compared with the servo valve, where the first stage torque motor requires constant flow. In addition, the spool stroke of proportional valves is large compared to that of most servo valves. Also, the electrical power input for single stage proportional valves is typically 100 times higher than that required for servo valves.

Servo and Proportional Valve Applications

Open Versus Closed Loop Configurations

Both servo and proportional valves can be used in either open loop or closed loop systems. A basic explanation of the difference between open and closed loop systems is provided in Figure 13. As an example of an open loop system we shall consider the often used analogy of an automobile on the highway. The position of the accelerator pedal controls the speed of the car. On level ground, the further the pedal is moved from its rest position, the faster the vehicle travels (Figure 13).

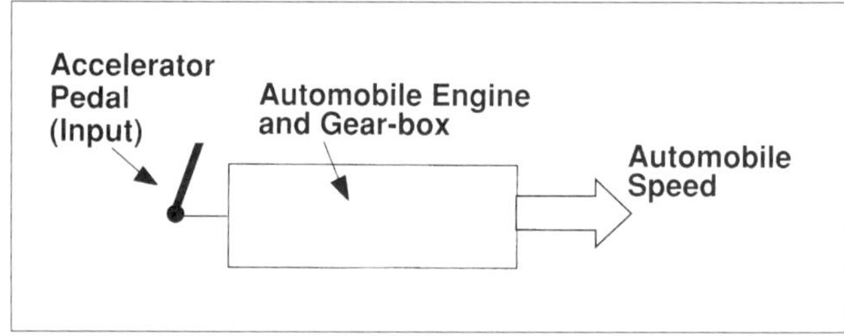

Figure 13 — Open loop automobile speed control

The relationship between pedal position and car velocity is not linear. Other factors that affect car speed include:

- The relative wind (i.e., headwind or tailwind)
- The condition of the road surface
- The condition of the engine air filter
- The condition of the fuel system
- The slope of the road
- The loaded weight of the vehicle

The alert and experienced driver can manipulate a constant speed by appropriate adjustments to the throttle pedal (Figure 14).

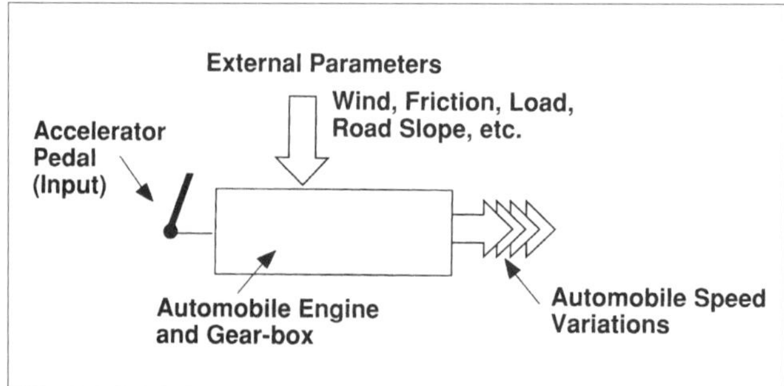

Figure 14 — Factors affecting car speed in open loop system

Open loop controls are commonplace in hydraulic systems. In many cases, the term open loop is technically a misnomer because it is the driver or operator who closes the loop. The driver/operator may, of course, be affected by such factors as the sight of a police cruiser in the rear view mirror, or, in the case of a front end loader, the proximity of the scoop boulder. As in the case of the automobile, other factors will affect motor speed, such as load, leakage, fluid temperature and system pressure.

For an automobile, the addition of a cruise control converts the car to a closed loop system. Now, after we set the desired speed on a potentiometer, output sensors provide an indication of actual speed. Actual speed as measured by the sensor is compared with the required speed to generate an error signal which changes the throttle setting in such a direction as to null itself.

As the speed drops, say because of an increasing headwind, an error signal is generated. The sign of this error signal is such to cause the throttle control to open, increasing the speed and decreasing the error signal (Figure 15).

Servo valves and high performance proportional valves are normally applied in closed loop applications. Low performance proportional valves are used in both open and closed loop applications. In some applications, servo and proportional valves are in direct competition with each other.

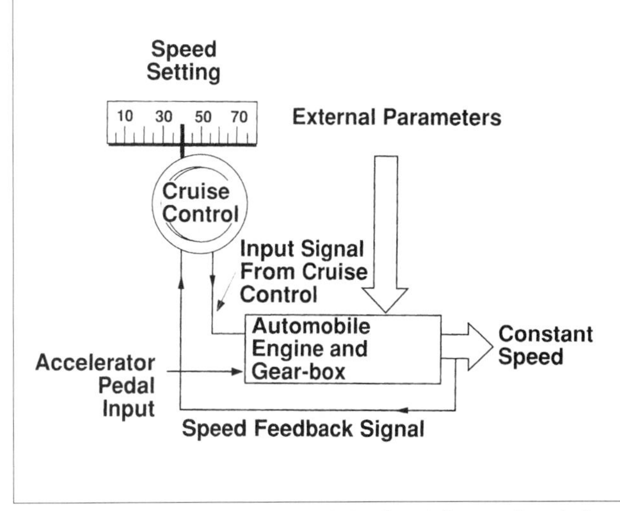

Figure 15 — Closed loop cruise control automobile speed control

VALVE FAILURE MODES AND THEIR CAUSES

The effects of contamination on servo and proportional valves have been studied by numerous investigators. Several different test techniques have been proposed; however, there is still no standard method for evaluating the sensitivity of servo or proportional valves to contamination. The primary modes of contaminant-induced failure include spool/sleeve degradation and first stage servo valve failures. The following sections describe how and why these failures occur in the presence of contamination.

Spool/Sleeve Failures

As previously discussed, all servo valves and almost all proportional valves use a spool/sleeve to control the final flow from the valve. In almost all of these valves, this spool/sleeve mechanism is the most sensitive to contamination-related failures. The spool/sleeve clearances of servo and proportional valves are very similar, and the failure modes are identical. These failure modes can be broken down into four categories:

1. Erosion
2. Silting
3. Stiction
4. Corrosion

Erosion Failures

Erosion failures are caused by particles harder than the surface of the spool or sleeve. The damage caused by particles is the result either of direct impact or of machining action. In all of these cases, the metering edges of the spool or sleeve are damaged, reducing the pressure gain, increasing null leakage, and making it more difficult for the valve to carry out its proper function.

Typical Industrial and Mobile Applications

Servo Valves and High Response Proportional Valves

- Injection molding machine shot size regulation
- Hydraulic presses
- Servo controlled variable displacement piston pumps
- Flight simulators and motion bases
- Painting, welding robots
- Some manipulators
- Automatic gage control in rolling mills
- Edge guide control in primary metal and paper plants
- Saw mills
- Precision control in plastic blow molding machines
- Electrode positioning in electric furnaces
- Steam and gas turbine electrohydraulic speed control
- Material and structures testing machines
- Metal extrusion press regulation
- Laser level control in road building and railroad maintenance equipment
- Roll stabilization for ships and trains

Proportional Valves

- Some injection molding machine models
- Some hydraulic press models
- Piston pump control with manual input
- Man lift platforms
- Fire fighting equipment
- Mobile equipment
- General machine tool positioning of fixtures
- Agricultural tractor plow position

Silting Failures

Silting between a spool and sleeve occurs when the valve is stationary and pressurized. Particles larger than the radial clearance are filtered out by the annular clearance space. As the contaminants collect, they cause an increase in break-out friction and stiction. Response time is increased; the valve can become unstable with a wide hysteresis. In severe cases, the valve can become jammed and inoperable. Increasing valve overlap increases the possibility of silting and stiction.

Stiction Failures

Stiction occurs when the valve becomes very unbalanced. The side loads cause microscopic adhesion (cold welding) to occur between the metal contact surfaces of the spool and sleeve. These severe side loads normally result from uneven silting of the annular clearance. Mild stiction causes an increase in break-out force resulting in jerky valve movement. Severe stiction can cause jamming failures in some valve designs.

Corrosion Failures

Corrosion of the spool/sleeve metering edges may be caused by water and/or chlorinated solvents in hydraulic oils. The damage to the valve is similar or that caused by erosion.

First Stage Failures

As mentioned earlier, servo valve designs now on the market include a first stage which drives the second stage spool/sleeve. When servo valves were first introduced, first stage contaminant-related catastrophic failures were often discussed and were considered common. As a result, most of the valves now on the market include an integral Last Chance Filter (LCF) chosen to prevent inlet contamination from reaching and causing catastrophic failures within the first stage. Thus, first stage failures which do occur are typically caused by a plugged LCF or erosive wear of a nozzle, a flapper, a jet pipe knife edge and target, a ball on end of feedback wire, etc. First stage failures are more rarely caused by built-in contamination. This built-in contamination can occur either in the original assembly by the valve manufacturer or during repairs if proper cleanliness is not observed. If there is adequate and properly maintained system filtration, the LCF should never plug and should function properly for the life of the system. Also, with proper system filtration, the erosive wear of the first stage should be minimized, if not completely eliminated. Proportional valve construction tolerates a much higher level of contamination.

How Are Failures Detected

Catastrophic failures can result in a hard-over signal in which the main stage spool is driven to its stop. In some valve applications, this type of hard-over failure can be very destructive to the equipment that the valve was intended to control. As mentioned above, the LCF installed within the valve should minimize this type of catastrophic failures.

A plugged LCF can result in sluggish valve response and poor system performance with respect to both the speed and accuracy of positioning. In severe cases of LCF plugging, the spool will become stuck in the sleeve because lack of driving force.

Any damage to the metering edges of a spool/sleeve will cause valve performance to deteriorate. Any increase in clearance of the spool/sleeve will also degrade valve performance. Such damage results in increased null leakage, hysteresis, and instability. Also, the valve will have poorer accuracy of both positioning and velocity control.

This is one of the common reasons for valve replacement and repair.

Causes of Valve Failures

Particulate Contaminants — Including Fibers

Particulate contaminants (including fibers) cause valve malfunctions and damage as a result of the size, population and/or physical properties of the particles. For example, in order for a particle to block an orifice or jam the flapper of a valve's first stage, the particle must be of the correct size to cause this type of malfunction. For the LCF that protects the first stage to become plugged, the hydraulic oil must be highly contaminated with a significant population of particles larger than the removal rating of the LCF. Typically, these types of first stage valve failures are associated with particles larger than twenty microns (20 µm).

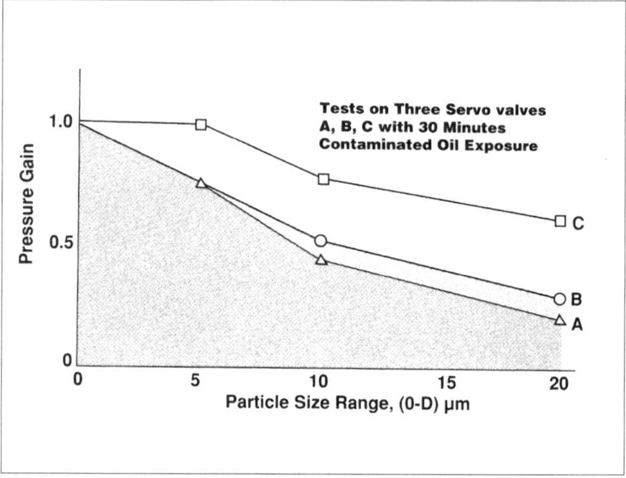

Figure 16 — Loss of pressure gain with erosion for three servo valves

Erosive wear can occur in the first stage of either flapper or jet pipe type valves. Any impingement surface will suffer from impact erosion in proportion to the number, size, velocity, and relative hardness of particles which impact that surface. This can cause wear to the control nozzles, the flapper, the jet pipe, etc. This erosive damage is most often caused by particles smaller than five microns since they are most numerous. Larger particles, if allowed to circulate, will also contribute to erosion.

In like manner, erosion of the metering edges of the valve spool will be proportional to the number, size, velocity, and relative hardness of the particles passing over the metering edges. Velocities in excess of 50 m/sec are common.

A study reported by Oklahoma State University[1], determined the deterioration in pressure gain for various servo valves (A, B, C) after thirty minutes exposure to 50 mg/l of various size ranges of contaminant. Figure 16 illustrates the high degree of deterioration after even brief exposure to contaminated fluid.

In contrast to erosion, silting of the spool/sleeve is in proportion to the number of particles of size similar to the radial valve clearance. Since the spool/sleeve clearances are typically very small, proper valve protection from silting requires fine filtration. For example, the typical servo valve has radial clearances ranging from 1-2 1/2 µm, and the typical proportional valve has clearances of 2-6 µm. The spool/sleeve is typically one of the most sensitive areas of an electrohydraulic valve and requires the finest filtration for protection against damage caused by particulate contamination. Proper filtration is needed to control particles smaller than one micron in order to minimize contaminant-related spool/sleeve silting, stiction and erosion damage.

Another study conducted by Oklahoma State University on proportional valves[2], revealed that valve hysteresis generally increases with exposure to contaminated fluid. Figure 17 shows the results of tests conducted to evaluate the contaminant sensitivity of the pilot stage of a proportional valve. It is seen that the valve is highly sensitive to dirt in the first stage. Figure 18 shows the results of tests on another valve which was highly sensitive in the second stage. In both cases, the result was wide hysteresis and a reduction in the maximum output flow.

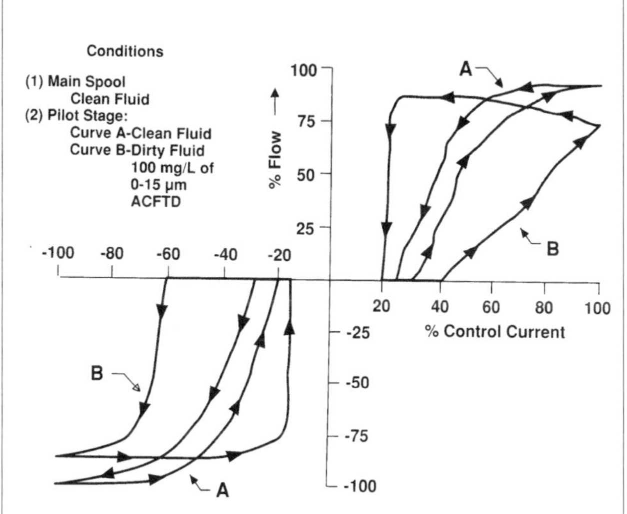

Figure 17 — Contaminant sensitivity of the pilot stage of a proportional valve

Water Contamination

Water is also a source of contamination and valve damage. Water occurs in hydraulic fluids in both free and dissolved states. Free water is the most troublesome as it can cause erosion. Also, free water collected in valves can freeze if the temperature drops below freezing. This can be particularly troublesome in equipment that must operate outdoors in cold weather. Water can also result in additives being

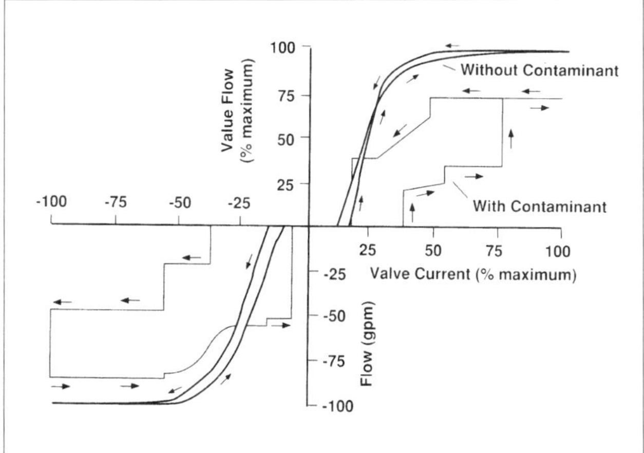

Figure 18 — Contaminant sensitivity of the second stage of a proportional valve

precipitated from some hydraulic fluids. These precipitates can contribute to valve failures as well as premature filter plugging.

Solvent Contamination

Solvent contamination can also be quite destructive in a hydraulic fluid. Chlorinated solvents, when allowed to combine with minute amounts of water, will hydrolyze to form hydrochloric acid. Acid will attack internal metallic surfaces in the system, particularly those that are ferrous, and produce a severe rust-like corrosion. There have also been cases where a hydraulic system was contaminated with a chlorinated solvent, and within hours the metering edges of the electrohydraulic valves were damaged beyond repair.

For long valve life with proper valve performance, care must be taken to maintain low levels of particulates, water and solvents.

[1] Izawa, K., et. al., **"Servovalve Contaminant Sensitivity - Parts 1-4,"** The BFPR Journal, Volume 17, No. 1, Oklahoma State University, 1983.

[2] Nair, K.S. et al., **"Electrohydraulic Proportional Flow Control Valves. How sensitive are they to contamination?"** The BFPR Journal, Volume 15, No. 1, Oklahoma State University, 1982.

Formulas For Accumulators

ISOTHERMAL EQUATIONS

VOLUME
Useable Working

$$V_u = V_1 \times \left(\frac{P_1}{P_2} - \frac{P_1}{P_3}\right)$$

V_u = Useable volume of an accumulator (in^3)
V_1 = Total accumulator capacity (in^3)
P_1 = Pre-charge pressure (psi)
(normally 85% of minimum working pressure)
P_2 = Minimum working pressure (psi)
P_3 = Maximum working pressure (psi)

VOLUME
Total Gas Capacity

$$V_1 = \frac{V_u \times P_2 \times P_3}{P_1 \times (P_3 - P_2)}$$

V_1 = Total accumulator gas capacity (in^3)
V_u = Useable volume of an accumulator (in^3)
P_2 = Minimum working pressure (psi)
P_3 = Maximum working pressure (psi)
P_1 = Pre-charge pressure (psi)
(normally 85% of minimum working pressure)

ADIABATIC EQUATIONS

VOLUME
Useable Working

$$V_u = V_1 \times \left[\left(\frac{P_1}{P_2}\right)^{\frac{1}{K}} - \left(\frac{P_1}{P_3}\right)^{\frac{1}{K}}\right]$$

V_u = Useable volume of an accumulator (in^3)
V_1 = Total accumulator capacity (in^3)
P_1 = Pre-charge pressure (psi)
(Normally 85% minimum working pressure)
P_2 = Minimum working pressure (psi)
K = Ratio of specific heats (no units)
K = 1.4 for Nitrogen
P_3 = Maximum working pressure (psi)

VOLUME
Total Gas Capacity

$$V_1 = \frac{V_u \times P_2^{1/K} \times P_3^{1/K}}{P_1^{1/K} \times (P_3^{1/K} - P_2^{1/K})}$$

V_1 = Total accumulator gas capacity (in^3)
V_u = Useable volume of an accumulator (in^3)
P_2 = Minimum working pressure (psi)
P_3 = Maximum working pressure (psi)
P_1 = Pre-charge pressure (psi)
(normally 85% of minimum working pressure)
K = Ratio of specific heats (no units)
K = 1.4 for Nitrogen

POLYTROPIC EQUATIONS

VOLUME
Useable Working

$$V_u = \frac{V_1 \times (P_1/P_2) \times (.95) \times ((P_3/P_2)^{1/n} - 1)}{(P_3/P_2)^{1/f}}$$

V_u = Useable volume of an accumulator (in^3)
V_1 = Total accumulator gas capacity (in^3)
P_1 = Pre-charge pressure (psi)
(normally 85% of minimum working pressure)
P_2 = Minimum working pressure (psi)
P_3 = Maximum working pressure (psi)
n = Discharge coefficient (no units)
f = Charge coefficient (no units)

VOLUME
Total Gas Capacity

$$V_1 = \frac{V_u \times (P_3/P_2)^{1/f}}{(P_1/P_2) \times (.95) \times ((P_3/P_2)^{1/n} - 1)}$$

V_1 = Total accumulator gas capacity (in^3)
V_u = Useable volume of an accumulator (in^3)
P_3 = Maximum working pressure (psi)
P_2 = Minimum working pressure (psi)
f = Charge coefficient (no units)
P_1 = Pre-charge pressure (psi)
(normally 85% of minimum working pressure)
n = Discharge coefficient (no units)

ISOTHERMAL EQUATIONS With Compressibility Factor, "Z"

VOLUME
Useable Working

$$V_u = V_1 \times \left(\frac{P_1 \times Z_2}{P_2 \times Z_1} - \frac{P_1 \times Z_3}{P_3 \times Z_1}\right)$$

V_u = Useable volume of an accumulator (in^3)
V_1 = Total accumulator capacity (in^3)
P_1 = Pre-charge pressure (psi)
(normally 85% of minimum working pressure)
Z = Coefficient of compressibility (no units)
$Z_1 = 1 + .000886 \times ((P_1 - 14.7)/477.8)^{2.48}$
$Z_2 = 1 + .000886 \times ((P_2 - 14.7)/477.8)^{2.48}$
$Z_3 = 1 + .000886 \times ((P_3 - 14.7)/477.8)^{2.48}$
P_2 = Minimum working pressure(psi)
P_3 = Maximum working pressure (psi)

VOLUME
Total Gas Capacity

$$V_1 = \frac{V_u \times P_2 \times P_3 \times Z_1}{P_1 \times (Z_2 \times P_3 - Z_3 \times P_2)}$$

V_1 = Total accumulator capacity (in^3)
V_u = Useable volume of an accumulator (in^3)
P_2 = Minimum working pressure(psi)
P_3 = Maximum working pressure (psi)
P_1 = Pre-charge pressure (psi)
(normally 85% of minimum working pressure)
Z = Coefficient of compressibility (no units)
$Z_1 = 1 + .000886 \times ((P_1 - 14.7)/477.8)^{2.48}$
$Z_2 = 1 + .000886 \times ((P_2 - 14.7)/477.8)^{2.48}$
$Z_3 = 1 + .000886 \times ((P_3 - 14.7)/477.8)^{2.48}$

Accumulators

An accumulator is a device that stores potential energy through the compression of a gas in a container open to a relatively incompressible fluid. There are two types of accumulators commonly used today: the first is the bladder type and the second is the piston type (Refer to Figures 1 & 2).

The bladder type uses a compressible gas contained in an elastic bladder contained inside of a shell. The shell provides a pressure container for both the gas (in the bladder) and fluid, the bladder provides the barrier between the gas and the fluid to prevent intermixing. The piston type uses a cylinder with a floating piston. The cylinder provides a pressure container for both the gas and fluid while the piston provides the barrier between the gas and the fluid to prevent intermixing.

The typical uses of an accumulator are:

- **Maintain Pressure** — An accumulator can be used to maintain pressure in a system due to leakage, thermal expansion, etc. It can also be used to provide a source of emergency power during pump failures, etc.
- **Hydraulic Line Shock Damper** — An accumulator can be used to cushion the pressure spike from sudden valve closure, the pulsation from pumps or the load reaction from sudden movement of parts connected to hydraulic cylinders.
- **Energy Conservation (Reduced pump/motor size)** — An accumulator can be used to supplement a pump during peak demand thereby reducing the size of the pump required. The accumulator is charged during the low demand portions of the pump cycle time and then discharges during the high demand portions of the system.
- **Source of Power** — An accumulator can be charged during a "run" period of equipment and then used as a power source for start up, such as a starter on diesel equipment.

Adjustment

Isothermal		**Adiabatic**
Slow ◄	Discharge Rate	► Fast
Large ◄	Surface Area to Volume Ratio	► Small

Selection

Bladder Accumulators

- Faster response times for very high speed applications
- Limited sizes

Piston Accumulators

- Various sizes

Maximum Recommended Accumulator Flow Rates, gpm

Piston Bore, in	Bladder Capacity	Piston[1]	Bladder	
			Standard	Hi-flow
2	1 qt	100	60	
4	1 ga	400	150	
6	2.5 gal+	800	220	600
7		1200		
9		2000		
12		3400		

[1] Limit piston velocity to 10 ft/sec

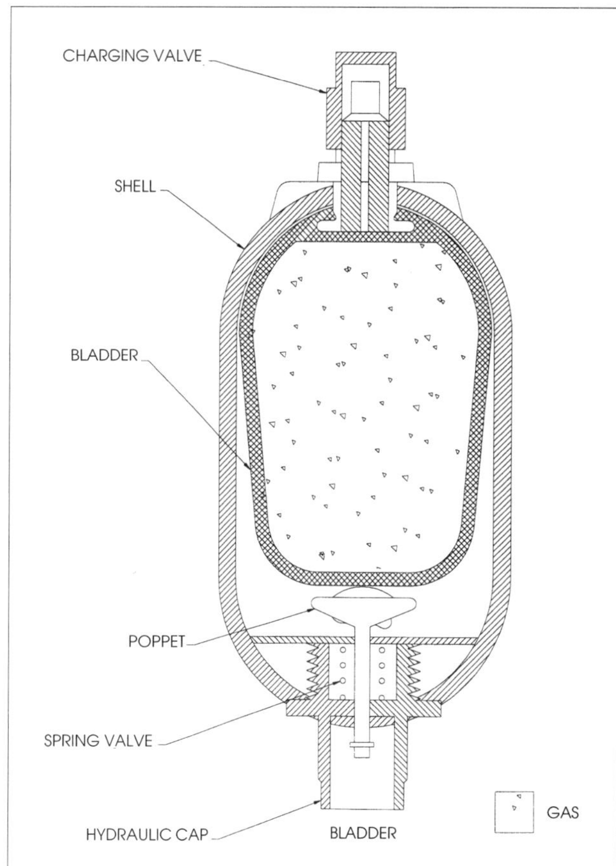

Figure 1 — Bladder accumulator

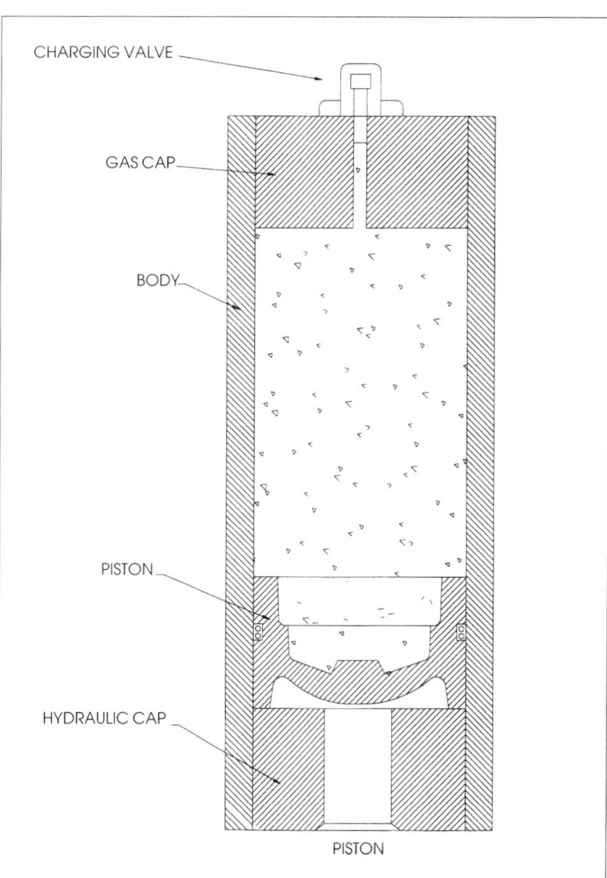

Figure 2 — Piston accumulator

Accumulator Sizing

There are four different formulas available for sizing a hydraulic accumulator. The first is an Isothermal equation which is the case where the gas is compressed or expanded at such a slow rate that the container can dissipate/collect the heat thereby maintaining the temperature of the gas as a constant. The second is an Adiabatic equation which is the case where the gas is compressed or expanded at such a fast rate that the container cannot dissipate/collect any of the heat so all the heat goes into changing the gases temperature. The third equation is a Polytropic equation with the exponents determined empirically, this equation takes into account the accumulator charge/discharge time and the average pressure and gas operating temperature. The forth equation is a modification to the idea gas law with the addition of a compressibility factor "Z" which is for practical purposes just an isothermal equation with a modification for gas compressibility. All four sets of equations are provided here for information along with a graph, and explanation of the calculation used to generate the graph on the following page, which is the recommended method of determining the accumulator size.

The graph was created by first generating the isothermal curves and then the adiabatic curves. A value that was 2/3 the way between the isothermal curve and the adiabatic curve (i.e. biased towards the adiabatic curve) was used for the graph.

Engineering judgement can be used to adjust the accumulator sizing graph for improved accuracy based on actual applications. At 0% volume, both the isothermal and adiabatic curves have the same value, at the high end of the curve (3000 psi working pressure) the values are per the chart on page 29-30.

Bladder and Diaphragm Accumulators

Bladder and Diaphragm accumulators are precharged to 80% of minimum hydraulic system pressure. Note: The first 50-100 psi of nitrogen should be introduced slowly into a bladder. Piston Accumulators are precharged to 100 psi below minimum hydraulic system pressure. The relative output of a piston accumulator is generally 10-15% higher by volume than a bladder. Due to the difference in allowable precharge.

Precharge is generally accomplished using Nitrogen (N_2), a dry inert gas.

Bladder accumulators are available up to a 40 gallon size, however, bladder accumulators are not used over 15 gallon size due to economics. Piston accumulators are available up to 200 gallon size. Using multiple components can cover most high volume requirements.

Bladder accumulators are more responsive to pressure variations than piston accumulators because of acceleration, deceleration required of the piston and the static friction of the piston seal.

A vertical mount with hydraulic port down is preferred for both piston and bladder accumulators. Mounting a piston accumulator horizontal is acceptable as long as the fluid is clean. With a bladder accumulator mounted horizontal, the bladder can see execrate wear from rubbing on top of the shell.[1]

[1] Additional information can be found in the Fluid Power Handbook & Directory, 1994-95.

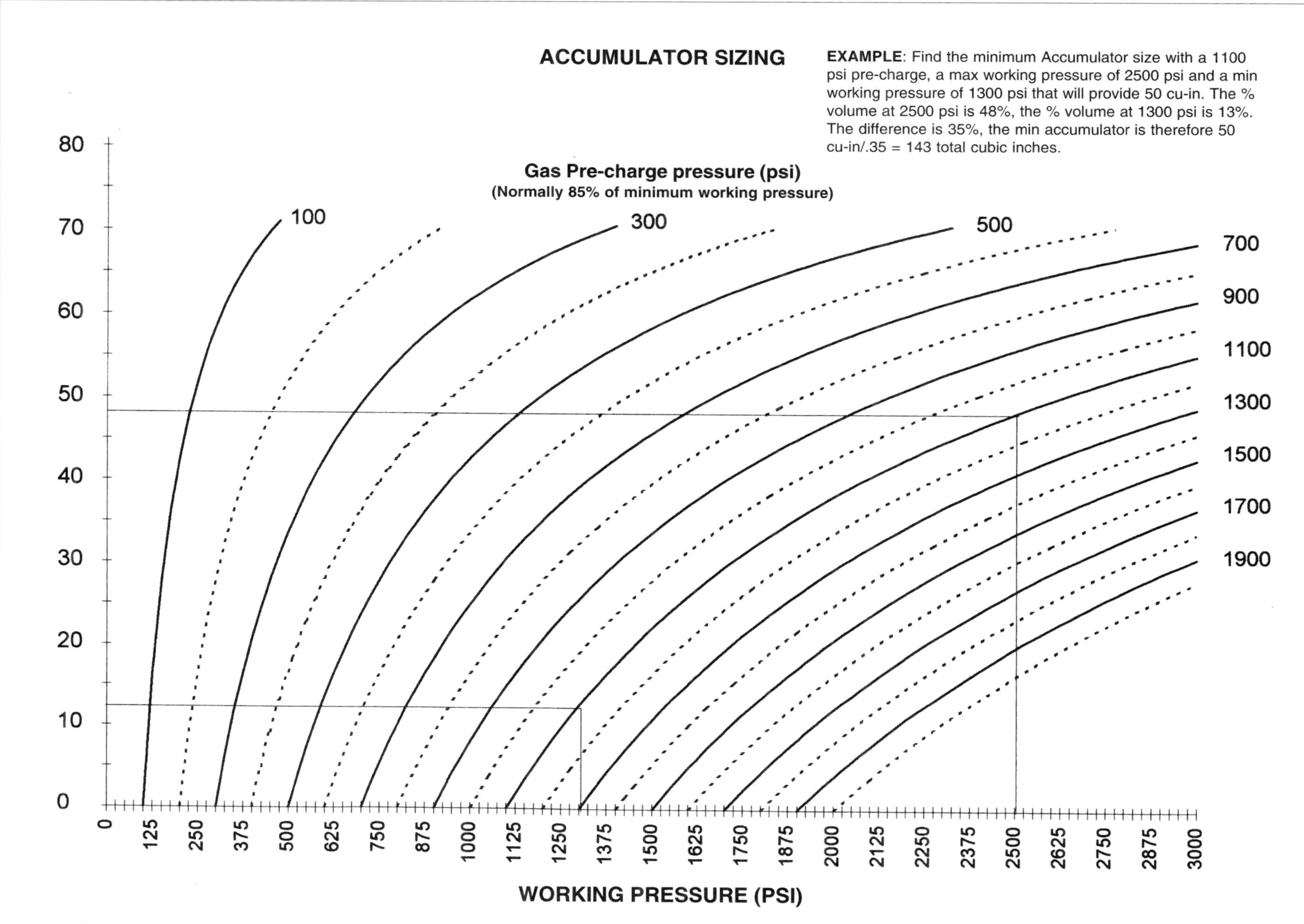
ACCUMULATOR SIZING
EXAMPLE: Find the minimum Accumulator size with a 1100 psi pre-charge, a max working pressure of 2500 psi and a min working pressure of 1300 psi that will provide 50 cu-in. The % volume at 2500 psi is 48%, the % volume at 1300 psi is 13%. The difference is 35%, the min accumulator is therefore 50 cu-in/.35 = 143 total cubic inches.
Gas Pre-charge pressure (psi)
(Normally 85% of minimum working pressure)
100
300
500
700
900
1100
1300
1500
1700
1900
80
70
60
50
40
30
20
10
0
0
125
250
375
500
625
750
875
1000
1125
1250
1375
1500
1625
1750
1875
2000
2125
2250
2375
2500
2625
2750
2875
3000
WORKING PRESSURE (PSI)

Accumulator Adjustment Chart

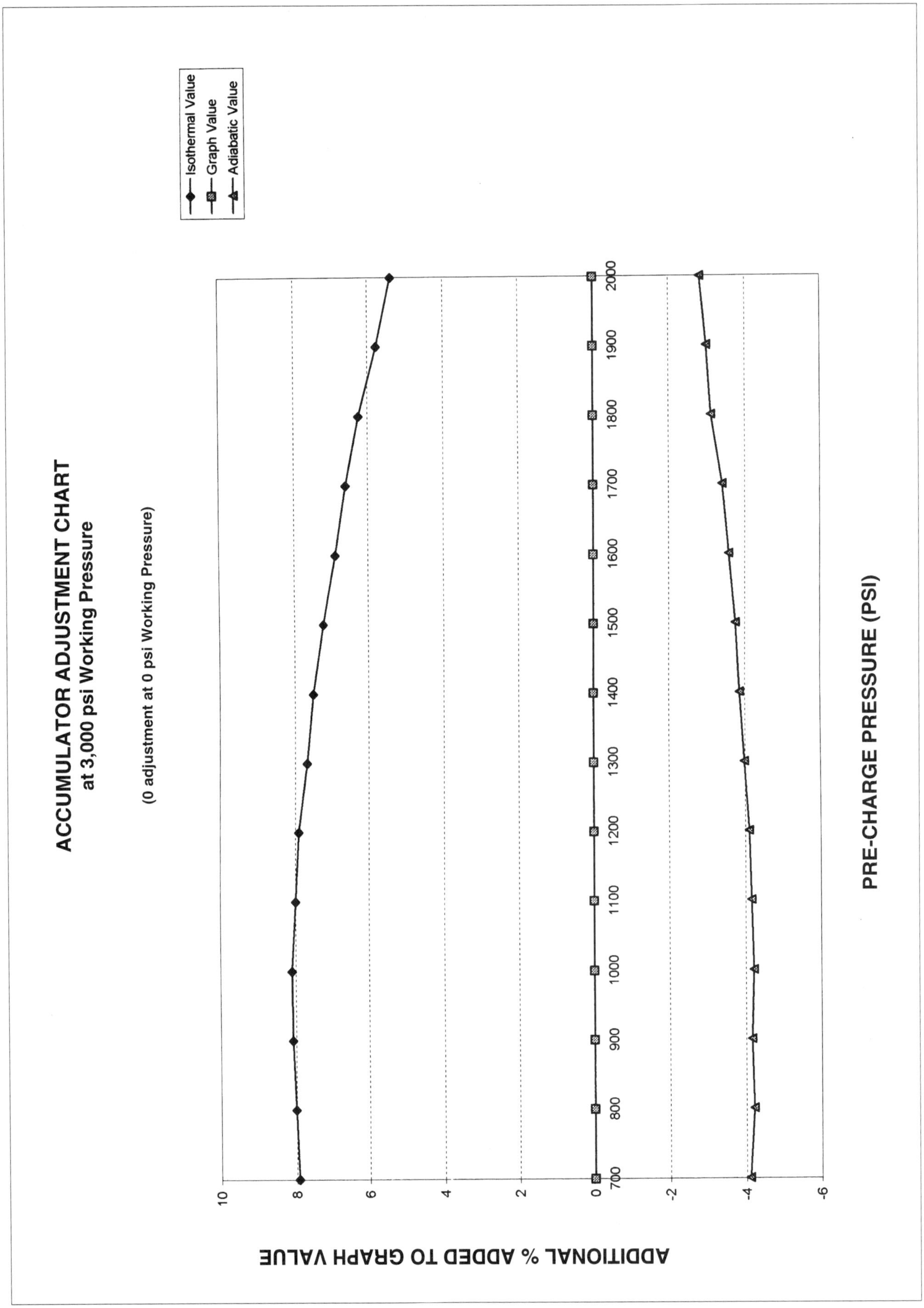

ACCUMULATOR ADJUSTMENT CHART
at 3,000 psi Working Pressure

(0 adjustment at 0 psi Working Pressure)

Isothermal Value
Graph Value
Adiabatic Value

ADDITIONAL % ADDED TO GRAPH VALUE

10
8
6
4
2
0
-2
-4
-6

700 800 900 1000 1100 1200 1300 1400 1500 1600 1700 1800 1900 2000

PRE-CHARGE PRESSURE (PSI)

Fluid Conditioning

A study by Dr. E. Rabinowicz was presented to the Society of Lubrication Engineers, which observed that 70% of component replacement (or "loss of usefulness") is due to surface degradation. In hydraulic and lubricating systems, 50% of these replacements result from mechanical wear, with 20% resulting from corrosion (water). Properly designed particulate removal (filtration) and water removal systems can eliminate a majority of these problems.

Common Results of Contaminant Control

- Decrease in premature component failures: 70% of surface degradation troubles stop
- Increase of system productivity
- Decrease in energy consumption costs: clearances stay the same contributing to higher efficiency
- General and dramatic system reliability improvements
- Reduced hazardous waste/oil disposal costs

Contaminant Sources

- Built-In Contaminants — Occur mostly from the components manufacturing processes. Built-in contaminants include weld splatter, core sand, metal chips, abrasive dust, test-stand dirt and lint.
- Introduced Contaminants — Enter the system through defective seals, open reservoirs, cylinder rods, filter tubes, top-off oil, breather caps, maintenance and defective coolers.
- System Generated Contaminants — Come from assembly, start-up, break-in and operation.
- Sludges and Acids — from reactions with water and heat.

Definitions of Filtration Terminology

ISO CODE

This quantifies your oil's overall cleanliness level with an internationally recognized standard. Adapted by the International Standards Organization (ISO) the ISO Cleanliness Code makes it quick and easy to judge the cleanliness of your fluid. Based on 1 milliliter (1 ml) sample of your oil, the ISO Code assigns a reference to the amount of 5 and 15 micron particles counted in the sample. One micron particle count will be available in the near future.

ISO Reference	# of Particles in 1 ml
10	5-10
11	10-20
12	20-40
13	40-80
14	80-160
15	160-320
16	320-640
17	640-1280
18	1280-2560
19	2560-5120

NOTE: As the Reference number increases, the number of particles doubles.

Example: ISO 17*/16/14

This number represents all 15 micron particles found in the same sample

This number represents all 5 micron particles found in a sample

* ISO will be adapting a reference code for two (2) micron particles. Currently only five (5) and fifteen (15) micron particles are referenced

For 1 ml of fluid, this example would have 640-1280 particles two (2) micron in size or greater, 320-640 particles 5 micron in size or greater (the 16 = 320-640 particles), while it would also have 80-160 particles 15 microns in size.

Definitions of Filtration Terminology

SERVICE LIFE VS. DIRT CAPACITY

Definitions of service life and dirt capacity are given below. Dirt capacity should not be used to predict filter service life due to the many variables that affect dirt capacity data.

SERVICE LIFE

Service life is the length of time that a filter will survive in an actual system before the minimum differential pressure (ΔP) is reached.

DIRT CAPACITY FROM THE MULTI-PASS TEST

(ISO 4572, ANSI/(NFPA) T3.10.8.8R1)

Test variables that affect capacity data include:

- Flow rate
- Contaminant
- Contaminant ingression rate
- Multi-pass vs. single pass
- Terminal pressure drop
- Filter integrity

Comparing dirt capacity of two elements:

- All variables above must be equal
- Elements must be of equivalent size
- Elements must be of equivalent efficiency
- Retained dirt capacity values must be compared.

Dirt capacity would appear to be an easy parameter to measure and understand; however, using dirt capacity to predict service life is quite difficult. Two different filters with the same dirt capacity will almost always have quite different service lives. For example, coarser filters with higher dirt capacities will allow more particle generation because of wear. They generally have shorter service lives than fine filters.

APPARENT DIRT CAPACITY

Apparent dirt capacity is the amount of dirt that can be added to the filter test system before the minimum differential pressure (ΔP) is reached.

RETAINED DIRT CAPACITY

Retained dirt capacity is the amount of dirt that is captured by the filter in a test system before the terminal ΔP is reached.

BETA RATIO

Developed by the International Standards Organization, the ISO 4572 Multi-Pass Efficiency Test (Beta Ratio) effectively compares the contaminant catching ability of "Filter A" to "Filter B". Contaminant (Air Cleaner Fine Test Dust or ACFTD) is put into a hydraulic test stand and per the ISO test standards, flow, pressure and ingression rates are controlled. The amount of contaminant before and after the filter is measured and a ratio is documented. The ratio of contaminant entering the element and leaving it is called the **Beta Ratio (β)**. The higher the Beta Ratio, the more dirt the element will capture on one pass.

Example: Beta (β) 6 = 1000

6 micron particles are being addressed

Ratio of particles In vs. Out. If 1 million particles pass through this filter, only 1000 would exit, one pass. If this number was 75, and 1 million particles pass through the filter, 13,333 would exit.

"Percentage of Efficiency" terminology, still used by some filter manufacturers, can be very misleading to the end-user. A filter that is 98% efficient versus a filter that is 99.2% efficient, may sound close to identical. However, the 98% filter actually allows almost 3 times more dirt downstream than the 99.2% filter.

BETA RATIO

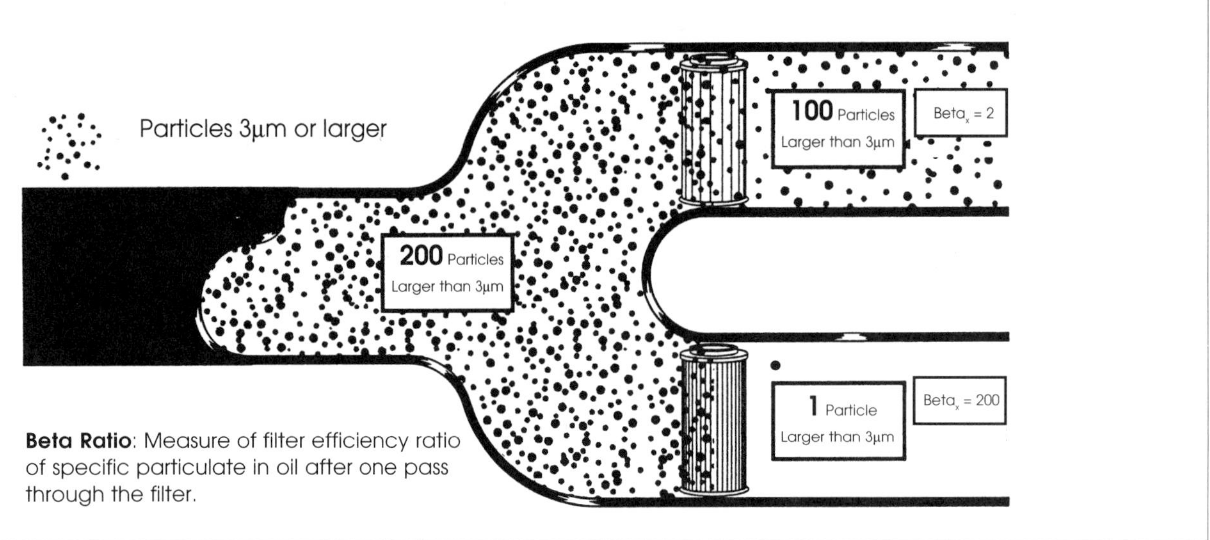

CONTAMINANT LOADING CURVE

As dirt is trapped by the filter, differential pressure (ΔP) increases. A differential pressure indicator is used to signal element change before the bypass relief valve opens. The bypass valve protects the filter and system from excessive differential pressure and/or element collapse.

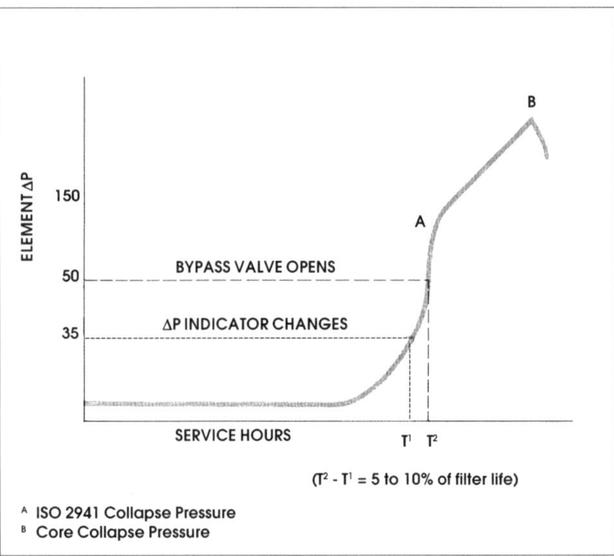

Effects of Contamination

Each microscopic particle inside the system acts as an "abrasive seed", resulting in eventual component failure. By bridging the dynamic (or running) clearances within the component (see table titled Typical Dynamic Clearances on page 33), the microscopic contaminant quickly grinds and erodes away pumps, valves, motors and cylinders. This creates more microscopic and highly abrasive contaminant, grinding away more and more of the components clearances. Component failure is inevitable. This is called the "Chain Reaction of Wear".

Abrasive Wear is the primary wear mechanism. Particles enter the clearance space between two moving surfaces, bury themselves in one of the surfaces, and act like cutting tools to remove material from the opposing surface. The particle sizes causing the most damage are those equal to and slightly larger than the clearance space. To protect opposing surfaces from abrasive wear, particles of approximately the dynamic clearance size range must be removed. This is called clearance protection filtration.

Abrasive wear effects:

- Dimensional changes
- Leakage
- Lower efficiency
- General particles = more wear

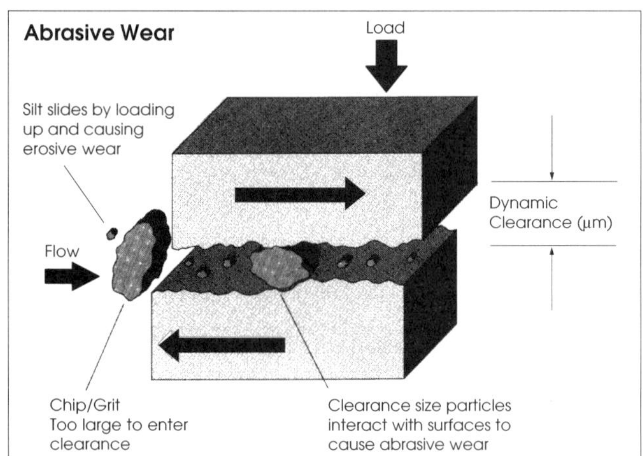

Sludge and Acids can also form within hydraulic and lubricating systems. This happens when the fluid chemically reacts to water, dirt, metal, air, heat, pressure and incompatible fluids. Sludge is not generally abrasive, but can generate heat due to loss of lubricating film. It is recognized as the gummy coating on moving parts, slowing or halting their operation. Acids corrode and pit the critical moving parts, destroying the running clearances and causing 20% of total component replacements.

By eliminating the water, dirt and heat within the system, assuming the oil has not been "burned" or improperly mixed, the oils additive package should remain consistent with new oil. The oil can last indefinitely.

Filter Selection Factors

Degree of Filtration

All hydraulic and lubrication systems should have filtration. Studies, such as those conducted by Oklahoma State University, the U.S. Navy, S.A.E., A.S.M.E., the Fluid Power Research Center and many others, conclude that the cleaner the system, the longer it will last. The degree of filtration your system should have is directly related to its operating characteristics.

In selecting filters, the following conditions need to be carefully evaluated:

- **Beta Ratio** - The ratio of contamination entering the filter vs. contamination passing through the filter (see page 31).
- **Pressure Drop** - All system components which have flow through them have pressure drop. This drop is the net pressure required for the fluid to flow from the inlet to the outlet. In all filters, this includes the pressure drop across the housing and the element. Pressure drop varies with flow rate, viscosity and specific gravity. When making a filter selection, the maximum allowable system pressure drop must be considered: operation temperature, highest and lowest temperature, plus dirty element bypass valve settings. Care must be taken when sizing a filter in a cylinder circuit or accumulator circuit to size for pressure drop at peak flows.
- **System Pressure** - Return line filters must withstand maximum return line pressure, while pressure line filters must withstand maximum system pressure, (including spikes). The housing, bypass valves and element must withstand fatigue from cycling and surges.
- **Temperature** - Operating temperature effects the viscosity (resistance to flow) of the fluid (see chart titled Correcting For Viscosity on page 35). Filters must be selected and sized for cold or ambient temperature, as well as system operating temperature.
- **Other Factors** - The environment in which the equipment operates is important, especially if the reservoir is vented to the atmosphere. It is becoming quite common to install Beta 3 ≥ 1000 air breathers on the system, removing any foreign, airborne particles from ever entering the reservoir.

  With increased emphasis today on preventing any machine downtime (Proactive Maintenance), the need for much higher standards of fluid cleanliness is evolving. Since pumps and valves have clearances on the order of 1-5 microns, it is easy to understand the need for cleaner oil. To meet this demand, filters are now capable of filtering down to the 1 micron level (Beta 1 ≥ 200), with long service lives and low pressure drops.

FILTER SELECTION HINTS

❶ The higher the Beta Ratio, the quicker the contaminant is removed.

❷ The lower the initial sized clean pressure drop, the longer the element will last in service.

With longer and longer "uptimes" being demanded by equipment operators, "downtime" for regular maintenance is becoming increasingly scarce. To allow continual equipment operation, while still being able to service a dirty filter, **fixed recirculation filtration** units are being installed on the machines. These low-cost units are bolted to the machine, continuously filtering the reservoir oil. If an element needs servicing, the production machine continues to run, while shutdown for service is done on the independent filtration system.

- **Filter Location** - See "Typical Filter Placement In Hydraulic Circuits" on page 37.

Typical Dynamic Clearances

Component	Clearance µm
Pump, Gear	
Tooth to Side Plate	0.5-5
Tooth Tip to Case	0.5-5
Pump, Vane	
Vane Sides	5-13
Vane Tip	0.5-1
Pump, Piston	
Piston to Bore	5-40
Valve Plate to Cylinder	0.5-5
Servo Valves	
Spool to Sleeve	1-4
Proportional Valves	1-6
Directional Valves	2-8
Roller Element Bearings	0.1-1
Journal Bearings	0.5-100
Hydrostatic Bearings	1-25
Gears	0.1-1
Dynamic Seal	0.05-0.5

Hydraulic System Filter Recommendations

System Pressure	Components	Fluid	ISO Rating
Above 3000 psi or lower pressure system with intermittent spikes ≥ 3000 psi	Servo/ Proportional	Petroleum Synthetic/HWBF	12/10 12/10
	Compensating Pumps	Petroleum Synthetic/HWBF	13/11 12/10
	Fixed Volume Pumps	Petroleum Synthetic/HWBF	14/12 13/11
Above 1000 psi and below 2000 psi or lower pressure system with intermittent spikes ≥ 1000 psi	Servo/ Proportional	Petroleum Synthetic/HWBF	12/10 12/10
	Compensating Pumps	Petroleum Synthetic/HWBF	13/11 13/11
	Fixed Volume Pumps	Petroleum Synthetic/HWBF	14/12 13/11
Above 2000 psi and below 3000 psi or lower pressure system with intermittent spikes ≥ 2000 psi	Servo/ Proportional	Petroleum Synthetic/HWBF	13/11 12/10
	Compensating Pumps	Petroleum Synthetic/HWBF	14/12 13/11
	Fixed Volume Pumps	Petroleum Synthetic/HWBF	15/13 14/12
Less than 1000 psi	Compensating Pumps	Petroleum Synthetic/HWBF	15/13 14/12
	Fixed Volume Pumps with Solenoid Valve	Petroleum Synthetic/HWBF	15/13 14/12

Lubrication System Filter Recommendations

System	Water	ISO Rating
Ball Bearings	<100 ppm	13/11
Roller Bearings	<100 ppm	14/12
Journal Bearings	<200 ppm	15/13
Gear Box/Industrial	<200 ppm	15/13
Gear Box/Mobile	<200 ppm	16/14
Diesel Engine	<200 ppm	17/15

Water and Air Contamination in Oil

Water Contamination in fluid systems causes:

- Fluid breakdown, such as additive precipitation and oil oxidation
- Reduced lubricating flim thickness
- Accelerated metal surface fatigue
- Corrosion
- Jamming of components due to ice crystals formed at low temperatures
- Loss of dielectric strength in insulating fluids

Dissolved air and other gases in oils cause:

- Foaming
- Slow system response with erratic action
- A reduction in system stiffness affecting servo/proportional valve response
- Higher fluid temperatures
- Pump damage due to cavitation
- Inability to develop full system pressure
- Acceleration of oil oxidation

Means of Water Removal

Type	Removal Capability	Operation
Absorption	Free and dissolved water	Absorbs and retains water. Good for very small amounts (1-3 quarts) of water. Must be disposed of as hazardous waste.
Gravity Precipitation Unit	Free water; nominal removal of particulate contamination	Settling out of gross contamination, water and particulate, by holding oil, water and particulate, in detention chambers. Also small water droplets are coalesced onto water-repelling wire mesh screens.
Centrifuge	Free water; nominal removal of particulate contamination	Cone shaped disks rapidly rotating inside a chamber generate centrifugal force to separate particles, water and oil based on their mass. Centrifuges are inefficient at silt size particulate removal, and often require frequent cleaning, maintenance and rebalancing.
Water Coalescer	Free water, removal of particulate contamination varies	Oil is passed through a prefilter to remove particulate, and then is passed through coalescing elements to combine water droplets. Combined water droplets settle out of the oil by gravity. Coalescer elements need periodic replacement.
Flash Distillation Purifier	Free and dissolved water; dissolved gases/solvents;	These large power consuming units pass heated oil through a vacuum chamber where water is boiled off. Because they use high vacuum and high heat to drive off water, flash distillation purifiers can alter the chemical or physical properties of the oil.
Vacuum Dehydration (mass transfer) purifier	Free and dissolved water; dissolved gases/solvents.	Employing the process of mass transfer, water, air and solvents are removed by exposing oil to low relative humidity air drawn up through a chamber by lower pressure maintained by a vacuum. Automatic, computer controlled units run 24 hours a day/7 days a week. Virtually maintenance free self diagnostics ease servicing if ever needed.

The following test data illustrates the dramatic effect that the vacuum dehydration method coupled with high performance filtration has on measured contamination:

Run Time (Min)	Initial	60	135	165
Water Content (ppm)	8,650	1,240	466	340
Particulate Level	22/20/16	16/14/11	14/13/11	14/13/10

Correcting For Viscosity

To determine unknown viscosity at system operating temperature, plot known temperature-viscosity points on the graph below and interpolate. Example: For a system operating at 130ºF using a fluid with a known viscosity of 200 SUS at 100ºF and 50 SUS at 210ºF, the viscosity is 130º found by plotting the two known points, draw a line between them, find the point where this line intersects the 130º line and read the corresponding viscosity. The result is a viscosity of 115 SUS at 130ºF.

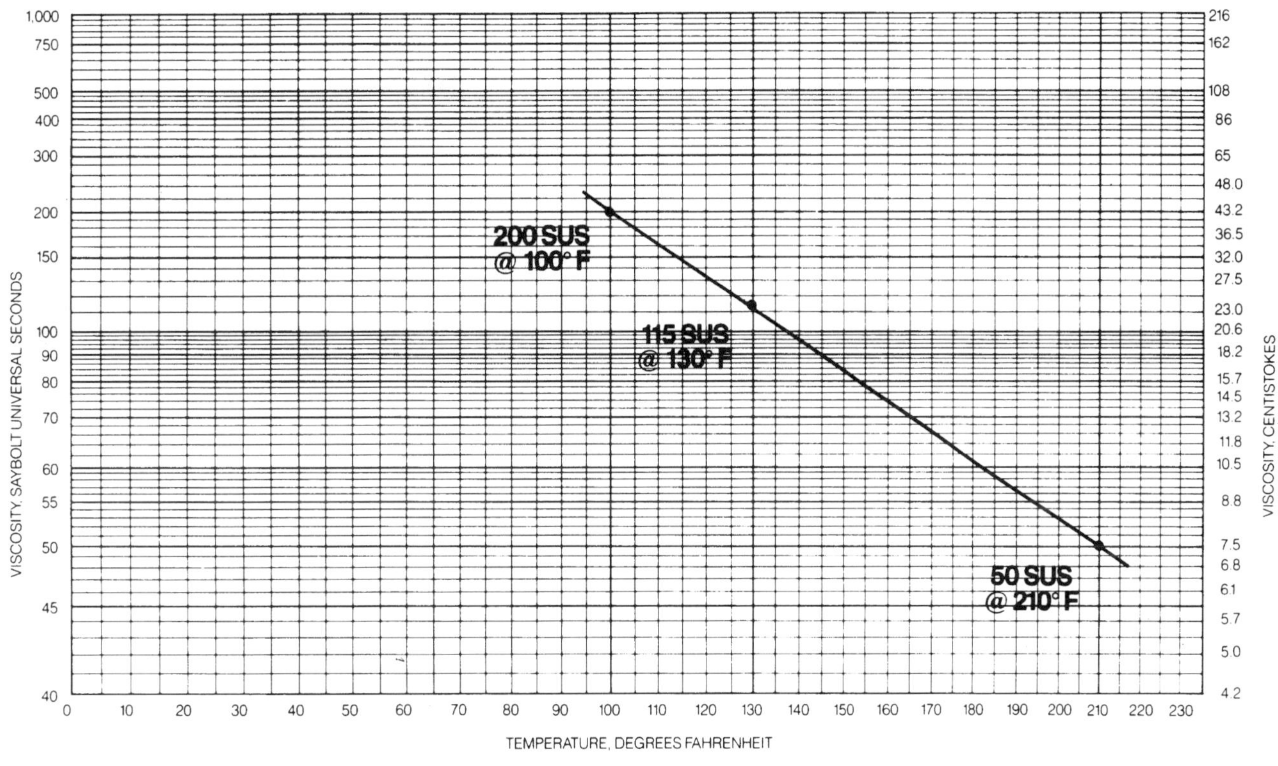

To determine actual total Δp through a filter assembly use manufacturer's data on housing Δp and element Δp and calculate as follows:

Assembly Δp = Housing Δp + Element Δp

Assembly Δp = Housing from mfg's data x $\frac{0.9}{\text{Actual SG}}$ + Element Δp from mfg's data x $\frac{0.9}{\text{Actual SG}}$ x $\frac{\text{Actual viscosity in SUS}}{\text{150 SUS}}$

VISCOSITY CONVERSION		
32 to 100 SUS	100 to 240 SUS	240 SUS+
$Cst = .2253 \times SUS - \frac{194.4}{SUS}$	$Cst = .2193 \times SUS - \frac{134.6}{SUS}$	$Cst = \frac{SUS}{4.635}$

EXAMPLE
Pressure Drop

From manufacturer's data:

Δp for housing: 32 psi
Δp for element: 14 psi
Specific gravity for fluid: .89
SUS at operating temperature: 130

$$32 \text{ psi} \times \frac{0.9}{.89} + 14 \text{ psi} \times \frac{0.9}{.89} \times \frac{130 \text{ SUS}}{150 \text{ SUS}} = 44.6 \text{ psi}$$

PROACTIVE MAINTENANCE
Methods of Analysis

There are two very common ways to know if your oil is clean:
- Constant breakdowns may indicate contaminated oil
- Independent oil analysis, comparing the results to industry standards (Refer to the table on page 33 For Hydraulic System Recommendations).

Constant problems keeping your equipment running, plus high repair and maintenance costs, can usually be traced back to component surface degradation (50% = mechanical wear, 20% = corrosive wear). If you are experiencing any of these difficulties, oil analysis may help you identify if your oil is the culprit.

If you are not experiencing these kinds of troubles, oil analysis may allow you to put into place some preventative methods, keeping you trouble free for many more years. There are many tests that can be performed on your fluids. Each of them will give you greater insight into the condition of your fluids.

- **Optical Particle Counts** — A particle counting procedure conducted by a trained technician. Your sample fluid is drawn down onto a 1.2 micron patch. Each particle that is seen under the 100x-500x microscope is tallied and reported on in the ISO Cleanliness Code. This test gives you very accurate counts on the amount and size of contaminant within your system. Unfortunately, it is very expensive.
- **Automatic Particle Counts** — Automatic particle counting procedure uses a machine to count your samples contaminant. It is very fast, (once the sample has arrived in the laboratory), inexpensive and mostly repeatable. The test can be inaccurate if subjected to high particle concentrations, water, air and gels. The report should be in an ISO Cleanliness Code.
- **Field Patch Testing** — A very fast and reliable, on-site measurement of your systems contaminant level. A fluid sample is taken, drawn down onto a 1.2 micron patch and viewed under a 100x microscope. It is then compared to a field comparator booklet, quickly giving you a general fluid contamination ISO Cleanliness Level. This test is the fastest field testing available and is accurate, although no actual report can be printed on-site.
- **Ferrography** — A laboratory test that provides basic information that may indicate the need for more sophisticated testing. It cannot detect nonferrous particles like brass, copper, silica, gels and water. It is usually used after a system failure.
- **Spectrometry** — An excellent test used to monitor the additive package within your system. It identifies and quantifies (by ppm and chemical makeup) additives and contaminants. Spectrometry cannot size the particles it "sees" and cannot read particles above 5 microns in size.
- **Gravimetric** — Only indicates the total amount of contaminant by weight. Gravimetric cannot distinguish particle sizes or quantities.
- **Viscosity** — Measures the actual viscosity of your oil, allowing you to compare it to new oil specifications.
- **Total Acid Number (T.A.N.)** — Measures the acidic properties of your oil, possibly indicating a water problem. Compare your T.A.N. to new oil specifications.
- **Water Content** — Usually measured by Karl Fischer test, in ppm. A ppm less than 100 is satisfactory for long equipment life.

How To Make Your Equipment Last Longer With Proactive Maintenance

Step One (1)

List all equipment you want to start monitoring
- Hydraulic systems
- Lube oil systems

Set and track fluid recommendations by individual systems
- **Contamination Level** - Pick a suitable ISO Cleanliness Level based on the information provided on page 33.
- **Water Level** - Measured in parts per million (ppm). Generally keeping your oil below the saturation levels, keeps corrosion at a minimum.

  Hydraulic 100-200 ppm (.01-.02%)
  Lubrication 200-400 ppm (.02-.04%)
  Transformer 10-50 ppm (.001-.005%)
- **Additives** - Take a Spectrometric Test of new oil as your "baseline" additive levels.
- **Viscosity, T.A.N.** - Consult your manufacturer.

Step Two (2)

Install a sample port on each machine, keeping the following in mind:
- One port per machine
- Port should be easily accessible
- Mark it "Sample Port"
- Port should be in the return line before the filter or at the "worst case" oil location

Develop a tracking method you are comfortable with which:
- Records all sample data
- Reminds you to take timely samples

Take reliable samples
- Open the sample valve and drain 1-2 cups of fluid (this flushes out the sample port)
- Open the sample bottle
- **DO NOT** touch the sample valve or port
- Put the sample bottle into the fluid stream
- Fill and swish the sample bottle 3 times then discard the oil (this removes contaminants from the bottle)
- Fill the bottle for the fourth time, and cap the bottle, making this your sample
- Turn off the sample valve and label the bottle
- Send in the bottle for analysis (ISO/Spectrometry/ Water Content/Viscosity/T.A.N.)

Step Three (3)

Compare your sample report to the established Fluid Recommendations. Immediately take corrective action if the sample fails to meet your recommendations

Problem	Solution
High contaminant or ISO levels	Upgrade existing filtration, flush the reservoir with Beta 1≥ 200 filters.
Additive levels outside target	Consult manufacturer.
A little water? **A lot of water?**	Use absorption methods. Vacuum Dehydration Unit.
Viscosity level not at "spec" System too hot Chemical reaction	Install or upgrade coolers. Consult oil manufacturer.
System acidic	Remove water & resample in 45 days.

Typical Filter Placement In Hydraulic Circuits

HYDRAULIC SYSTEM WITH SOLENOID VALVES

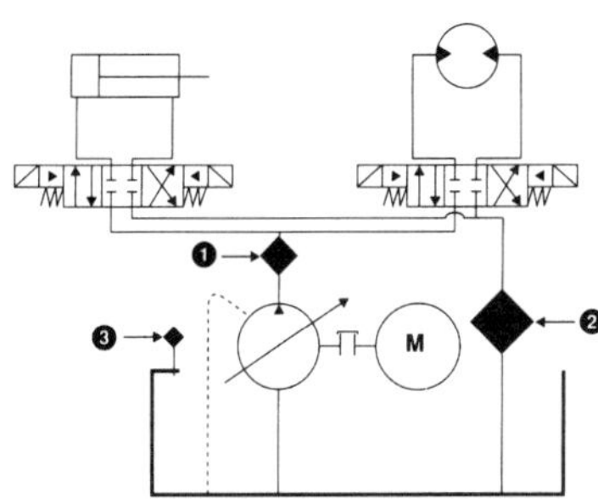

① Pressure Line Filter — Beta 3≥200
② Return Line Filter — Beta 3≥200
③ Air Breather Filter — Beta 3≥200

System Parameters:
Pump Flow: 30 gpm
Max. Operating Pressure: 1800 psi
Fluid Type: Petroleum based oil

HYDRAULIC SYSTEM WITH KIDNEY LOOP FILTRATION

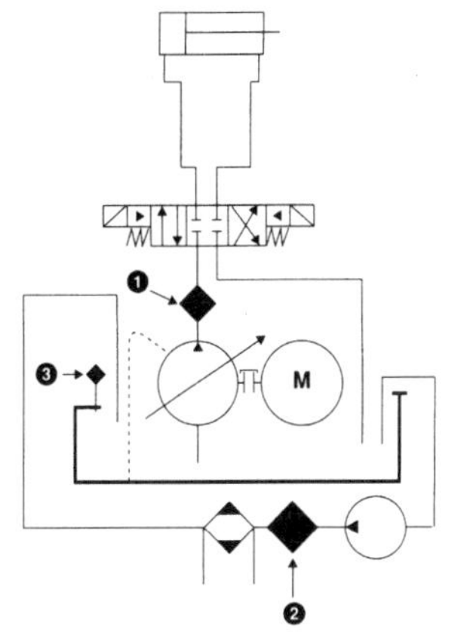

① Pressure Line Filter
② Kidney Loop Filter
③ Air Breather Filter

System Parameters:
Pump Type: Piston
Variable displacement
Pressure compensated
Pump Flow: 50 gpm (intermittent)
Max. Operating Pressure: 2800 psi
Fluid Type: Water glycol
Total System Fluid Volume: 300 gal

HYDRAULIC SYSTEM WITH SOLENOID VALVES AND A SERVO VALVE

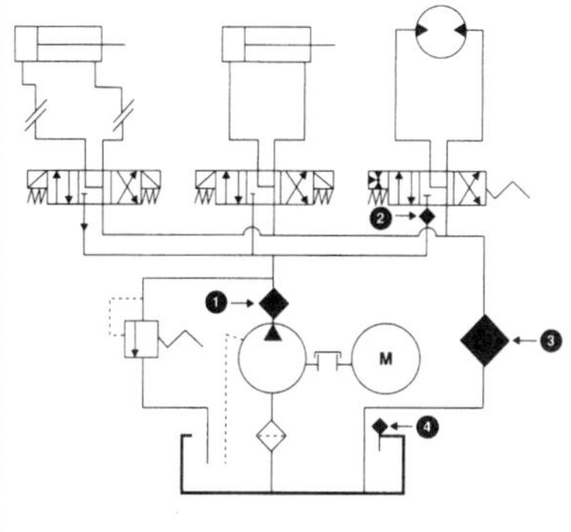

① Pressure Line Filter
② Remote Mounted Non-Bypass Filter
③ Return Line Filter
④ Air Breather Filter

CLOSED LOOP HYDROSTATIC TRANSMISSION CIRCUIT

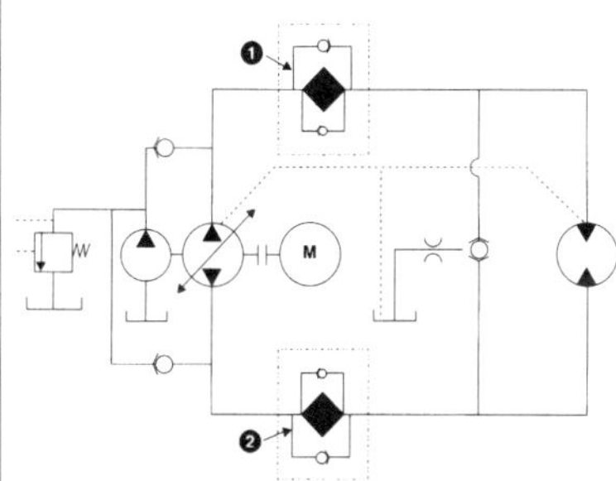

① & ② In-loop filters with heavy duty reverse flow valve

System Parameters:
Change Pressure: 125 psi
Change Flow: 4 gpm
Loop Pressure: 3200 psi
Loop Flow: 32 gpm
Loop Flow Direction: Bi-directional
Fluid Type: Petroleum based oil

BEARING LUBE CIRCUIT PAPER MACHINE MAIN LUBE SYSTEM

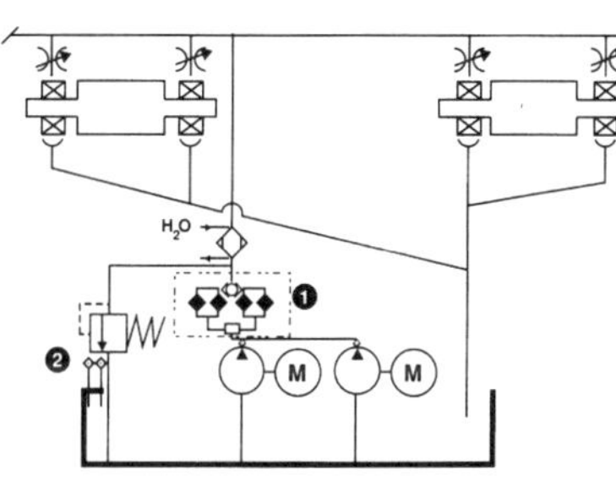

① Duplex filter in Presssure Line
② Air Breathers

System Parameters:
Pump Flow: 120 gpm
System Pressure: 45 psi
Max. Operating Pressure: 120 psi
Lubrication Oil: Petroleum based oil

TURBINE KIDNEY LOOP FILTER LARGE STEAM TURBINE

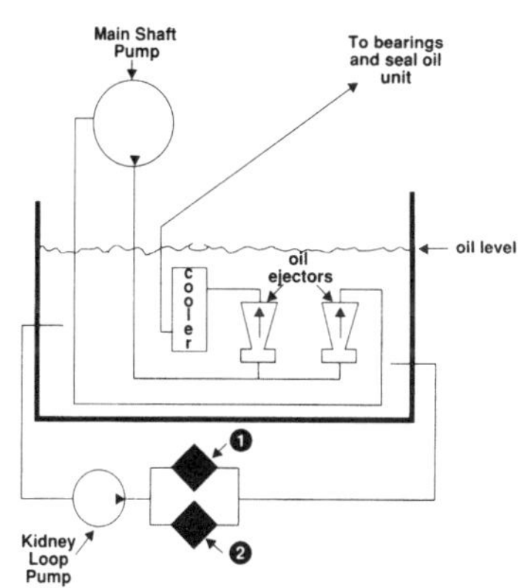

① & ② Kidney Loop Filters

System Parameters:
Main reservoir Volume: 6000 gal
Operating Pressure: 15 psi
Kidney Loop Flow (added): 100 gpm
Fluid Type: Petroleum based-turbine oil

NOTE: The complexity of main bearing lube oil supply circuit, combined with the high flow and low available pressure, means that installing a kidney loop is most often the best approach to upgrading to high efficiency filtration.

Pipe Size

The original system for classifying pipe size was based on a named wall thickness: Standard, Extra Heavy and Double Extra Heavy. The additional wall thickness is achieved by the reduction of the inside diameter, while the outside diameter remains the same for all weights of the same nominal pipe size.

The current system used in classification is the Schedule system. The schedule numbers are specified by the American Standards Institute (ANSI) and range from 10 to 160, giving the system the ability to define 10 different wall thicknesses. Like the original system, the schedule system reduces the inside diameter to gain wall thickness, thereby retaining the same outside diameter for the nominal pipe size.

For reference, Schedule 40 is the same as Standard for $^1/_8$ to 6 inch nominal pipe, Schedule 80 is the same as Extra Heavy for $^1/_8$ to 8 inch nominal pipe. Schedule 160 is in between Schedule 80 and Double Extra Heavy.

Pressure Ratings For Tube and Pipe

The pressure ratings tabulated are for plain end carbon steel tubing, the calculation method is per the Code for Pressure Piping, ANSI/ASME B31.1. No allowances have been included in these ratings for fabrication tolerances such as bending allowances. Plain end piping is piping that is joined by a method that does not reduce the wall thickness at the joint. If the joining method reduces the wall thickness the maximum pressure will be reduced from the tabulated values.

The following formula was used for the calculation of the tabulated values for plain end pipe:

$$P = \frac{2 \times S \times E \times (Tm - A)}{Do - 2 \times Y \times (Tm - A)}$$

Definition	Value Used In Tabulated Rating Calculation
P = Maximum internal pressure in pounds per square inch gage	
S = Allowable stress in material for internal pressure, at the design temperature, in pounds per square inch	15,000 psi at -20 to 650° F for ASTM A-53 Grade B and ASTM A-106 Grade B
E = Efficiency factor for longitudinal welded pipe. See code for specific values for other material	1 for ASTM A-53 Grade B and ASTM A-106 Grade B
Tm = Minimum pipe wall thickness in inches	87.5% of nominal wall thickness per ASTM A-53 per Table X2.2
A = Allowance for threading, corrosion, etc. in inches	.0015 for corrosion and erosion
Do = Outside diameter of pipe in inches	Per ASTM A-53 Table X2.2
Y = Temperature adjusting factor	.4 for Ferritic steel for <900° F

Sample Calculation

- Schedule 80
- $^1/_2$" pipe
- Straight pipe
- Plain end
- 0.0015 inch allowance for corrosion and erosion

$$P = 5{,}165 = \frac{2 \times 15{,}000 \times 1 \times ((.875 \times .147) - .0015)}{.840 - 2 \times .4 \times ((.875 \times .147) - .0015)}$$

The following formula is used to adjust the plain end maximum pressure ratings of pipe for system application including threading and water hammer.

$$Pa = \frac{2 \times S \times E \times [Tm - (Aa + Ab)]}{Do - 2 \times Y \times [Tm - (Aa + Ab)]} - wh$$

Definition	Value Used In Tabulated Rating Calculation
Pa = Adjusted maximum internal pressure in pounds per square inch gage	
S = Allowable stress in material for internal pressure, at the design temperature, in pounds per square inch	15,000 at -20 to 650° F for ASTM A-53 Grade B and ASTM A-106 Grade B
E = Efficiency factor for longitudinal welded pipe. See code for specific values for other material	1 for ASTM A-53 Grade B and ASTM A-106 Grade B
Tm = Minimum pipe wall thickness in inches	87.5% of nominal wall thickness per ASTM A-53, Table X2.2
Aa = Allowance for erosion, corrosion, etc. in inches	.0015 for corrosion and erosion
Ab = Threading factor	.8 x 1/thread pitch
Do = Outside diameter of pipe in inches	Per ASTM A-53 Table X2.2
Y = Temperature adjusting factor	.4 for Ferritic steel for <900° F
Wh = Water Hammer Spike, psi $= \frac{F \times (SG \times B)^{.5}}{20 \times Di^2}$	
F = Flow rate in gpm	Velocity (ft/sec) x Area (in²)/.3208
SG = Specific Gravity	1.00 (water)
B = Bulk Modulus	320,000 psi (water)
Di = Inside diameter	Per ASTM Table X2.2

Sample Calculation

- Schedule 80
- $^1/_2$" pipe
- Straight pipe
- Bulk modulus = 320,000 psi (water)
- Adjusted for water hammer
- Adjusted for threaded fittings (14 TPI)
- 10.9 gpm flow (15 foot/sec velocity)
- 0.0015 inches allowance for corrosion and erosion

$$Wh = 1034\text{ psi} = \frac{10.9 \times (1 \times 320{,}000)^{.5}}{20 \times .546^2}$$

Pa = 1,650 psi

$$= \frac{2 \times 15{,}000 \times 1 \times [(.875 \times .147) - (.057 + .0015)]}{.840 - 2 \times .4 [(.875 \times .147) - (.057 + .0015)]} - 1034\text{ psi}$$

Water Hammer Factor

The value "Water Hammer Factor" shown on the following pages is: $\Delta p/Q$. To determine the increased pressure due to water hammer (psi) multiply the value shown in the table by the flow in the pipe (gpm). The Specific Gravity and Bulk Modulus of Elasticity used in the calculation for the table was for water, 1 and 320,000 (psi) respectively. To correct for different fluids, multiply the "Water Hammer Factor" by the square root of the ratio of the Specific Gravity x Bulk Modulus of Elasticity of the particular fluid used to that of water (1 x 320,000).

Example

Determine the added pressure created when a valve is closed quickly in a water system that is flowing at 12 gpm in a 1/2" Schedule 40 pipe using the table.

12 = Flow (gpm)
69.63 = Water Hammer Factor from chart (psi/gpm)

12 x 69.63 = 836 (psi)

The same problem but with a fluid other than water:

12 = Flow (gpm)
69.63 = Water Hammer Factor from chart (psi/gpm)
SG = .88 (no units)
Bm = 280,000 (psi)

$$12 \times 69.63 \times \sqrt{\frac{.88 \times 280{,}000}{1 \times 320{,}000}} = 733 \text{ (psi)}$$

Water Hammer

Whenever a valve is closed quickly in a hydraulic system, a positive pressure wave is created in the pipe that travels through the pipe at the speed of sound. The additional pressure created by this wave, known as water hammer, can reach several times the normal pressure of the system causing catastrophic failure. If you assume that the pipe is rigid, the equation for the increase pressure can be determined by a momentum equation. The classical form of which is:

$$\Delta p = \rho \times V \times c$$

Δp = Increased pressure due to water hammer (psi)
ρ = Fluid density (slugs/ft^3)
V = Fluid velocity (ft/sec)
c = Speed of sound in the fluid (ft/sec)

A more useful form of this equation can be obtained by manipulation of the variables and use of values such as the density of water (1.94 slugs/ft^3). The manipulation is in three parts: (1) get the density into a function of specific gravity (2) get the fluid velocity as a function of flow (3) get the speed of sound in the fluid as a function of Bulk Modulus and density. The "conversion" is fairly lengthy and is outside of the scope of this document and is therefore not included. The final configuration is as follows:

$$\Delta p = \frac{Q}{21 \times D^2} \times \sqrt{SG \times Bm}$$

Δp = Increased pressure due to Water Hammer (psi)
Q = Flow (gpm)
SG = Specific Gravity (no units)
Bm = Bulk Modulus (psi)
21 = conversion factor
D = Pipe inside diameter (in)

Example

Determine the added pressure created when a valve is closed quickly in a water system that is flowing at 12 gpm in a 1/2" Schedule 40 pipe:

Q = 12 (gpm)
SG = 1 (no units)
Bm = 320,000 (psi)
D = .622 in

$$\Delta p = 836 = \frac{12}{21 \times .622^2} \times \sqrt{1 \times 320{,}000}$$

The same problem but with a 1" Schedule 40 pipe is:

Q = 12 (gpm)
SG = 1 (no units)
Bm = 320,000 (psi)
D = 1.049 in

$$\Delta p = 293 = \frac{12}{21 \times 1.049^2} \times \sqrt{1 \times 320{,}000}$$

Critical Time For Valve Closure

The following formula is not provided for calculation but to indicate that the time for valve closure is a function of the pipe run length and the type of fluid used in the system. The longer the pipe run and the faster the valve acts, the greater the chance that water hammer will be a problem.

$$t_c = \frac{2 \times L}{c}$$

t_c = Maximum time of closure to produce maximum pressure (sec)
2 = Indicates the total length of travel of the wave (no units)
L = Length of the pipe run (ft)
c = Speed of sound in the fluid (ft/sec) where:

$$c = \sqrt{\frac{Bm}{\rho}}$$

Bm = Bulk Modulus of Elasticity (psi)
ρ = Fluid density (slugs/ft^3)

Recommended Maximum Flows For Pressure Lines

Pump Suction
Size to maintain the velocity in the range of 2 to 4 feet per second.

Oil Return Lines
Size to maintain the velocity in the range of 10 to 15 feet per second.

Medium Pressure Lines
Size to maintain the velocity in the range of 15 to 20 feet per second.

High Pressure Lines
Size to maintain the velocity at a maximum of 30 feet per second.

Safe Pump Inlet Vacuum

	Gear Pumps	Vane Pumps	Piston Pumps
Max. Safe Inlet Vacuum, psi	3 to 5	2 to 3	2
Max. Safe Inlet Vacuum, "Hg	6 to 10	4 to 6	4

NOTE: The suction strainer should be cleaned or replaced when inlet vacuum on a hydraulic pump reaches these values. Sustained operation at these vacuums may damage the pump. When suction strainer is clean, inlet vacuum should not be more than 1/3 of these values.

Hose Dash Numbers For Approximately Equivalent or Greater Flow

Pipe Size	Pipe Type 40	80	160	XXS
1/4	-6	-6		
3/8	-8	-8		
1/2	-10	-10	-8	-4
3/4	-16	-12	-10	-8
1	-16	-16	-16	-10
1 1/4	-24	-20	-20	-16
1 1/2	-32	-24	-24	-20
2	-32	-32	-32	-24
2 1/2	-40	-40	-40	-32

Spacing of Piping Supports

Tubing Outside Diameter in Inches	Pipe, Nominal Size	Length between Supports in Feet
1/4, 3/8	1/8, 1/4	3
1/2, 5/8, 3/4, 7/8, 1	3/8, 1/2, 3/4	5
1 1/4 and larger	1 and larger	7

Size, Schedule, Pressure Ratings[1] & Flow Capacity of Steel Pipe

Schedule 40 — Standard Steel Pipe

Pipe Size	I.D.	O.D.	Wall	Max. Pressure	Bursting Pressure	Water Hammer[6]	Int Area	Wt/ft	gpm @ Velocity (ft/sec) 2	5	10	15	20	25
1/8	.269	.405	.068	4900	20100		.057	.245	0.4	0.9	1.8	2.7	3.5	4.4
1/4	.364	.540	.088	4700	19600		.104	.425	0.6	1.6	3.2	4.9	6.5	8.1
3/8	.493	.675	.091	3800	16200		.191	.567	1.2	3.0	5.9	8.9	11.9	14.9
1/2	.622	.840	.109	3700	15600	69.63	.304	.852	1.9	4.7	9.5	14.2	18.9	23.7
3/4	.824	1.050	.113	3000	12900	39.67	.533	1.132	3.3	8.3	16.6	24.9	33.2	41.6
1	1.049	1.315	.133	2800	12100	24.48	.864	1.679	5.4	13.5	26.9	40.4	53.9	67.3
1 1/4	1.380	1.660	.140	2300	10100	14.14	1.495	2.273	9.3	23.3	46.6	69.9	93.2	116.5
1 1/2	1.610	1.900	.145	2100	9200	10.39	2.036	2.718	12.7	31.7	63.5	95.2	126.9	158.6
2	2.067	2.375	.154	1800	7800	6.3	3.356	3.653	20.9	52.3	104.6	156.9	209.2	261.5
2 1/2	2.469	2.875	.203	1900	8500	4.42	4.788	5.793	29.8	74.6	149.2	223.8	298.5	373.1
3	3.068	3.500	.216	1700	7400	2.86	7.393	7.575	46.1	115.2	230.4	345.6	460.8	576.1
3 1/2	3.548	4.000	.226	1500	6800	2.14	9.886	9.109	61.6	154.1	308.2	462.2	616.3	770.4
4	4.026	4.500	.237	1400	6300	1.66	12.73	10.79	79.4	198.4	396.8	595.2	793.6	992.0
5	5.047	5.563	.258	1200	5600	1.06	20.01	14.62	124.7	311.8	623.6	935.3	1247.1	1558.9
6	6.065	6.625	.280	1100	5100	0.73	28.89	18.97	180.1	450.2	900.5	1350.7	1801.0	2251.2
8	7.981	8.625	.322	1000	4500	0.42	50.03	28.55	311.9	779.6	1559.3	2338.9	3118.6	3898.2
10	10.02	10.75	.365	900	4100	0.27	78.85	40.48	491.6	1228.9	2457.8	3686.7	4915.6	6144.5
12[4]	11.94	12.75	.405	800	3800	0.19	111.9	53.56	698.0	1745.0	3489.9	5234.9	6979.9	8724.9

Schedule 80 — Standard Steel Pipe

Pipe Size	I.D.	O.D.	Wall	Max Pressure	Bursting Pressure	Water Hammer[6]	Int Area	Wt/ft	gpm @ Velocity (ft/sec) 2	5	10	15	20	25
1/8	.215	.405	.095	7200	28100		.036	.314	0.2	0.6	1.1	1.7	2.3	2.8
1/4	.302	.540	.119	6700	26400		.072	.535	0.4	1.1	2.2	3.3	4.5	5.6
3/8	.423	.675	.126	5500	22400		.141	.738	0.9	2.2	4.4	6.6	8.8	11.0
1/2	.546	.840	.147	5200	21000	90.36	.234	1.087	1.5	3.6	7.3	10.9	14.6	18.2
3/4	.742	1.050	.154	4200	17600	48.93	.433	1.473	2.7	6.7	13.5	20.2	27.0	33.7
1	.957	1.315	.179	3900	16300	29.41	.719	2.171	4.5	11.2	22.4	33.6	44.8	56.0
1 1/4	1.278	1.660	.191	3300	13800	16.49	1.283	2.996	8.0	20.0	40.0	60.0	80.0	100.0
1 1/2	1.500	1.900	.200	3000	12600	11.97	1.767	3.631	11.0	27.5	55.1	82.6	110.2	137.7
2	1.939	2.375	.218	2600	11000	7.16	2.953	5.022	18.4	46.0	92.0	138.1	184.1	230.1
2 1/2	2.323	2.875	.276	2700	11500	4.99	4.238	7.661	26.4	66.1	132.1	198.2	264.2	330.3
3	2.900	3.500	.300	2400	10300	3.2	6.605	10.25	41.2	102.9	205.9	308.8	411.8	514.7
3 1/2	3.364	4.000	.318	2200	9500	2.38	8.888	12.50	55.4	138.5	277.0	415.5	554.1	692.6
4	3.826	4.500	.337	2100	9000	1.84	11.50	14.98	71.7	179.2	358.3	537.5	716.7	895.9
5	4.813	5.563	.375	1800	8100	1.16	18.19	20.78	113.4	283.5	567.1	850.6	1134.2	1417.7
6	5.761	6.625	.432	1800	7800	0.81	26.07	28.57	162.5	406.2	812.5	1218.7	1624.9	2031.2
8	7.625	8.625	.500	1600	7000	0.46	45.66	43.39	284.7	711.6	1423.3	2134.9	2846.6	3558.2
10[2]	9.562	10.75	.594	1500	6600	0.29	71.81	64.40	447.6	1119.1	2238.2	3357.4	4476.5	5595.6
12[3]	11.37	12.75	.688	1500	6500	0.21	101.61	88.57	632.9	1582.3	3164.7	4747.0	6329.4	7911.7

Schedule 160

Pipe Size	I.D.	O.D.	Wall	Max. Pressure	Bursting Pressure	Water Hammer[5]	Int Area	Wt/ft	gpm @ Velocity (ft/sec) 2	5	10	15	20	25
1/2	.466	.840	.187	6800	26700		1.71	1.310	1.1	2.7	5.3	8.0	10.6	13.3
3/4	.612	1.050	.219	6400	25000	71.92	.294	1.940	1.8	4.6	9.2	13.8	18.3	22.9
1	.815	1.315	.250	5700	22800	40.55	.522	2.850	3.3	8.1	16.3	24.4	32.5	40.7
1 1/4	1.160	1.660	.250	4400	18100	20.02	1.060	3.764	6.6	16.5	32.9	49.4	65.9	82.4
1 1/2	1.338	1.900	.281	4300	17700	15.05	1.410	4.862	8.8	21.9	43.8	65.7	87.7	109.6
2	1.687	2.375	.344	4200	17400	9.47	2.235	7.46	13.9	34.8	69.7	104.5	139.3	174.2
2 1/2	2.125	2.875	.375	3700	15700	5.97	3.542	10.01	22.1	55.3	110.5	165.8	221.1	276.4
3	2.624	3.500	.438	3600	15000	3.91	5.408	14.32	33.7	84.3	168.6	252.8	337.1	421.4
4	3.438	4.500	.531	3400	14200	2.28	9.283	22.52	57.9	144.7	289.3	434.0	578.7	723.4
5	4.313	5.563	.625	3200	13500	1.45	14.61	33.0	91.1	227.7	455.4	683.1	910.8	1138.4
6	5.187	6.625	.719	3100	13000	1	21.13	45.35	131.7	329.3	658.6	987.9	1317.3	1646.6
8	6.813	8.625	.906	3000	12600	0.58	36.44	74.70	227.3	568.1	1136.3	1704.4	2272.6	2840.7
10	8.500	10.75	1.125	3000	12600	0.37	56.75	115.64	353.7	884.3	1768.7	2653.0	3537.4	4421.7
12	10.126	12.75	1.312	2900	12300	0.26	80.53	160.33	502.0	1255.0	2510.1	3765.1	5020.1	6275.2

Double Extra Strong Pipe — XXS

Pipe Size	I.D.	O.D.	Wall	Max Pressure	Bursting Pressure	Water Hammer[5]	Int Area	Wt/ft	gpm @ Velocity (ft/sec) 2	5	10	15	20	25
1/2	.252	.840	.294	12100	42000		.050	1.71	0.3	0.8	1.6	2.3	3.1	3.9
3/4	.434	1.050	.308	9600	35200		.148	2.440	0.9	2.3	4.6	6.9	9.2	11.5
1	.599	1.315	.358	8800	32700	75.08	.282	3.659	1.8	4.4	8.8	13.2	17.6	22.0
1 1/4	.896	1.660	.382	7200	27600	33.55	.630	5.214	3.9	9.8	19.7	29.5	39.3	49.1
1 1/2	1.100	1.900	.400	6400	25300	22.26	.950	6.408	5.9	14.8	29.6	44.4	59.2	74.1
2	1.503	2.375	.436	5500	22000	11.92	1.774	9.029	11.1	27.7	55.3	83.0	110.6	138.3
2 1/2	1.771	2.875	.552	5800	23000	8.59	2.463	13.70	15.4	38.4	76.8	115.2	153.6	191.9
3	2.300	3.500	.600	5100	20600	5.09	4.154	18.58	25.9	64.7	129.5	194.2	259.0	323.7
4	3.152	4.500	.674	4400	18000	2.71	7.803	27.54	48.6	121.6	243.2	364.8	486.4	608.0
5	4.063	5.563	.750	3900	16200	1.63	12.97	38.55	80.8	202.1	404.1	606.2	808.2	1010.3
6	4.897	6.625	.864	3800	15600	1.12	18.83	53.16	117.4	293.5	587.0	880.6	1174.1	1467.6
8	6.875	8.625	.875	2900	12200	0.57	37.12	72.42	231.4	578.5	1157.1	1735.6	2314.1	2892.6
10	8.750	10.75	1.000	2600	11200	0.35	60.13	104.1	374.8	937.1	1874.2	2811.4	3748.5	4685.6
12	10.75	12.75	1.000	2200	9400	0.23	90.76	125.5	565.8	1414.5	2829.0	4243.4	5657.9	7072.4

NOTES APPLY TO ALL SCHEDULES (pp. 40 & 41)

[1] Based on ASTM A53 Grade B or A106 Grade B Seamless ANSI.B31.1. 1992, allowances for connections and fittings reduces these working pressures approximately 25%.
[2] Not extra strong. Schedule 60 is extra strong in this size.
[3] Not extra strong. Extra strong does not have a schedule number in this size (I.D. of 12" xs is 11.75 inches).
[4] Not standard weight. Standard weight does not have a schedule number in this size (I.D. of 12" standard is 12.00 inches).
[5] Water Hammer Factor, see page 39 for application.

SAE Pressure Rating of Hose

SAE Hose #	Pressure Range PSI	I.D."	PSI	I.D."
100R1	3000	3/16	375	2
100R2	5000	3/16	375	2
100R3	1500	3/16	375	1 1/4
100R4	300	3/4	35	4
100R5	3000	3/16	350	1 13/16
100R6	500	3/16	350	5/8
100R7	3000	3/16	1000	1
100R8	5000	3/16	2000	1
100R9	4500	3/8	2000	2
100R10	10,000	3/16	2500	2
100R11	12,500	3/16	3000	2
100R12	4000	3/8	2500	2

Temperatures -40 to 200°F for thermoplastic

Hose Pressure Drop

Hose Dash Size	-04		-05		-06		-08		-10		-12		-16		-20		-24		-32		-40	-48
Hose I.D. (Inches)	.19	.25	.25	.31	.31	.38	.41	.50	.50	.63	.63	.75	.88	1.00	1.13	1.25	1.38	1.50	1.81	2.00	2.38	3.00
Working Pressure, psi .25	10	3.1	3.1																			
.50	19	6	6	2.7	2.7																	
1	40	12	12	5.5	5.5	2.4																
2	95	24	24	10	10	4.8	3.5															
3	185	46	46	17	17	7	5	2.2	2.2													
4		78	78	29	29	12	8	3	3	1.2	1.2											
5		120	120	44	44	18	12	4.5	4.5	1.6	1.6	7.2										
8				95	95	39	26	10	10	3.6	3.6	1.4	.60									
10						59	40	15	15	5.7	5.7	2	1	.55								
12						80	52	20	20	7.2	7.2	2.6	1.5	7.5	.43							
15							75	30	30	10	10	4.2	2.2	1.2	.67	.38						
18							107	40	40	15	15	6.3	3	1.5	.70	.55	.35					
20								49	49	19	19	8	3.4	2	1.1	.65	.43	.27				
25								72	72	26	26	11	5.5	3	1.6	1	.64	.40	.17			
30										34	34	14	7	3.6	2.2	1.3	.80	.52	.22	.14		
35										47	47	19	9.5	5	2.8	1.7	1.1	.70	.27	.18		
40												25	12	6.5	3.4	2.2	1.4	.90	.38	.24		
50												36	17	9	5.3	3.3	2	1.3	.54	.35	.15	
60												50	23	12	7.5	4.4	2.8	1.8	.75	.45	.20	
70													31	17	9.3	6	3.8	2.4	1	.65	.30	
80													38	21	12	7.1	4.6	3	1.2	.76	.34	.11
90													49	27	15	9	5.9	3.8	1.5	1	.45	.13
100														33	19	12	7	4.7	1.9	1.3	.55	.18
150														60	36	22	13	8.5	3.4	2.2	1	.33
200																36	23	15	6	3.9	1.7	.55
250																54	33	22	8.5	5.3	2.5	.75
300																	45	29	12	7.5	4	1.1
400																		51	21	14	6.5	2.2
500																			32	20	10	3
800																					18	5
1000																						10

NOTE: Pressure drop in psi (pounds per square inch)/gpm (gallons per minute)/for 10 feet of hose (smooth bore) without fittings. Fluid specification: Specify gravity = .85; Viscosity = v = 20 centistokes (Cst), (20 Cst = 97 SSU).

Pressure drop values listed are typical of many petroleum based hydraulic oils at approximately +100°F (+38°C). Differences in fluids, fluid temperature and viscosity can increase or decrease actual pressure drop compared to the values listed.

Friction Loss in Standard Valves and Fittings, Schedule 40

Table gives equivalent lengths of straight pipe in feet (laminar flow)

Type of Fitting	Nominal Pipe Diameter 1/2	3/4	1	1 1/4	1 1/2	2	2 1/2	3	4	5	8	10	12
Gate Valve (open)	0.7	0.9	1.1	1.5	1.7	2.2	2.7	3.3	4.4	5.5	8.6	10.9	12.9
Angle Valve (open)	7.5	10.0	12.7	16.7	19.5	25.0	29.8	37.1	48.6	61.0	96.4	121.1	144.3
Ball Valve (open)	0.2	0.2	0.3	0.3	0.4	0.5	0.6	0.8	1.0	1.3	2.0	2.5	3.0
Square Corner Elbow	3.0	3.9	5.0	6.6	7.6	9.8	11.7	14.6	19.1	24.0	37.9	47.6	56.7
90° Street Elbow	2.6	3.4	4.4	5.8	6.7	8.6	10.3	12.8	16.8	21.0	33.3	41.8	49.8
90° Standard Elbow	1.6	2.1	2.6	3.5	4.0	5.2	6.2	7.7	10.1	12.6	20.0	25.1	29.9
90° Long Radius Elbow	1.0	1.4	1.7	2.3	2.7	3.4	4.1	5.1	6.7	8.4	13.3	16.7	19.9
45° Street Elbow	1.3	1.8	2.3	3.0	3.5	4.5	5.3	6.6	8.7	10.9	17.3	21.7	25.9
45° Standard Elbow	0.8	1.1	1.4	1.8	2.1	2.8	3.3	4.1	5.4	6.7	10.6	13.4	15.9
Tee (Straight Thru)	1.0	1.4	1.7	2.3	2.7	3.4	4.1	5.1	6.7	8.4	13.3	16.7	19.9
Tee (Thru branch)	3.1	4.1	5.2	6.9	8.1	10.3	12.3	15.3	20.1	25.2	39.9	50.1	59.7

Schedule 40 Pipe
Flow & Velocity Chart

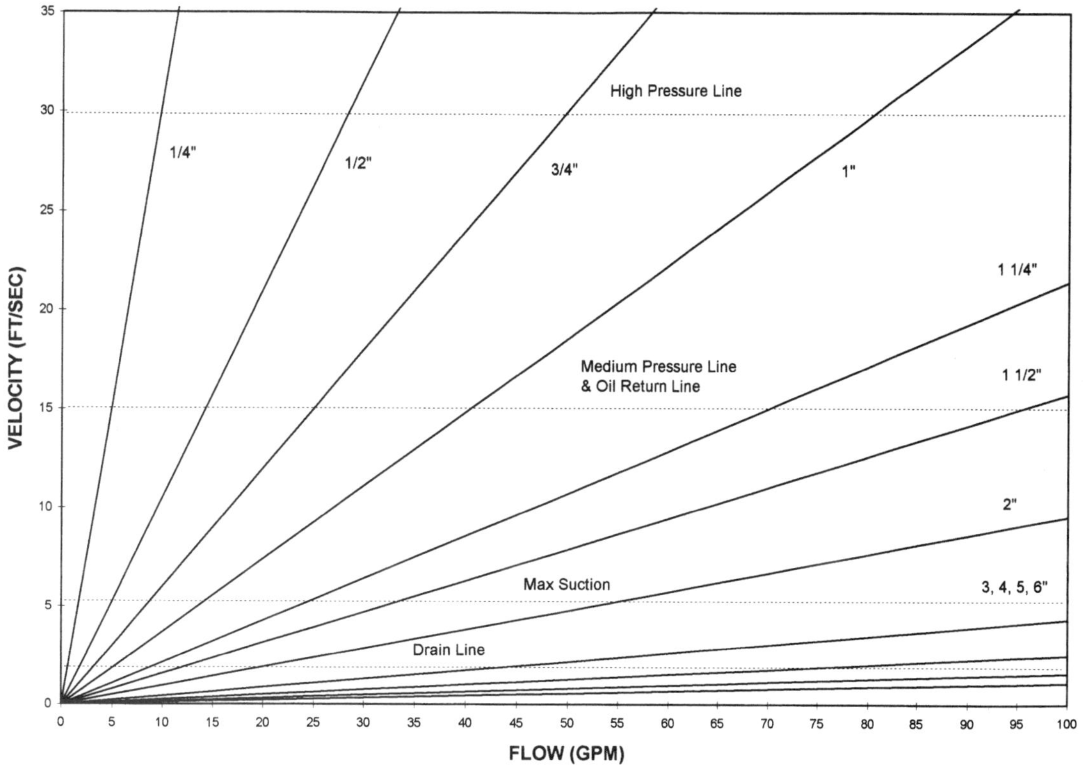

Schedule 80 Pipe
Flow & Velocity Chart

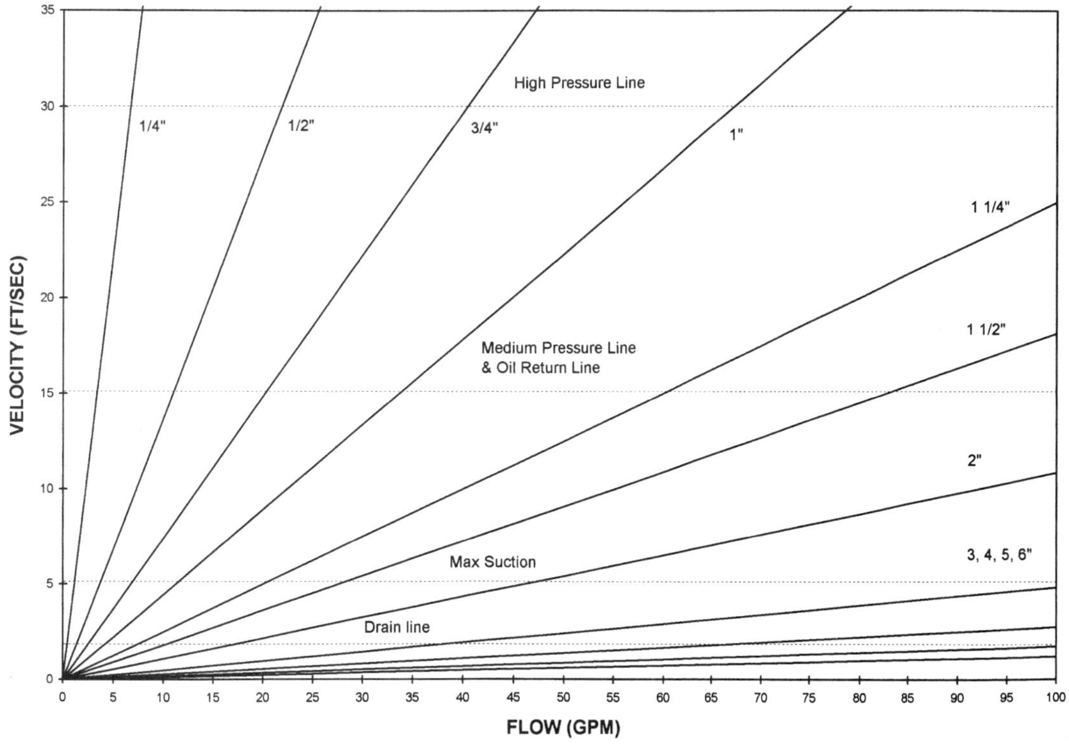

Schedule 160 Pipe Flow & Velocity Chart

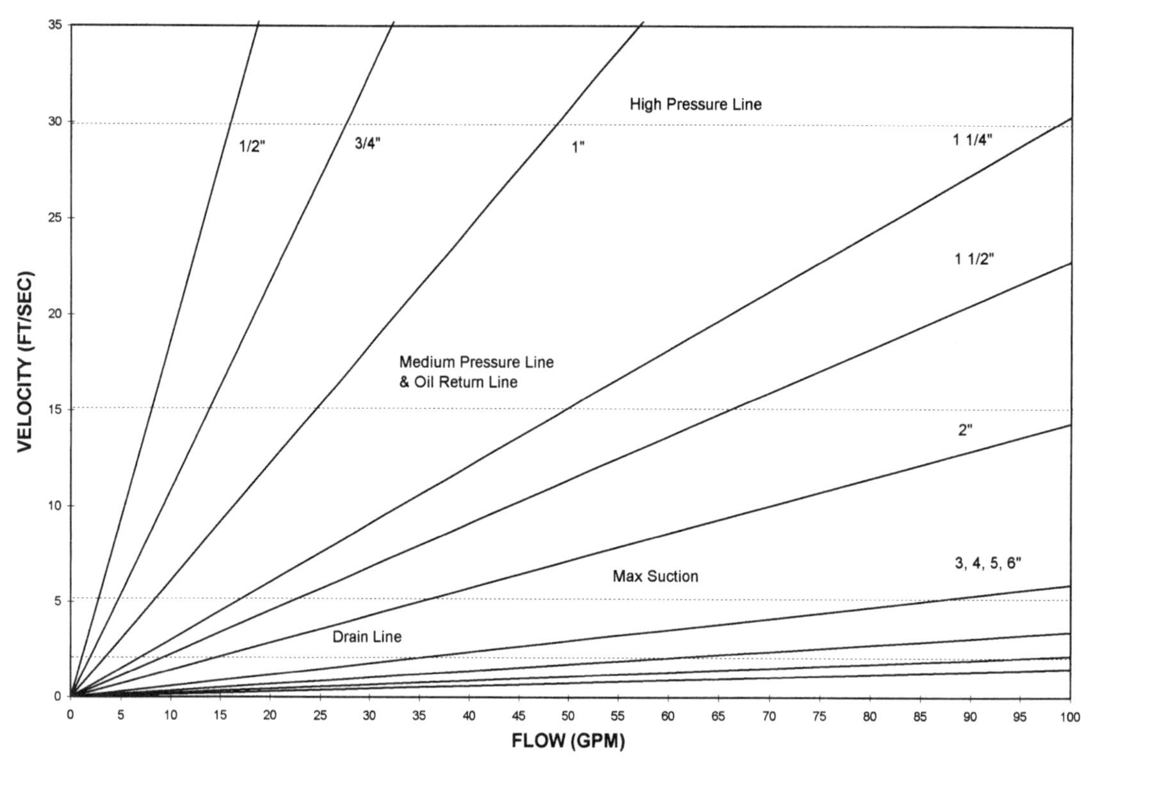

Schedule XXS Pipe Flow & Velocity Chart

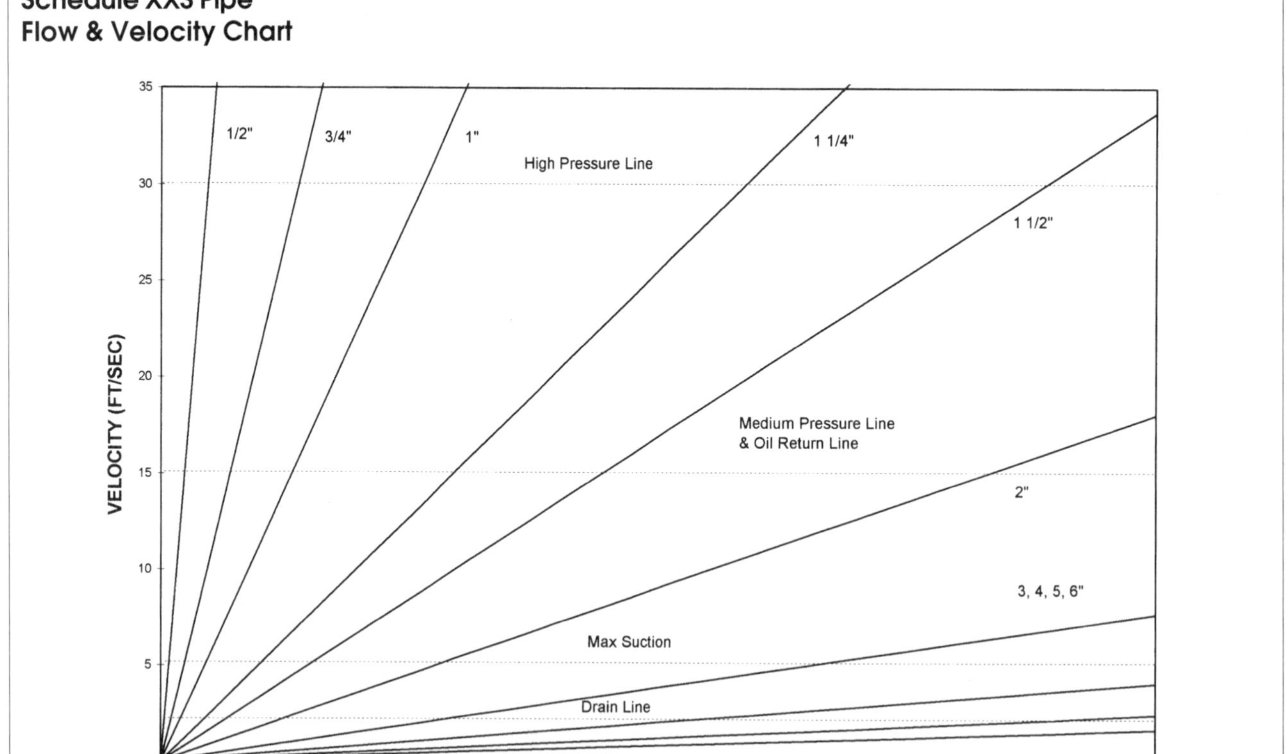

Steel Hydraulic Tubing Data

25,000 psi yield — 350° F limit — Safety Factor of 6.0										
Nominal Tube Size	I.D.	Wall Thickness	Maximum Pressure	Bursting Pressure	Flow gpm (ft/sec) 2	5	10	15	20	25
1/8	.055	.035	2390	9330						
	.061	.032	2130	8530						
	.065	.030	1960	8000						
	.069	.028	1800	7470						
3/16	.117	.035	1480	6240						
	.123	.032	1330	5700						
	.127	.030	1230	5350						
1/4	.120	.065	2240	8670						
	.134	.058	1950	7730						
	.152	.049	1590	6530						
	.166	.042	1320	5600						
	.180	.035	1070	4670						
	.190	.032	950	4200						
5/16	.182	.065	1720	6940						
	.196	.058	1510	6200						
	.214	.049	1240	5240						
	.228	.042	1040	4490						
	.242	.035	840	3740						
	.248	.032	760	3420						
3/8	.245	.065	1400	5780	0.3	0.7	1.5	2.2	2.9	3.7
	.259	.058	1220	5160	0.3	0.8	1.6	2.5	3.3	4.1
	.277	.049	1010	4360	0.4	0.9	1.9	2.8	3.8	4.7
	.291	.042	850	3730	0.4	1.0	2.1	3.1	4.1	5.2
	.305	.035	690	3110	0.5	1.1	2.3	3.4	4.6	5.7
	.311	.032	620	2840	0.5	1.2	2.4	3.6	4.7	5.9
1/2	.310	.095	1560	6330	0.5	1.2	2.4	3.5	4.7	5.9
	.334	.083	1340	5530	0.5	1.4	2.7	4.1	5.5	6.8
	.358	.072	1130	4780	0.6	1.6	3.1	4.7	6.3	7.8
	.370	.065	1010	4330	0.7	1.7	3.4	5.0	6.7	8.4
	.384	.058	890	3870	0.7	1.8	3.6	5.4	7.2	9.0
	.402	.049	740	3270	0.8	2.0	4.0	5.9	7.9	9.9
	.416	.042	620	2800	0.8	2.1	4.2	6.4	8.5	10.6
	.430	.035	510	2330	0.9	2.3	4.5	6.8	9.1	11.3
	.436	.032	460	2130	0.9	2.3	4.7	7.0	9.3	11.6
5/8	.435	.095	1220	5070	0.9	2.3	4.6	6.9	9.3	11.6
	.459	.083	1040	4430	1.0	2.6	5.2	7.7	10.3	12.9
	.481	.072	890	3840	1.1	2.8	5.7	8.5	11.3	14.2
	.495	.065	790	3470	1.2	3.0	6.0	9.0	12.0	15.0
	.509	.058	700	3090	1.3	3.2	6.3	9.5	12.7	15.9
	.527	.049	580	2610	1.4	3.4	6.8	10.2	13.6	17.0
	.541	.042	490	2240	1.4	3.6	7.2	10.7	14.3	17.9
	.555	.035	400	1870	1.5	3.8	7.5	11.3	15.1	18.9
3/4	.532	.109	1160	4840	1.4	3.5	6.9	10.4	13.9	17.3
	.560	.095	990	4220	1.5	3.8	7.7	11.5	15.4	19.2
	.584	.083	860	3690	1.7	4.2	8.3	12.5	16.7	20.9
	.606	.072	730	3200	1.8	4.5	9.0	13.5	18.0	22.5
	.620	.065	650	2890	1.9	4.7	9.4	14.1	18.8	23.5
	.634	.058	580	2580	2.0	4.9	9.8	14.8	19.7	24.6
	.652	.049	480	2180	2.1	5.2	10.4	15.6	20.8	26.0
	.666	.042	410	1870	2.2	5.4	10.9	16.3	21.7	27.1
	.680	.035	330	1560	2.3	5.7	11.3	17.0	22.6	28.3

Calculated flows in these pipe sizes are not consistent with actual flows. Testing is required.

Steel Hydraulic Tubing Data

25,000 psi yield — 350° F limit — Safety Factor of 6.0										
Nominal Tube Size	I.D.	Wall Thickness	Maximum Pressure	Bursting Pressure	Flow gpm (ft/sec) 2	5	10	15	20	25
7/8	.657	.109	980	4150	2.1	5.3	10.6	15.9	21.1	26.4
	.685	.095	840	3620	2.3	5.7	11.5	17.2	23.0	28.7
	.709	.083	720	3160	2.5	6.2	12.3	18.5	24.6	30.8
	.731	.072	620	2740	2.6	6.5	13.1	19.6	26.2	32.7
	.745	.065	560	2480	2.7	6.8	13.6	20.4	27.2	34.0
	.759	.058	490	2210	2.8	7.1	14.1	21.2	28.2	35.3
	.777	.049	410	1870	3.0	7.4	14.8	22.2	29.6	36.9
1	.760	.120	940	4000	2.8	7.1	14.1	21.2	28.3	35.3
	.782	.109	850	3630	3.0	7.5	15.0	22.5	29.9	37.4
	.810	.095	730	3170	3.2	8.0	16.1	24.1	32.1	40.2
	.834	.083	630	2770	3.4	8.5	17.0	25.5	34.1	42.6
	.856	.072	540	2400	3.6	9.0	17.9	26.9	35.9	44.8
	.870	.065	480	2170	3.7	9.3	18.5	27.8	37.1	46.3
	.884	.058	430	1930	3.8	9.6	19.1	28.7	38.3	47.8
	.902	.049	360	1630	4.0	10.0	19.9	29.9	39.8	49.8
1 1/4	.982	.134	830	3570	4.7	11.8	23.6	35.4	47.2	59.0
	1.010	.120	740	3200	5.0	12.5	25.0	37.5	49.9	62.4
	1.032	.109	670	2910	5.2	13.0	26.1	39.1	52.1	65.2
	1.060	.095	570	2530	5.5	13.8	27.5	41.3	55.0	68.8
	1.084	.083	500	2210	5.8	14.4	28.8	43.1	57.5	71.9
	1.106	.072	430	1920	6.0	15.0	29.9	44.9	59.9	74.9
	1.120	.065	380	1730	6.1	15.4	30.7	46.1	61.4	76.8
	1.134	.058	340	1550	6.3	15.7	31.5	47.2	63.0	78.7
	1.152	.049	280	1310	6.5	16.2	32.5	48.7	65.0	81.2
1 1/2	1.232	.134	690	2980	7.4	18.6	37.2	55.7	74.3	92.9
	1.260	.120	610	2670	7.8	19.4	38.9	58.3	77.7	97.2
	1.282	.109	550	2420	8.0	20.1	40.2	60.4	80.5	100.6
	1.310	.095	470	2110	8.4	21.0	42.0	63.0	84.0	105.0
	1.334	.083	410	1840	8.7	21.8	43.6	65.3	87.1	108.9
	1.356	.072	350	1600	9.0	22.5	45.0	67.5	90.0	112.5
	1.370	.065	320	1440	9.2	23.0	45.9	68.9	91.9	114.9
2	1.732	.134	510	2230	14.7	36.7	73.4	110.2	146.9	183.6
	1.760	.120	450	2000	15.2	37.9	75.8	113.7	151.7	189.6
	1.782	.109	410	1820	15.5	38.9	77.7	116.6	155.5	194.3
	1.810	.095	350	1580	16.0	40.1	80.2	120.3	160.4	200.5
	1.834	.083	310	1380	16.5	41.2	82.3	123.5	164.7	205.8
	1.856	.072	260	1200	16.9	42.2	84.3	126.5	168.7	210.8
	1.870	.065	240	1080	17.1	42.8	85.6	128.4	171.2	214.0

Steel Hydraulic Tubing Data

55,000 psi yield — 350° F limit — Safety Factor of 6.0										
Nominal Tube Size	I.D.	Wall Thickness	Maximum Pressure	Bursting Pressure	Flow gpm (ft/sec) 2	5	10	15	20	25
1/8	.055	.035	5250	20530						
	.061	.032	4680	18770						
	.065	.030	4310	17600						
	.069	.028	3960	16430						
3/16	.117	.035	3260	13730						
	.123	.032	2930	12550						
	.127	.030	2710	11760						
1/4	.120	.065	4940	19070		Calculated flows in these pipe sizes are not consistent with actual flows. Testing is required.				
	.134	.058	4290	17010						
	.152	.049	3500	14370						
	.166	.042	2910	12320						
	.180	.035	2360	10270						
	.190	.032	2090	9240						
5/16	.182	.065	3790	15280						
	.196	.058	3310	13630						
	.214	.049	2720	11520						
	.228	.042	2280	9870						
	.242	.035	1850	8230						
	.248	.032	1670	7520						
3/8	.245	.065	3070	12710	0.3	0.7	1.5	2.2	2.9	3.7
	.259	.058	2690	11340	0.3	0.8	1.6	2.5	3.3	4.1
	.277	.049	2220	9580	0.4	0.9	1.9	2.8	3.8	4.7
	.291	.042	1860	8210	0.4	1.0	2.1	3.1	4.1	5.2
	.305	.035	1520	6840	0.5	1.1	2.3	3.4	4.6	5.7
	.311	.032	1370	6260	0.5	1.2	2.4	3.6	4.7	5.9
1/2	.310	.095	3440	13930	0.5	1.2	2.4	3.5	4.7	5.9
	.334	.083	2940	12170	0.5	1.4	2.7	4.1	5.5	6.8
	.358	.072	2490	10520	0.6	1.6	3.1	4.7	6.3	7.8
	.370	.065	2230	9530	0.7	1.7	3.4	5.0	6.7	8.4
	.384	.058	1960	8510	0.7	1.8	3.6	5.4	7.2	9.0
	.402	.049	1620	7190	0.8	2.0	4.0	5.9	7.9	9.9
	.416	.042	1370	6160	0.8	2.1	4.2	6.4	8.5	10.6
	.430	.035	1120	5130	0.9	2.3	4.5	6.8	9.1	11.3
	.436	.032	1010	4690	0.9	2.3	4.7	7.0	9.3	11.6
5/8	.435	.095	2670	11150	0.9	2.3	4.6	6.9	9.3	11.6
	.459	.083	2300	9740	1.0	2.6	5.2	7.7	10.3	12.9
	.481	.072	1960	8450	1.1	2.8	5.7	8.5	11.3	14.2
	.495	.065	1750	7630	1.2	3.0	6.0	9.0	12.0	15.0
	.509	.058	1540	6810	1.3	3.2	6.3	9.5	12.7	15.9
	.527	.049	1280	5750	1.4	3.4	6.8	10.2	13.6	17.0
	.541	.042	1080	4930	1.4	3.6	7.2	10.7	14.3	17.9
	.555	.035	890	4110	1.5	3.8	7.5	11.3	15.1	18.9
3/4	.532	.109	2550	10660	1.4	3.5	6.9	10.4	13.9	17.3
	.560	.095	2190	9290	1.5	3.8	7.7	11.5	15.4	19.2
	.584	.083	1880	8120	1.7	4.2	8.3	12.5	16.7	20.9
	.606	.072	1610	7040	1.8	4.5	9.0	13.5	18.0	22.5
	.620	.065	1440	6360	1.9	4.7	9.4	14.1	18.8	23.5
	.634	.058	1270	5670	2.0	4.9	9.8	14.8	19.7	24.6
	.652	.049	1060	4790	2.1	5.2	10.4	15.6	20.8	26.0
	.666	.042	900	4110	2.2	5.4	10.9	16.3	21.7	27.1
	.680	.035	730	3420	2.3	5.7	11.3	17.0	22.6	28.3

Steel Hydraulic Tubing Data

55,000 psi yield — 350⁰ F limit — Safety Factor of 6.0										
Nominal Tube Size	I.D.	Wall Thickness	Maximum Pressure	Bursting Pressure	Flow gpm (ft/sec) 2	5	10	15	20	25
7/8	.657	.109	2150	9140	2.1	5.3	10.6	15.9	21.1	26.4
	.685	.095	1850	7960	2.3	5.7	11.5	17.2	23.0	28.7
	.709	.083	1590	6960	2.5	6.2	12.3	18.5	24.6	30.8
	.731	.072	1370	6030	2.6	6.5	13.1	19.6	26.2	32.7
	.745	.065	1220	5450	2.7	6.8	13.6	20.4	27.2	34.0
	.759	.058	1080	4860	2.8	7.1	14.1	21.2	28.2	35.3
	.777	.049	900	4110	3.0	7.4	14.8	22.2	29.6	36.9
1	.760	.120	2070	8800	2.8	7.1	14.1	21.2	28.3	35.3
	.782	.109	1860	7990	3.0	7.5	15.0	22.5	29.9	37.4
	.810	.095	1600	6970	3.2	8.0	16.1	24.1	32.1	40.2
	.834	.083	1380	6090	3.4	8.5	17.0	25.5	34.1	42.6
	.856	.072	1190	5280	3.6	9.0	17.9	26.9	35.9	44.8
	.870	.065	1060	4770	3.7	9.3	18.5	27.8	37.1	46.3
	.884	.058	940	4250	3.8	9.6	19.1	28.7	38.3	47.8
	.902	.049	780	3590	4.0	10.0	19.9	29.9	39.8	49.8
1 1/4	.982	.134	1830	7860	4.7	11.8	23.6	35.4	47.2	59.0
	1.010	.120	1630	7040	5.0	12.5	25.0	37.5	49.9	62.4
	1.032	.109	1460	6390	5.2	13.0	26.1	39.1	52.1	65.2
	1.060	.095	1260	5570	5.5	13.8	27.5	41.3	55.0	68.8
	1.084	.083	1090	4870	5.8	14.4	28.8	43.1	57.5	71.9
	1.106	.072	940	4220	6.0	15.0	29.9	44.9	59.9	74.9
	1.120	.065	840	3810	6.1	15.4	30.7	46.1	61.4	76.8
	1.134	.058	750	3400	6.3	15.7	31.5	47.2	63.0	78.7
	1.152	.049	620	2870	6.5	16.2	32.5	48.7	65.0	81.2
1 1/2	1.232	.134	1510	6550	7.4	18.6	37.2	55.7	74.3	92.9
	1.260	.120	1340	5870	7.8	19.4	38.9	58.3	77.7	97.2
	1.282	.109	1210	5330	8.0	20.1	40.2	60.4	80.5	100.6
	1.310	.095	1040	4640	8.4	21.0	42.0	63.0	84.0	105.0
	1.334	.083	900	4060	8.7	21.8	43.6	65.3	87.1	108.9
	1.356	.072	780	3250	9.0	22.5	45.0	67.5	90.0	112.5
	1.370	.065	700	3180	9.2	23.0	45.9	68.9	91.9	114.9
2	1.732	.134	1110	4910	14.7	36.7	73.4	110.2	146.9	183.6
	1.760	.120	990	4400	15.2	37.9	75.8	113.7	151.7	189.6
	1.782	.109	890	4000	15.5	38.9	77.7	116.6	155.5	194.3
	1.810	.095	770	3480	16.0	40.1	80.2	120.3	160.4	200.5
	1.834	.083	670	3040	16.5	41.2	82.3	123.5	164.7	205.8
	1.856	.072	580	2640	16.9	42.2	84.3	126.5	168.7	210.8
	1.870	.065	520	2380	17.1	42.8	85.6	128.4	171.2	214.0

Stainless Steel Hydraulic Tubing Data

70,000 psi yield — 350° F limit — Safety Factor of 6.0										
Nominal Tube Size	I.D.	Wall Thickness	Maximum Pressure	Bursting Pressure	Flow gpm (ft/sec) 2	5	10	15	20	25
1/8	.055	.035	6680	26130	Calculated flows in these pipe sizes are not consistent with actual flows. Testing is required.					
	.061	.032	5960	23890						
	.065	.030	5490	22400						
	.069	.028	5030	20910						
3/16	.117	.035	4150	17470						
	.123	.032	3730	15970						
	.127	.030	3450	14970						
1/4	.120	.065	6280	24270						
	.134	.058	5460	21650						
	.152	.049	4450	18290						
	.166	.042	3710	15680						
	.180	.035	3000	13070						
	.190	.032	2660	11760						
5/16	.182	.065	4830	19440						
	.196	.058	4220	17350						
	.214	.049	3460	14660						
	.228	.042	2900	12560						
	.242	.035	2350	10470						
	.248	.032	2130	9570						
3/8	.245	.065	3910	16180	0.3	0.7	1.5	2.2	2.9	3.7
	.259	.058	3420	14440	0.3	0.8	1.6	2.5	3.3	4.1
	.277	.049	2820	12200	0.4	0.9	1.9	2.8	3.8	4.7
	.291	.042	2370	10450	0.4	1.0	2.1	3.1	4.1	5.2
	.305	.035	1930	8710	0.5	1.1	2.3	3.4	4.6	5.7
	.311	.032	1750	7960	0.5	1.2	2.4	3.6	4.7	5.9
1/2	.310	.095	4380	17730	0.5	1.2	2.4	3.5	4.7	5.9
	.334	.083	3750	15490	0.5	1.4	2.7	4.1	5.5	6.8
	.358	.072	3170	13390	0.6	1.6	3.1	4.7	6.3	7.8
	.370	.065	2840	12130	0.7	1.7	3.4	5.0	6.7	8.4
	.384	.058	2490	10830	0.7	1.8	3.6	5.4	7.2	9.0
	.402	.049	2070	9150	0.8	2.0	4.0	5.9	7.9	9.9
	.416	.042	1740	7840	0.8	2.1	4.2	6.4	8.5	10.6
	.430	.035	1430	6530	0.9	2.3	4.5	6.8	9.1	11.3
	.436	.032	1290	5970	0.9	2.3	4.7	7.0	9.3	11.6
5/8	.435	.095	3400	14190	0.9	2.3	4.6	6.9	9.3	11.6
	.459	.083	2920	12390	1.0	2.6	5.2	7.7	10.3	12.9
	.481	.072	2490	10750	1.1	2.8	5.7	8.5	11.3	14.2
	.495	.065	2230	9710	1.2	3.0	6.0	9.0	12.0	15.0
	.509	.058	1960	8660	1.3	3.2	6.3	9.5	12.7	15.9
	.527	.049	1630	7320	1.4	3.4	6.8	10.2	13.6	17.0
	.541	.042	1380	6270	1.4	3.6	7.2	10.7	14.3	17.9
	.555	.035	1130	5230	1.5	3.8	7.5	11.3	15.1	18.9
3/4	.532	.109	3250	13560	1.4	3.5	6.9	10.4	13.9	17.3
	.560	.095	2780	11820	1.5	3.8	7.7	11.5	15.4	19.2
	.584	.083	2390	10330	1.7	4.2	8.3	12.5	16.7	20.9
	.606	.072	2050	8960	1.8	4.5	9.0	13.5	18.0	22.5
	.620	.065	1830	8090	1.9	4.7	9.4	14.1	18.8	23.5
	.634	.058	1620	7220	2.0	4.9	9.8	14.8	19.7	24.6
	.652	.049	1350	6100	2.1	5.2	10.4	15.6	20.8	26.0
	.666	.042	1140	5230	2.2	5.4	10.9	16.3	21.7	27.1
	.680	.035	940	4360	2.3	5.7	11.3	17.0	22.6	28.3

Stainless Steel Hydraulic Tubing Data

70,000 psi yield — 350° F limit — Safety Factor of 6.0										
Nominal Tube Size	I.D.	Wall Thickness	Maximum Pressure	Bursting Pressure	Flow gpm (ft/sec)					
					2	5	10	15	20	25
7/8	.657	.109	2740	11630	2.1	5.3	10.6	15.9	21.1	26.4
	.685	.095	2350	10130	2.3	5.7	11.5	17.2	23.0	28.7
	.709	.083	2030	8850	2.5	6.2	12.3	18.5	24.6	30.8
	.731	.072	1740	7680	2.6	6.5	13.1	19.6	26.2	32.7
	.745	.065	1560	6930	2.7	6.8	13.6	20.4	27.2	34.0
	.759	.058	1380	6190	2.8	7.1	14.1	21.2	28.2	35.3
	.777	.049	1150	5230	3.0	7.4	14.8	22.2	29.6	36.9
1	.760	.120	2630	11200	2.8	7.1	14.1	21.2	28.3	35.3
	.782	.109	2370	10170	3.0	7.5	15.0	22.5	29.9	37.4
	.810	.095	2040	8870	3.2	8.0	16.1	24.1	32.1	40.2
	.834	.083	1760	7750	3.4	8.5	17.0	25.5	34.1	42.6
	.856	.072	1510	6720	3.6	9.0	17.9	26.9	35.9	44.8
	.870	.065	1350	6070	3.7	9.3	18.5	27.8	37.1	46.3
	.884	.058	1200	5410	3.8	9.6	19.1	28.7	38.3	47.8
	.902	.049	1000	4570	4.0	10.0	19.9	29.9	39.8	49.8
1 1/4	.982	.134	2330	10010	4.7	11.8	23.6	35.4	47.2	59.0
	1.010	.120	2070	8960	5.0	12.5	25.0	37.5	49.9	62.4
	1.032	.109	1860	8140	5.2	13.0	26.1	39.1	52.1	65.2
	1.060	.095	1610	7090	5.5	13.8	27.5	41.3	55.0	68.8
	1.084	.083	1390	6200	5.8	14.4	28.8	43.1	57.5	71.9
	1.106	.072	1200	5380	6.0	15.0	29.9	44.9	59.9	74.9
	1.120	.065	1070	4850	6.1	15.4	30.7	46.1	61.4	76.8
	1.134	.058	950	4330	6.3	15.7	31.5	47.2	63.0	78.7
	1.152	.049	790	3660	6.5	16.2	32.5	48.7	65.0	81.2
1 1/2	1.232	.134	1920	8340	7.4	18.6	37.2	55.7	74.3	92.9
	1.260	.120	1700	7470	7.8	19.4	38.9	58.3	77.7	97.2
	1.282	.109	1540	6780	8.0	20.1	40.2	60.4	80.5	100.6
	1.310	.095	1330	5910	8.4	21.0	42.0	63.0	84.0	105.0
	1.334	.083	1150	5160	8.7	21.8	43.6	65.3	87.1	108.9
	1.356	.072	990	4480	9.0	22.5	45.0	67.5	90.0	112.5
	1.370	.065	890	4040	9.2	23.0	45.9	68.9	91.9	114.9
2	1.732	.134	1420	6250	14.7	36.7	73.4	110.2	146.9	183.6
	1.760	.120	1260	5600	15.2	37.9	75.8	113.7	151.7	189.6
	1.782	.109	1140	5090	15.5	38.9	77.7	116.6	155.5	194.3
	1.810	.095	980	4430	16.0	40.1	80.2	120.3	160.4	200.5
	1.834	.083	850	3870	16.5	41.2	82.3	123.5	164.7	205.8
	1.856	.072	740	3360	16.9	42.2	84.3	126.5	168.7	210.8
	1.870	.065	660	3030	17.1	42.8	85.6	128.4	171.2	214.0

Copper Tubing Data

Nominal Tube Size	I.D.	Wall Thickness	Maximum Pressure	Bursting Pressure	Flow gpm (ft/sec) 2	5	10	15	20	25
30,000 psi yield — 350° F limit — Safety Factor of 6.0										
1/8	.055	.035	2860	11200	Calculated flows in these pipe sizes are not consistent with actual flows. Testing is required.					
	.061	.032	2550	10240						
	.065	.030	2350	9600						
	.069	.028	2160	8960						
3/16	.117	.035	1780	7490						
	.123	.032	1600	6840						
	.127	.030	1480	6420						
1/4	.120	.065	2690	10400						
	.134	.058	2340	9280						
	.152	.049	1910	7840						
	.166	.042	1590	6720						
	.180	.035	1280	5600						
	.190	.032	1140	5040						
5/16	.182	.065	2070	8330						
	.196	.058	1810	7440						
	.214	.049	1480	6280						
	.228	.042	1240	5380						
	.242	.035	1010	4490						
	.248	.032	910	4100						
3/8	.245	.065	1670	6930	0.3	0.7	1.5	2.2	2.9	3.7
	.259	.058	1470	6190	0.3	0.8	1.6	2.5	3.3	4.1
	.277	.049	1210	5230	0.4	0.9	1.9	2.8	3.8	4.7
	.291	.042	1020	4480	0.4	1.0	2.1	3.1	4.1	5.2
	.305	.035	830	3730	0.5	1.1	2.3	3.4	4.6	5.7
	.311	.032	750	3410	0.5	1.2	2.4	3.6	4.7	5.9
1/2	.310	.095	1880	7600	0.5	1.2	2.4	3.5	4.7	5.9
	.334	.083	1610	6640	0.5	1.4	2.7	4.1	5.5	6.8
	.358	.072	1360	5740	0.6	1.6	3.1	4.7	6.3	7.8
	.370	.065	1220	5200	0.7	1.7	3.4	5.0	6.7	8.4
	.384	.058	1070	4640	0.7	1.8	3.6	5.4	7.2	9.0
	.402	.049	890	3920	0.8	2.0	4.0	5.9	7.9	9.9
	.416	.042	750	3360	0.8	2.1	4.2	6.4	8.5	10.6
	.430	.035	610	2800	0.9	2.3	4.5	6.8	9.1	11.3
	.436	.032	550	2560	0.9	2.3	4.7	7.0	9.3	11.6
5/8	.435	.095	1460	6080	0.9	2.3	4.6	6.9	9.3	11.6
	.459	.083	1250	5310	1.0	2.6	5.2	7.7	10.3	12.9
	.481	.072	1070	4610	1.1	2.8	5.7	8.5	11.3	14.2
	.495	.065	950	4160	1.2	3.0	6.0	9.0	12.0	15.0
	.509	.058	840	3710	1.3	3.2	6.3	9.5	12.7	15.9
	.527	.049	700	3140	1.4	3.4	6.8	10.2	13.6	17.0
	.541	.042	590	2690	1.4	3.6	7.2	10.7	14.3	17.9
	.555	.035	480	2240	1.5	3.8	7.5	11.3	15.1	18.9
3/4	.532	.109	1390	5810	1.4	3.5	6.9	10.4	13.9	17.3
	.560	.095	1190	5070	1.5	3.8	7.7	11.5	15.4	19.2
	.584	.083	1030	4430	1.7	4.2	8.3	12.5	16.7	20.9
	.606	.072	880	3840	1.8	4.5	9.0	13.5	18.0	22.5
	.620	.065	780	3470	1.9	4.7	9.4	14.1	18.8	23.5
	.634	.058	690	3090	2.0	4.9	9.8	14.8	19.7	24.6
	.652	.049	580	2610	2.1	5.2	10.4	15.6	20.8	26.0
	.666	.042	490	2240	2.2	5.4	10.9	16.3	21.7	27.1
	.680	.035	400	1870	2.3	5.7	11.3	17.0	22.6	28.3

Copper Tubing Data

30,000 psi yield — 350° F limit — Safety Factor of 6.0										
Nominal Tube Size	I.D.	Wall Thickness	Maximum Pressure	Bursting Pressure	Flow gpm (ft/sec) 2	5	10	15	20	25
7/8	.657	.109	1170	4980	2.1	5.3	10.6	15.9	21.1	26.4
	.685	.095	1010	4340	2.3	5.7	11.5	17.2	23.0	28.7
	.709	.083	870	3790	2.5	6.2	12.3	18.5	24.6	30.8
	.731	.072	740	3290	2.6	6.5	13.1	19.6	26.2	32.7
	.745	.065	670	2970	2.7	6.8	13.6	20.4	27.2	34.0
	.759	.058	590	2650	2.8	7.1	14.1	21.2	28.2	35.3
	.777	.049	490	2240	3.0	7.4	14.8	22.2	29.6	36.9
1	.760	.120	1130	4800	2.8	7.1	14.1	21.2	28.3	35.3
	.782	.109	1010	4360	3.0	7.5	15.0	22.5	29.9	37.4
	.810	.095	870	3800	3.2	8.0	16.1	24.1	32.1	40.2
	.834	.083	750	3320	3.4	8.5	17.0	25.5	34.1	42.6
	.856	.072	650	2880	3.6	9.0	17.9	26.9	35.9	44.8
	.870	.065	580	2600	3.7	9.3	18.5	27.8	37.1	46.3
	.884	.058	510	2320	3.8	9.6	19.1	28.7	38.3	47.8
	.902	.049	430	1960	4.0	10.0	19.9	29.9	39.8	49.8
1 1/4	.982	.134	1000	4290	4.7	11.8	23.6	35.4	47.2	59.0
	1.010	.120	890	3840	5.0	12.5	25.0	37.5	49.9	62.4
	1.032	.109	800	3490	5.2	13.0	26.1	39.1	52.1	65.2
	1.060	.095	690	3040	5.5	13.8	27.5	41.3	55.0	68.8
	1.084	.083	600	2660	5.8	14.4	28.8	43.1	57.5	71.9
	1.106	.072	510	2300	6.0	15.0	29.9	44.9	59.9	74.9
	1.120	.065	460	2080	6.1	15.4	30.7	46.1	61.4	76.8
	1.134	.058	410	1860	6.3	15.7	31.5	47.2	63.0	78.7
	1.152	.049	340	1570	6.5	16.2	32.5	48.7	65.0	81.2
1 1/2	1.232	.134	820	3570	7.4	18.6	37.2	55.7	74.3	92.9
	1.260	.120	730	3200	7.8	19.4	38.9	58.3	77.7	97.2
	1.282	.109	660	2910	8.0	20.1	40.2	60.4	80.5	100.6
	1.310	.095	570	2530	8.4	21.0	42.0	63.0	84.0	105.0
	1.334	.083	490	2210	8.7	21.8	43.6	65.3	87.1	108.9
	1.356	.072	420	1920	9.0	22.5	45.0	67.5	90.0	112.5
	1.370	.065	380	1730	9.2	23.0	45.9	68.9	91.9	114.9
2	1.732	.134	610	2680	14.7	36.7	73.4	110.2	146.9	183.6
	1.760	.120	540	2400	15.2	37.9	75.8	113.7	151.7	189.6
	1.782	.109	490	2180	15.5	38.9	77.7	116.6	155.5	194.3
	1.810	.095	420	1900	16.0	40.1	80.2	120.3	160.4	200.5
	1.834	.083	370	1660	16.5	41.2	82.3	123.5	164.7	205.8
	1.856	.072	320	1440	16.9	42.2	84.3	126.5	168.7	210.8
	1.870	.065	280	1300	17.1	42.8	85.6	128.4	171.2	214.0

Tube Fittings

2-piece Assembly

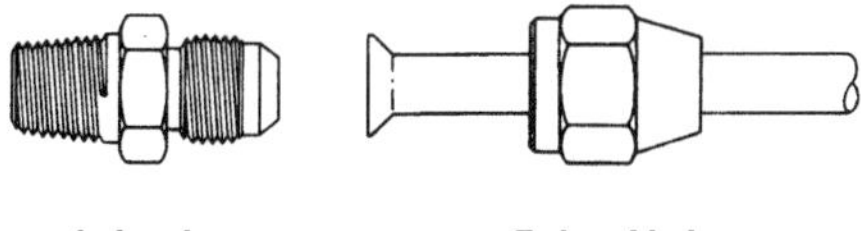

Adapter   Tube Nut

3-piece Assembly

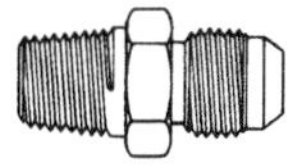

Adapter

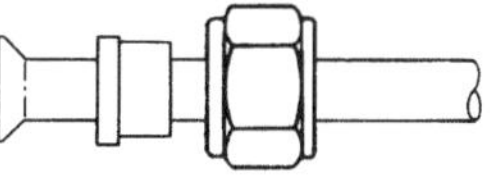

Sleeve   Nut

Thread Size Chart

The following chart is intended as a quick reference guide for thread size by dash size.

Dash Size	N.P.T.F.	N.P.S.M.	SAE 45⁰ auto refrig.	SAE 37⁰ (JIC) hydraulic	SAE O-Ring boss	P.T.T. 300 automotive	SAE invert flare	ORS
-02	1/8-27	1/8-27	5/16-24	5/16-24	5/16-24		5/16-24	
-03			3/8-24	3/8-24	3/8-24		3/8-24	
-04	1/4-18	1/4-18	7/16-20	7/16-20	7/16-20		7/16-24	9/16-18
-05			1/2-20	1/2-20	1/2-20		1/2-20	
-06	3/8-18	3/8-18	5/8-18	9/16-18	9/16-18		5/8-18	11/16-16
-07			11/16-24				11/16-18	
-08	1/2-14	1/2-14	3/4-16	3/4-16	3/4-16		3/4-18	13/16-16
-10			7/8-14	7/8-14	7/8-14		7/8-18	1-14
-12	3/4-14	3/4-14	11/16-14	1 1/16-12	1 1/16-12		1 1/16-16	1 3/16-12
-14				1 3/16-12	1 3/16-12			
-16	1-11 1/2	1-11 1/2	15/16-12	1 5/16-12	1 5/16-12			1 7/16-12
-20	1 1/4-11 1/2	1 1/4-11 1/2	1 5/8-12	1 5/8-12	1 5/8-12			1 11/16-12
-24	1 1/2-11 1/2	1 1/2-11 1/2		1 7/8-12	1 7/8-12	1 7/8-14		2-12
-32	2-11 1/2	2-11 1/2		2 1/2-12	2 1/2-12	2 1/2-12		
-40	2 1/2-8			3-12	3-12			
-48	3-8			3 1/2-12	3 1/2-12			

Tube/Pipe Conversions

Tube O.D.	Dash Number	SAE Thread	Pipe Thread	Tube O.D.	Metric Thread	BSPP Thread
1/4	-4	7/16-20	1/4-18	6 mm	M10 x 1.0	1/4-19
5/16	-5	1/2-20	N/A	8 mm	M12 x 1.5	N/A
3/8	-6	9/16-18	3/8-18	10 mm	M14 x 1.5	3/8-19
1/2	-8	3/4-16	1/2-14	12 mm	M18 x 1.5	1/2-14
5/8	-10	7/8-14	1/2-14	12 mm	M18 x 1.5	1/2-14
3/4	-12	1 1/16-12	3/4-14	18 mm	M26 x 1.5	N/A
7/8	-14	1 3/16-12	N/A	22 mm	M26 x 1.5	N/A
1	-16	1 5/16-12	1-11 1/2	25 mm	M33 x 2.0	1-11
1 1/4	-20	1 5/8-12	1 1/4-11 1/2	32 mm	M42 x 2.0	1 1/4-11
1 1/2	-24	1 7/8-12	1 1/2-11 1/2	8 mm	M42 x 2.0	1 1/2-11
2	-32	2 1/2-12	2-11 1/2	50 mm	M48 x 2.0	2-11

Recommended Parallel Connection Assembly Torque

A torque wrench should be used to assure proper fitting assembly of these connections.

	Dash Size	Thread Size (Inches)	Swivel Nut Torque	
			Lb-Ft	Newton Meters
ORS	-04	9/16-18	10-12	14-16
	-06	11/16-16	18-20	24-27
	-08	13/16-16	32-35	43-47
	-10	1-14	46-50	62-68
	-12	1 3/16-12	65-70	88-95
	-16	1 7/16-12	92-100	125-136
	-20	1 11/16-12	125-140	170-190
	-24	2-12	150-165	204-224

	Dash Size	Thread Size (Inches)	Jam Nut or Straight Fitting Torque	
			Lb-Ft	Newton Meters
Straight Thread O-Ring Boss Low Pressure with 37° (SAEJ514)	-03	3/8-24	8-9	12-13
	-04	7/16-20	13-15	18-20
	-05	1/2-20	14-15	19-21
	-06	9/16-18	23-24	32-33
	-08	3/4-16	40-43	55-57
	-10	7/8-14	43-48	59-64
	-12	1 1/16-12	68-75	93-101
	-14	1 3/16-12	83-90	113-122
	-16	1 5/16-12	112-123	152-166
	-20	1 5/8-12	146-161	198-218
	-24	1 7/8-12	154-170	209-230
	-32	2 1/2-12	218-240	296-325

	Dash Size	Thread Size (Inches)	Jam Nut or Straight Fitting Torque	
			Lb-Ft	Newton Meters
Straight Thread O-Ring Boss High Pressure with ORS (J1453)	-03	3/8-24	8-10	11-13
	-04	7/16-20	14-16	20-22
	-05	1/2-20	18-20	24-27
	-06	9/16-18	24-26	33-35
	-08	3/4-16	50-60	68-78
	-10	7/8-14	72-80	98-110
	-12	1 1/16-12	125-135	170-183
	-14	1 3/16-12	160-180	215-245
	-16	1 5/16-12	200-220	270-300
	-20	1 5/8-12	210-280	285-380
	-24	1 7/8-12	270-360	370-490

Recommended Parallel Connection Assembly Torque

A torque wrench should be used to assure proper fitting assembly of these connections.

	Dash Size	Thread Size (Inches)	Swivel Nut Torque	
			Lb-Ft	Newton Meters
SAE 37° (JIC)	-04	$^{7}/_{16}$-20	11-12	15-16
	-05	$^{1}/_{2}$-20	15-16	20-22
	-06	$^{9}/_{16}$-18	18-20	24-28
	-08	$^{3}/_{4}$-16	38-42	52-58
	-10	$^{7}/_{8}$-14	57-62	77-85
	-12	1 $^{1}/_{16}$-12	79-87	108-119
	-16	1 $^{5}/_{16}$-12	108-113	148-154
	-20	1 $^{5}/_{8}$-12	127-133	173-182
	-24	1 $^{7}/_{8}$-12	158-167	216-227
	-32	2 $^{1}/_{2}$-12	245-258	334-352

	Thread Size (mm)	Straight Adapter or Locknut Torque	
		Lb-Ft	Newton Meters
Metric	M10 x 1	13-15	18-20
	M12 x 1.5	15-19	20-25
	M14 x 1.5	19-23	25-30
	M16 x 1.5	33-40	45-55
	M18 x 1.5	37-44	50-60
	M20 x 1.5	52-66	70-90
	M22 x 1.5	55-70	75-95
	M26 x 1.5	81-96	110-130
	M27 x 2	96-111	130-150
	M33 x 2	162-184	220-250
	M42 x 2	170-192	230-260
	M48 x 2	258-347	350-470

	Thread Size (mm)	Straight Adapter or Locknut Torque	
		Lb-Ft	Newton Meters
BSPP (Port connection only) "G" denotes parallel threads	G$^{1}/_{8}$-28	13-15	18-20
	G$^{1}/_{4}$-19	19-23	25-30
	G$^{3}/_{8}$-19	33-40	45-55
	G$^{1}/_{2}$-14	55-70	75-95
	G$^{3}/_{4}$-14	103-118	140-160
	G1-11	162-184	220-250
	G1 $^{1}/_{4}$-11	170-192	230-260
	G1 $^{1}/_{2}$-11	258-347	350-470

Basic Dimensions,

American National Standard Taper Pipe Threads[1], NPT (ANSI B2.1 — 1968)

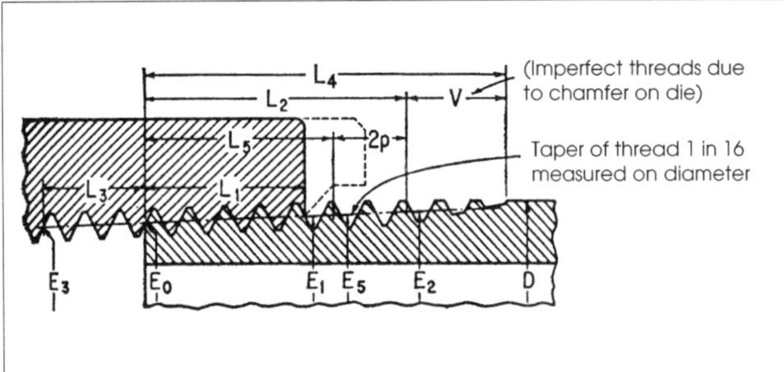

For all dimensions see corresponding reference letters in table. Angle between sides of thread is 60 degrees. Taper of thread, on diameter is 3/4 inch per foot. Angle of taper with centerline is 1°47'. The basic maximum thread height, h, of the truncated thread is 0.8 x pitch of thread. The crest and root are truncated a minimum of 0.033 x pitch for all pitches. For maximum depth of truncation see the table below.

American National Standards Internal Threads in Pipe Couplings,

NPC for Pressuretight Joints with Lubricant or Sealer (ANSI B2.1 — 1968)

Nom. Pipe Size	Thds. per Inch	Min. Dia[2] Min.	Pitch Diameter[1] Min.	Max.
1/8	27	0.342	0.3701	0.3771
1/4	18	0.440	0.4864	0.4968
3/8	18	0.577	0.6218	0.6322
1/2	14	0.715	0.7717	0.7851
3/4	14	0.925	0.9822	0.9956
1	11 1/2	1.161	1.2305	1.2468
1 1/4	11 1/2	1.506	1.5752	1.5915
1 1/2	11 1/2	1.745	1.7142	1.8305
2	11 1/2	2.219	2.2881	2.3044
2 1/2	8	2.650	2.7504	2.7739
3	8	3.277	3.3768	3.4002
3 1/2	8	3.777	3.8771	3.9005
4	8	4.275	4.3754	4.3988

1 The actual pitch diameter of the straight tapped hole will be slightly smaller than the value given when gaged with a taper plug gage as called for in ANSI B2.1.
2 As the ANSI Standard Pipe Thread form is maintained, the major and minor diameters of the internal thread vary with the pitch diameter. All dimensions are given in inches.

Nominal Pipe Size	Outside Diameter of Pipe D	Threads per Inch n	Pitch of Thread p	Pitch Diameter At Beginning of External Thread $E0$	Handtight Engagement: Length,[2] L_1 In.	Handtight Engagement: Diameter[3] E_1	Effective Thread, External: Length[4] L_2 In.	Effective Thread, External: Diameter E_2
1/16	0.3125	27	0.03704	0.27118	0.160	0.28118	0.2611	0.28750
1/8	0.405	27	0.03704	0.36351	0.1615	0.37360	0.2639	0.38000
1/4	0.540	18	0.05556	0.47739	0.2278	0.49163	0.4018	0.50250
3/8	0.675	18	0.05556	0.61201	0.240	0.62701	0.4078	0.63750
1/2	0.840	14	0.07143	0.75843	0.320	0.77843	0.5337	0.79179
3/4	1.050	14	0.07143	0.96768	0.339	0.98887	0.5457	1.00179
1	1.315	11 1/2	0.08696	1.21363	0.400	1.23863	0.6828	1.25630
1 1/4	1.660	11 1/2	0.08696	1.55713	0.420	1.58338	0.7068	1.60130
1 1/2	1.900	11 1/2	0.08696	1.79609	0.420	1.82234	0.7235	1.84130
2	2.375	11 1/2	0.08696	2.26902	0.436	2.29627	0.7505	2.31030
2 1/2	2.875	8	0.12500	2.71953	0.682	2.76216	1.1375	2.79062
3	3.500	8	0.12500	3.34062	0.766	3.38850	1.2000	3.41562
3 1/2	4.000	8	0.12500	3.83750	0.821	3.88881	1.2500	3.91562
4	4.500	8	0.12500	4.33438	0.844	4.38712	1.3000	4.41502
5	5.503	8	0.12500	5.39073	0.937	5.44929	1.4063	5.47862
6	6.625	8	0.12500	6.44609	0.958	6.50597	1.5125	6.54002
8	8.625	8	0.12500	8.43359	1.063	8.50003	1.7125	8.54062
10	10.750	8	0.12500	10.54531	1.210	10.62094	1.9250	10.66562
12	12.750	8	0.12500	12.53281	1.360	12.61781	2.1250	12.66562

All dimensions given in inches.
1 The basic dimensions of the ANSI Standard Taper Pipe Thread are given in inches to four or five decimal places. While this implies a greater degree of precision than is ordinarily attained, these dimensions are the basis of gage dimensions and are so expressed for the purpose of eliminating errors in computations.
2 Also length of thin ring gage and length from gaging notch to small end of plug gage.
3 Also pitch diameter at gaging notch (handtight plane).
4 Also length of plug gage.

SAE 4 Bolt Flange Dimensions

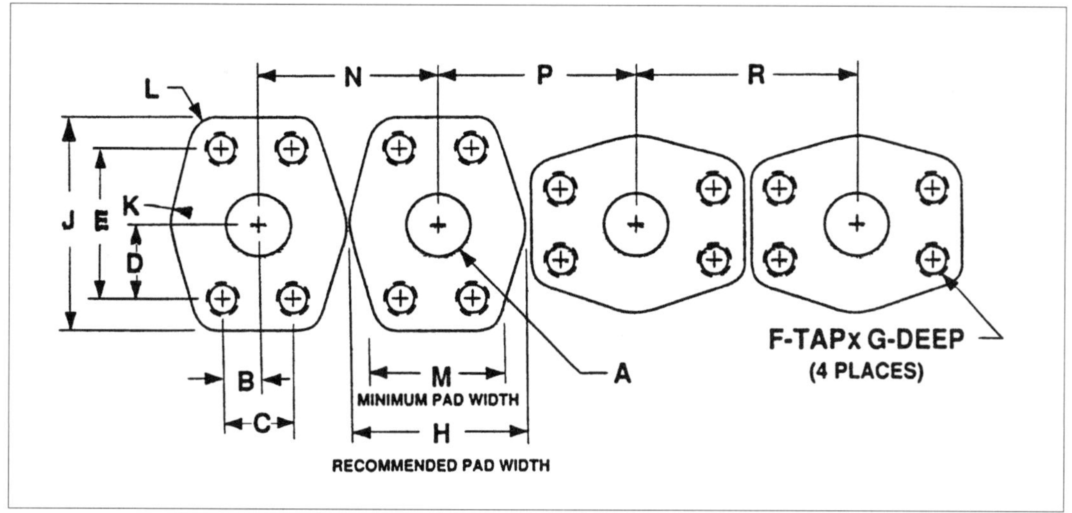

SAE Code 61

Size Code	Max Press	A Dia Max	B ±.01	C ±.01	D ±.01	E ±.01	F UNC-2B	G Min	H ±.01	J ±.03	K Rad	L Rad	M Min	N Min	P Min	R Min
08	5000	.50	.344	.688	.750	1.500	5/16-18	.94	1.81	2.12	.91	.31	1.31	1.91	2.06	2.22
12	5000	.75	.438	.875	.938	1.875	3/8-16	.88	2.06	2.56	1.03	.34	1.62	2.16	2.41	2.66
16	5000	1.00	.515	1.031	1.031	2.062	3/8-16	.88	2.31	2.75	1.16	.34	1.88	2.41	2.62	2.84
20	4000	1.25	.594	1.188	1.156	2.312	7/16-14	1.12	2.88	3.12	1.44	.41	2.12	2.97	3.09	3.22
24	3000	1.50	.703	1.406	1.375	2.750	1/2-13	1.06	3.25	3.69	1.62	.47	2.50	3.34	3.56	3.78
32	3000	2.00	.844	1.688	1.531	3.062	1/2-13	1.06	3.81	4.00	1.91	.47	3.00	3.91	4.00	4.09
40	2500	2.50	1.000	2.000	1.750	3.500	1/2-13	1.19	4.28	4.50	2.14	.50	3.50	4.38	4.50	4.59
48	2000	3.00	1.219	2.438	2.094	4.188	5/8-11	1.19	5.16	5.31	2.58	.56	4.19	5.25	5.34	5.41
56	500	3.50	1.375	2.75	2.375	4.750	5/8-11	1.31	5.50	6.00	2.75	.62	4.69	5.59	5.84	6.09
64	500	4.00	1.531	3.062	2.562	5.125	5/8-11	1.19	6.00	6.38	3.00	.62	5.19	6.09	6.28	6.47
80	500	5.00	1.812	3.625	3.000	6.000	5/8-11	1.31	7.12	7.25	3.56	.62	6.19	7.22	7.28	7.34

SAE Code 62

Size Code	Max Press	A Dia Max	B ±.01	C ±.01	D ±.01	E ±.01	F UNC-2B	G Min	H ±.01	J ±.03	K Rad	L Rad	M Min	N Min	P Min	R Min
08	6000	.50	.359	.718	.797	1.594	5/16-18	.81	1.88	2.22	.94	.31	1.50	2.09	2.22	2.34
12	6000	.75	.469	.937	1.000	2.000	3/8-16	.94	2.38	2.81	1.19	.41	1.88	2.59	2.75	2.94
16	6000	1.00	.547	1.093	1.125	2.250	7/16-14	1.06	2.75	3.19	1.38	.47	2.12	2.97	3.16	3.31
20	6000	1.25	.625	1.250	1.312	2.625	1/2-13	1.00	3.06	3.75	1.53	.56	2.38	3.25	3.56	3.88
24	6000	1.50	.719	1.437	1.562	3.125	5/8-11	1.38	3.75	4.44	1.88	.66	2.75	3.97	4.25	4.56
32	6000	2.00	.875	1.750	1.906	3.812	3/4-10	1.50	4.50	5.25	2.25	.72	3.38	4.72	5.03	5.38
40[1]	6000	2.50	1.156	2.312	2.437	4.875	7/8-9	1.81	5.87	6.87	2.94	1.00	4.38	6.09	6.54	7.00
48[1]	6000	3.00	1.406	2.812	3.000	6.000	1 1/8-7	2.31	7.00	8.50	3.50	1.25	5.38	7.22	7.92	8.62

[1] Industry standard, not an SAE standard

SAE Port Dimensions

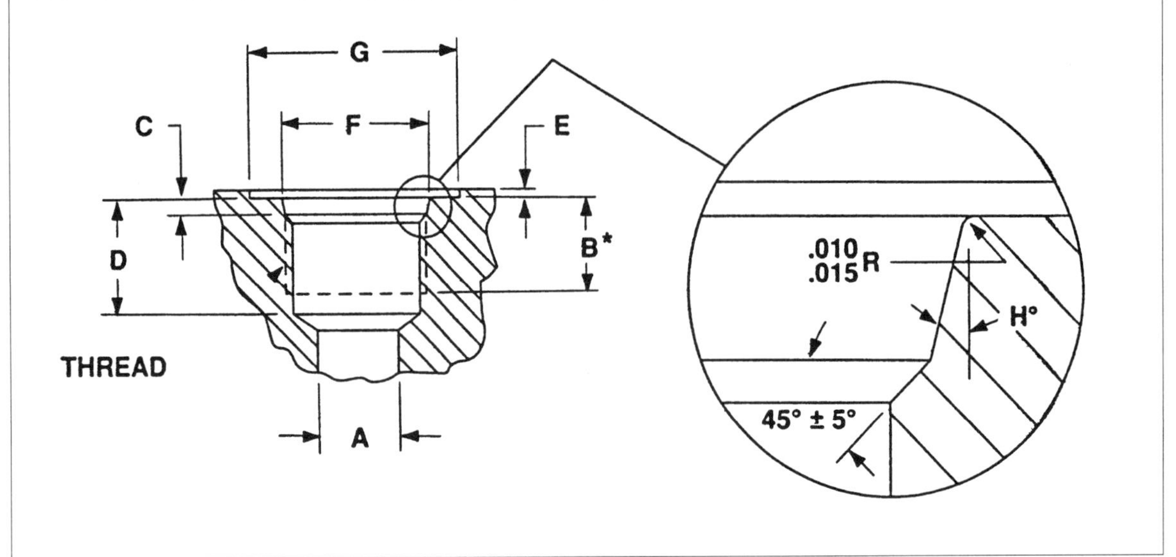

Size Code	Nominal Tube Size	Thread Size	A Min.	B[1] Min.	C +.015 -.000	D Min.	E Max.	F +.005 -.000	G Min.	H ±1°
-2	1/8	5/16-24	.062	.390	.074	.468	.067	.358	.672	12
-3	3/16	3/8-24	.125	.390	.074	.468	.062	.421	.750	12
-4	1/4	7/16-20	.172	.454	.093	.547	.062	.487	.828	12
-5	5/16	1/2-20	.234	.454	.093	.547	.062	.550	.906	12
-6	3/8	9/16-18	.297	.500	.097	.609	.062	.616	.969	12
-8	1/2	3/4-16	.391	.562	.100	.688	.094	.811	1.188	15
-10	5/8	7/8-14	.484	.656	.100	.781	.094	.942	1.344	15
-12	3/4	1 1/16-12	.609	.750	.130	.906	.094	1.148	1.625	15
-14	7/8	1 3/16-12	.719	.750	.130	.906	.094	1.273	1.765	15
-16	1	1 5/16-12	.844	.750	.130	.906	.125	1.398	1.910	15
-20	1 1/4	1 5/8-12	1.078	.750	.132	.906	.125	1.713	2.270	15
-24	1 1/2	1 7/8-12	1.312	.750	.132	.906	.125	1.962	2.560	15
-32	2	2 1/2-12	1.781	.750	.132	.906	.125	2.587	3.480	15

[1] Requires using a bottom tap

Torque Values For Pipe Plugs

Dryseal Type — 3/4 inch Taper/ft			
Nominal Thread Size		Threads Per Inch	Max. Torque Inch-Pounds
1/16	0.062	27	150
1/8	0.125	27	250
1/4	0.250	18	600
3/8	0.375	18	1,200
1/2	0.500	14	1,800
3/4	0.750	14	3,000
1	1.000	11 1/2	4,200
1 1/4	1.250	11 1/2	5,400
1 1/2	1.500	11 1/2	6,900
2	2.000	11 1/2	8,500

ANSI Standard Steel Pipe Flanges

Nominal Pipe Size	Flange OD	Flange Thick-ness	Bolt Circle Dia.	Bore Weld Neck Socket Weld	No. of Bolts	Bolt Dia.
150 lb. Standard*						
1/2	3 1/2	7/16	2 3/8	0.62	4	1/2
3/4	3 7/8	1/2	2 3/4	0.82	4	1/2
1	4 1/4	9/16	3 1/8	1.05	4	1/2
1 1/4	4 5/8	3 1/2	5/8	1.38	4	1/2
1 1/2	5	11-16	3 7/8	1.61	4	1/2
2	6	3/4	4 3/4	2.07	4	5/8
2 1/2	7	7/8	5 1/2	2.47	4	5/8
3	7 1/2	15/16	6	3.07	4	5/8
3 1/2	8 1/2	15/16	7	3.55	8	5/8
4	9	15/16	7 1/2	4.03	8	5/8
5	10	15/16	8 1/2	5.05	8	3/4
6	11	1	9 1/2	6.07	8	3/4
8	13 1/2	1 1/8	11 3/4	7.98	8	3/4
10	16	1 3/16	14 1/4	10.02	12	7/8
12	19	1 1/4	17	12.00	12	7/8
300 lb. Standard *						
1/2	3 3/4	9/16	2 5/8	0.62	4	1/2
3/4	4 5/8	5/8	3 1/4	0.82	4	5/8
1	4 7/8	11/16	3 1/2	1.05	4	5/8
1 1/4	5 1/4	3/4	3 7/8	1.38	4	5/8
1 1/2	6 1/8	13/16	4 1/2	1.61	4	3/4
2	6 1/2	7/8	5	2.07	8	5/8
2 1/2	7 1/2	1	5 7/8	2.47	8	3/
3	8 1/4	1 1/8	6 5/8	3.07	8	3/4
3 1/2	9	1 3/16	7 1/4	3.55	8	3/4
4	10	1 1/4	7 7/8	4.03	8	3/4
5	11	1 3/8	9 1/4	5.05	8	3/4
6	12 1/2	1 7/16	10 5/8	6.07	12	3/4
8	15	1 5/8	13	7.98	12	7/8
10	17 1/2	1 7/8	15 1/4	10.02	16	1
12	20 1/2	2	17 3/4	12.00	16	1 1/8
600 lb. Standard §						
1/2	3 3/4	9/16	2 5/8	†	4	1/2
3/4	4 5/8	5/8	3 1/4	†	4	5/8
1	4 7/8	11/16	3 1/2	†	4	5/8
1 1/4	5 1/4	13/16	3 7/8	†	4	5/8
1 1/2	6 1/8	7/8	4 1/2	†	4	3/4
2	6 1/2	1	5	†	8	5/8
2 1/2	7 1/2	1 1/8	5 7/8	†	8	3/4
3	8 1/4	1 1/4	6 5/8	†	8	3/4
3 1/2	9	1 3/8	7 1/4	†	8	7/8
4	10 3/4	1 1/2	8 1/2	†	8	7/8
5	13	1 3/4	10 1/2	†	8	1
6	14	1 7/8	11 1/2	†	12	1
8	16 1/2	2 3/16	13 3/4	†	12	1 1/8
10	20	2 1/2	17	†	16	1 1/4
12	22	2 5/8	19 1/4	†	20	1 1/4

Nominal Pipe Size	Flange OD OD	Flange Thick-ness	Bolt Circle Dia.	Bore Weld Neck Socket Weld	No. of Bolts	Bolt Dia.
900 lb. Standard						
1/2	4 3/4	7/8	3 1/4	†	4	3/4
3/4	5 1/8	1	3 1/2	†	4	3/4
1	5 7/8	1 1/8	4	†	4	7/8
1 1/4	6 1/4	1 1/8	4 3/8	†	4	7/8
1 1/2	7	1 1/4	4 7/8	†	4	1
2	8 1/2	1 1/2	6 1/2	†	8	7/8
2 1/2	9 5/8	1 5/8	7 1/2	†	8	1
3	10 1/2	1 1/2	7 1/2	†	8	7/8
4	11 1/2	1 3/4	9 1/4	†	8	1 1/8
5	13 3/4	2	11	†	8	1 1/4
6	15	2 3/16	12 1/2	†	12	1 1/8
8	18 1/2	2 1/2	15 1/2	†	12	1 3/8
10	21 1/2	2 3/4	18 1/2	†	16	1 3/8
12	24	3 1/8	21	†	20	1 3/8
500 lb. Standard						
1/2	4 3/4	7/8	3 1/4	†	4	3/4
3/4	5 1/8	1	3 1/2	†	4	3/4
1	5 7/8	1 1/8	4	†	4	7/8
1 1/4	6 1/4	1 1/8	4 3/8	†	4	7/8
1 1/2	7	1 1/4	4 7/8	†	4	1
2	8 1/2	1 1/2	6 1/2	†	8	7/8
2 1/2	9 5/8	1 5/8	7 1/2	†	8	1
3	10 1/2	1 7/8	8	†	8	1 1/8
4	12 1/4	2 1/8	9 1/2	†	8	1 1/4
5	14 3/4	2 7/8	11 1/2	†	8	1 1/2
6	15 1/2	3 1/4	12 1/2	†	12	1 3/8
8	19	3 5/8	15 1/2	†	12	1 5/8
10	23	4 1/4	19	†	12	1 7/8
12	26 1/2	4 7/8	22 1/2	†	16	2
500 lb. Standard §						
1/2	5 1/4	1 3/16	3 1/2	†	4	3/4
3/4	5 1/2	1 1/4	3 3/4	†	4	3/4
1	6 1/4	1 3/8	4 1/4	†	4	7/8
1 1/4	7 1/4	1 1/2	5 1/8	†	4	1
1 1/2	8	1 3/4	5 3/4	†	4	1 1/8
2	9 1/4	2	6 3/4	†	8	1
2 1/2	10 1/2	2 1/4	7 3/4	†	8	1 1/8
3	12	2 5/8	9	†	8	1 1/4
4	14	3	10 3/4	†	8	1 1/2
5	16 1/2	3 5/8	12 3/4	†	8	1 3/4
6	19	4 1/4	12 1/2	†	8	2
8	21 3/4	5	17 1/4	†	12	2
10	26 1/2	6 1/2	21 1/4	†	12	2 3/4
12	30	7 1/4	24 3/8	†	12	2 3/4

From ANSI Standard, Steel Pipe Flanges and Flanges Fittings, B16.5—1968
*Flange thickness includes 1/16 inch raised face.
†To be specified by manufacturer.
§Flange thickness does not include 1/4 in. raised face.

ANSI Standard Cast-Iron Pipe Flanges

Nominal Pipe Size	Flange OD	Flange Thickness	Bolt Circle Dia.	No. of Bolts	Bolt Dia.
125 lb. Standard					
1	4 1/4	7/16	3 1/8	4	1/2
1 1/4	4 5/8	1/2	3 1/2	4	1/2
1 1/2	5	9/16	3 7/8	4	1/2
2	6	5/8	4 3/4	4	5/8
2 1/2	7	11/16	5 1/2	4	5/8
3	7 1/2	3/4	6	4	5/8
3 1/2	8 1/2	13/16	7	8	5/8
4	9	15/16	7 1/2	8	5/8
5	10	15/16	8 1/2	8	3/4
6	11	1	9 1/2	8	3/4
8	13 1/2	1 1/8	11 3/4	8	3/4
10	16	1 3/16	14 1/4	12	7/8
12	19	1 1/4	17	12	7/8
250 lb. Standard *					
1	4 7/8	11/16	3 1/2	4	5/8
1 1/4	5 1/4	3/4	3 7/8	4	5/8
1 1/2	6 1/8	13/16	4 1/2	4	3/4
2	6 1/2	7/8	5	8	5/8
2 1/2	7 1/2	1	5	8	3/4
3	8 1/4	1 1/8	6 5/8	8	3/4
3 1/2	9	1 3/16	7 1/4	8	3/4
4	10	1 1/4	7 7/8	8	3/4
5	11	1 3/8	9 1/4	8	3/4
6	12 1/2	1 7/16	10 5/8	12	3/4
8	15	1 5/8	13	12	7/8
10	17 1/2	1 7/8	15 1/4	16	1
12	20 1/2	2	17 3/4	16	1 1/8

From ANSI Standard, Cast-Iron Pipe Flange Fittings B16.1—1967.

* Flange thickness includes 1/16 inch raised face.

ANSI Standard Pipe Threads

Nominal Pipe Size	Pipe OD	Threads Per Inch	Length of Effective Threads	Length of Hand-Tight Engage-ment	Total Thread Length
900 lb. Standard					
1/16	0.3125	27	0.26	0.16	0.39
1/8	0.405	27	0.26	0.16	.039
1/4	0.540	18	0.40	0.23	0.60
3/8	0.675	18	0.41	0.24	0.60
1/2	0.840	14	0.53	0.32	0.78
3/4	1.050	14	0.55	0.34	0.79
1	1.315	11 1/2	0.68	0.40	0.99
1 1/4	1.660	11 1/2	0.71	0.42	1.01
1 1/2	1.900	11 1/2	0.72	0.42	1.03
2	2.375	11 1/2	0.76	0.44	1.06
2 1/2	2.875	8	1.14	0.68	1.57
3	3.500	8	1.20	0.77	1.63
3 1/2	4.000	8	1.25	0.82	1.68
4	4.500	8	1.30	0.84	1.73
5	5.563	8	1.41	0.94	1.84
6	6.625	8	1.51	0.96	1.95
8	8.625	8	1.71	10.06	2.15
10	10.750	8	1.93	1.21	2.36
12	12.750	8	2.13	1.36	2.56

Table abridged from ANSI Standard B2.1—1968.

Maximum Non-Shock Hydraulic Working Pressure

Temp °F	Working Presure PSI							
	Cast Iron		Steel					
	125#	250#	150#	500#	600#	900#	1500#	2500#
-20 to 100	200	500	285	740	1480	2220	3705	6170
200	190	460	260	675	1350	2025	3375	5625
300	165	375	250	655	1310	1970	3280	5470
400	140	290	245	635	1265	1900	3170	5280
500	—	—	230	600	1195	1795	2995	4990
600	—	—	210	545	1095	1640	2735	4560

Notes: 125# and 250# from ANSI B16.1-1975 for ASTM A126, Class B, 1-12" size.
Others from ANSI B16.1-5—1973 for ASTM A105.

Thread Engagement Dimensions — Nominal

Dimensions may vary due to tolerance conditions.

Listed below are the thread engagement dimensions (B) which must be taken into consideration when making connection with ports or appropriate female adapters. The "B" dimension must be subtracted from the overall length (A) to insure proper connection.

Dash size	Male Pipe Straight and Angled Dimension "B"	SAE O-Ring boss SAE J514 Straight and Adjustable Dimension "B"	SAE O-Ring boss SAE J1453 Straight and Adjustable Dimension "B"
-02	.25	—	—
-04	.38	.36	.43
-05	—	.36	.43
-06	.38	.39	.47
-08	.50	.43	.55
-10	—	.50	.63
-12	.62	.59	.73
-14	—	.59	—
-16	.69	.59	.73
-20	.69	.59	.73
-24	.69	.59	.73
-32	.75	.59	—

All dimensions in inches. Dimensions may vary due to tolerance conditions.

Thru Hole Dimensions

All dimensions are nominal. In jump size bodies, the minimum thru hole dimensions will correspond to the smallest dash size.

Dash	E thru Hole SAE 37°
-03	.12
-04	.17
-05	.23
-06	.30
-08	.39
-10	.48
-12	.61
-16	.84
-20	1.08
-24	1.31
-32	1.78

SAE[1], BSP and DIN[2] Recommended Maximum Operating Pressures (psi)[3]

Sealing is achieved by means of an o-ring, retaining washer and a properly machined port. The o-ring is "captured" by the i.d. of the retaining washer. The port may be spot faced or a flat machined surface as long as the D6 dimension is met.

Dash Size	Inch Size	NPTF/ NPSM (thru -32)	J514 37° (thru -32)	SAE J514 ORB	SAE J1453 ORB/ STR	SAE J1453 ORB/ ADJ	SAE J518 Code 61	SAE J518 Code 62	BSPP Port	BSPT & BSPP 600
02	1/8	5000								5000
03	3/16									
04	1/4	5000	4500	4500	6000	6000			4500	5000
05	5/16		4000	4000						
06	3/8	4000	4000	4000	6000	6000			4000	4000
08	1/2	3000	4000	4000	6000	6000	5000	6000	4000	3000
10	5/8		3000	3000	6000	6000			3000	2500
12	3/4	2500	3000	3000	6000	6000	5000	6000	3000	2500
14	7/8		2500	2500						
16	1	2000	2500	2500	6000	5000	5000	6000	2500	2000
20	1 1/4	1150	2000	2000	4000	4000	4000	6000	2000	1150
24	1 1/2	1000	1500	1500	4000	3000	3000	6000	1500	1000
32	2	1000	1125	1125			3000	6000		
40	2 1/2	200	200				2500			
48	3	200	200				2000			

Metric Threads	DIN 3852 x to A, B & E Heavy Series	DIN 3902 24° (l.Rh) Light Series	DIN 3902 24° (s.Rh) Heavy Series	DIN 7631 60°
M8x1.0				
M10x1.0				
M12x1.5	5700 (9135)	2250 (3625)		2250 (3625)
M14x1.5	5700 (9135)	2250 (3625)	5700 (9135)	2250 (3625)
M16x1.5	5700 (9135)	2250 (3625)	5700 (9135)	2250 (3625)
M18x1.5	5700 (9135)	2250 (3625)	5700 (9135)	2250 (3625)
M20x1.5	5700 (9135)		5700 (9135)	
M22x1.5	3600 (5800)	2250 (3625)	5700 (9135)	2250 (3625)
M24x1.5	5700 (9135)		3600 (5800)	
M26x1.5		1450 (2320)		1450 (2320)
M27x1.5	3600 (5800)			
M30x1.5				900 (1450)
M33x2.0	3600 (5800)	900 (1450)		
M36x2.0		900 (1450)	3600 (5800)	
M38x1.5				900 (1450)
M42x2.0	2250 (3625)		2250 (3625)	
M45x1.5				900 (1450)
M45x2.0		900 (1450)		
M48x2.0	2250 (3625)			
M50x2.0				
M52x1.5				900 (1450)
M52x2.0		900 (1450)	2250 (3625)	
M60x2.0				

[1] SAE pressures are based on a 4:1 safety/design factor.
[2] DIN pressures in parentheses are based on a 2.5:1 safety/design factor.
[3] For applications with higher operating pressures, contact manufacturer.

How to Identify Fluid Connectors

Measuring Tools

A seat angle gage and an I.D./O.D. caliper are necessary to make accurate measurements of commonly used connectors.

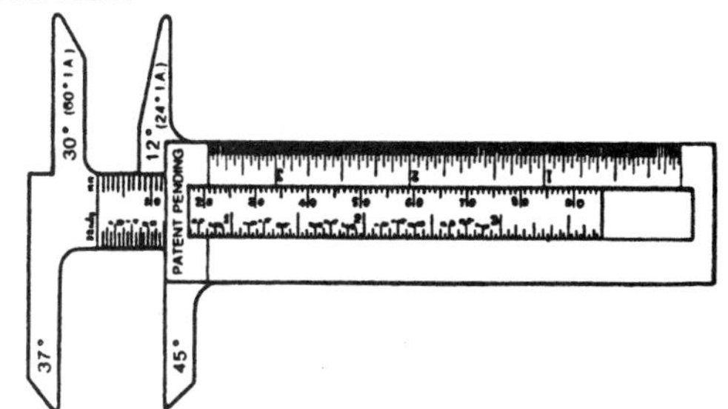

I.D./O.D. Angle Gage Caliper

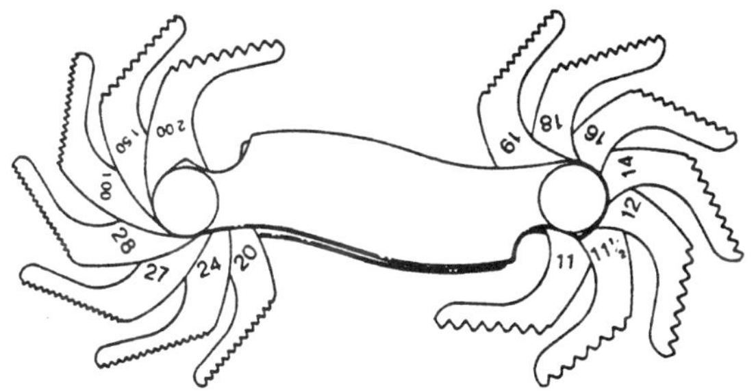

Thread Pitch Gage

How to Measure Threads

Use a thread pitch gage to determine the number of threads per inch or the distance betweeen threads in metric connections. Place the gage on the threads until the fit is snug. Match the measurement to the charts.

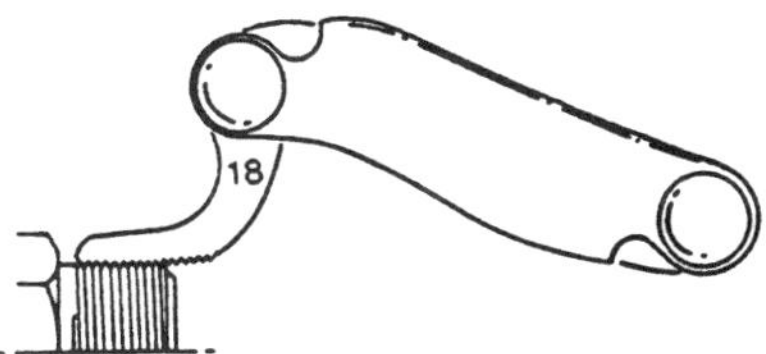

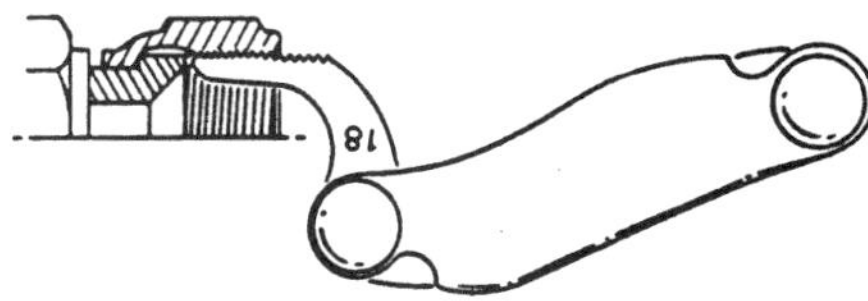

Measure the thread diameter with an I.D./O.D. caliper as shown. Match the measurements to the charts.

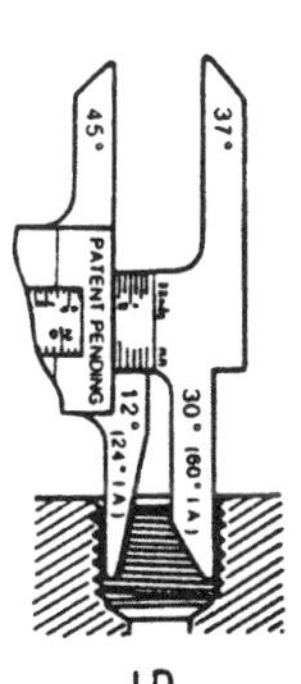

I.D.

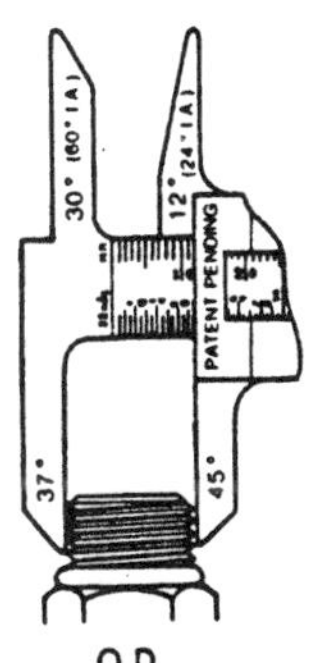

O.D.

How to Measure Sealing Surface Angles

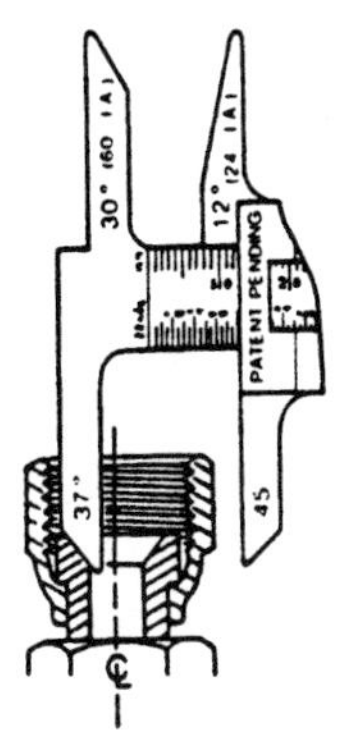

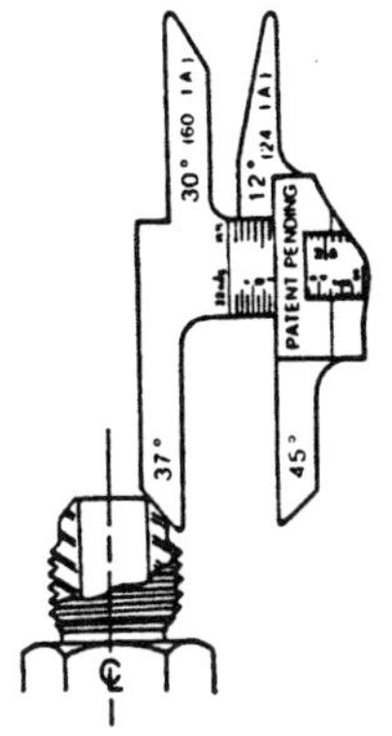

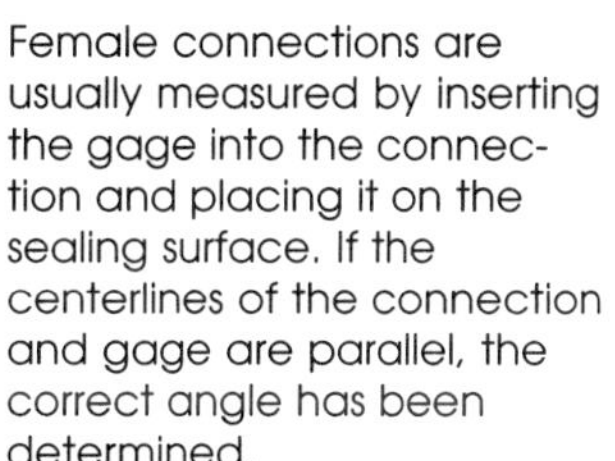

Female connections are usually measured by inserting the gage into the connection and placing it on the sealing surface. If the centerlines of the connection and gage are parallel, the correct angle has been determined.

Male flare type connectors are usually measured by placing the gage on the sealing surface. If the center-lines of the connection and gage are parallel, the correct angle has been determined.

How to Measure Non-Threaded Connections

Four Bolt Flange — First measure the port hole diameter using the caliper. Next, measure the longest bolt hole spacing from center-to-center or measure the flange head diameter.

Staplok® — Measure the male diameter with the O.D. portion of the caliper. Measure the female half by inserting the I.D. portion of the caliper into the thru hole.

Dash Numbers

Most fluid piping system sizes in the United States are measured by dash numbers. These are universally used abbreviations for the size of the component expressed as the numerator of the fraction with the denominator always being 16. For example, a -04 port is $^4/_{16}$ or $^1/_4$-inch. Dash numbers are usually nominal (in name only) and are abbreviations that make ordering of components easier.

Torque Values For Pipe Plugs

Dryseal Type — $^3/_4$ inch Taper/ft			
Nominal Thread Size		Threads Per Inch	Max. Torque Inch-Pounds
$^1/_{16}$	0.062	27	150
$^1/_8$	0.125	27	250
$^1/_4$	0.250	18	600
$^3/_8$	0.375	18	1,200
$^1/_2$	0.500	14	1,800
$^3/_4$	0.750	14	3,000
1	1.000	11 $^1/_2$	4,200
1 $^1/_4$	1.250	11 $^1/_2$	5,400
1 $^1/_2$	1.500	11 $^1/_2$	6,900
2	2.000	11 $^1/_2$	8,500

American Connections

NPTF
National Pipe Tapered Fuel

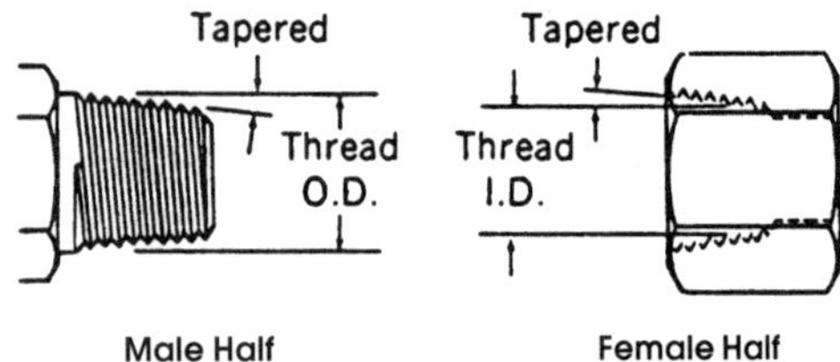

This connection is still widely used in fluid power systems, even though it is not recommended by the National Fluid Power Association (NFPA) for use in hydraulic applications. The thread is tapered and the seal takes place by deformation of the threads.

NPSM
National Pipe Straight Mechanical

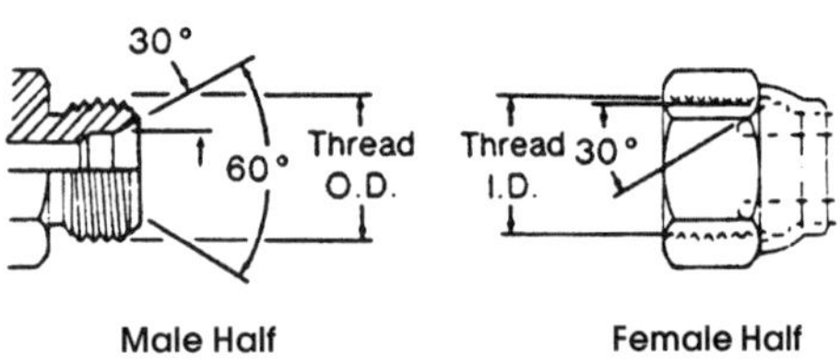

This connection is sometimes used in fluid power systems. The female half has a straight thread and an inverted 30° seat. The male half of the connection has a straight thread and a 30° internal chamfer. The seal takes place by compression of the 30° seat on the chamfer. The threads holds the connection mechanically.

NOTE: A properly chamfered NPTF male will also seal with the NPSM female.

Inch Size	Dash Size	Nominal Thread Size	Male O.D. (Inch) Fraction	Male O.D. (Inch) Decimal	Female I.D. (Inch) Fraction	Female I.D. (Inch) Decimal
1/8	02	1/8-27	13/32	.41	3/8	.38
1/4	04	1/4-18	17/32	.54	1/2	.49
3/8	06	3/8-18	11/16	.68	5/8	.63
1/2	08	1/2-14	27/32	.84	25/32	.77
3/4	12	3/4-14	1 1/16	1.05	1	.98
1	16	1-11 1/2	1 5/16	1.32	1 1/4	1.24
1 1/4	20	1 1/4-11 1/2	1 21/32	1.66	1 19/32	1.58
1 1/2	24	1 1/2-11 1/2	1 29/32	1.90	1 13/16	1.82
2	32	2-11 1/2	2 3/8	2.38	2 5/16	2.30

NOTE: For NPTF threads, measure thread diameter and subtract 1/4-inch to find the nominal pipe size.

SAE J514 Straight Thread
O-Ring Boss (ORB)

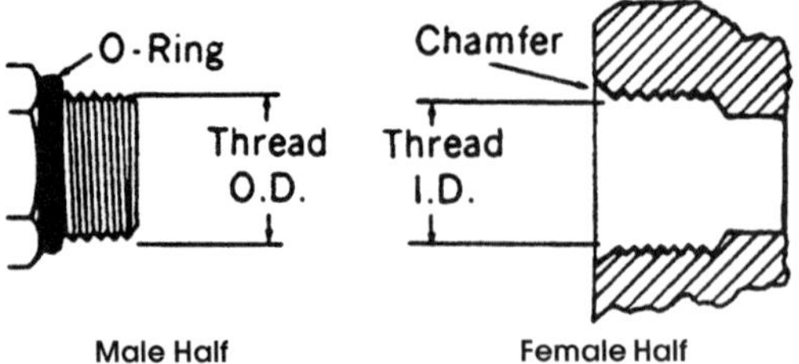

This port connection is recommended by the NFPA for optimum leakage control in medium and high pressure hydraulic systems. The male connector has a straight thread and an o-ring. The female port has a straight thread, a machined surface (minimum spotface) and a chamfer to accept the o-ring. The seal takes place by compressing the o-ring into the chamfer. The threads hold the connection mechanically.

SAE J514 37°
Hydraulic (Formerly JIC)

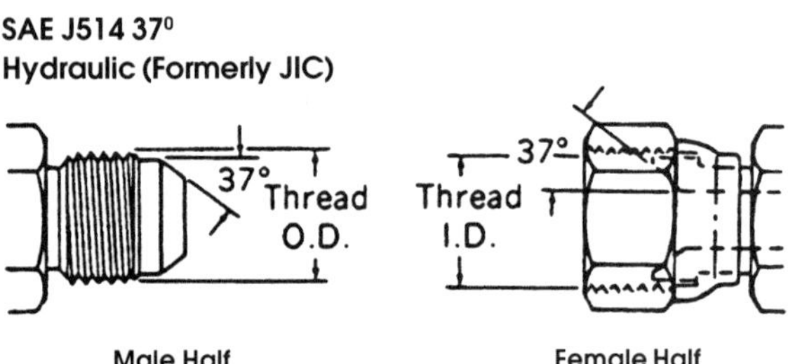

This connection is very common in fluid power systems. Both the male and female halves of the connections have 37° seats. The seal takes place by establishing a line contact between the male flare and the female cone seat. The threads hold the connection mechanically.

CAUTION: In the -02, -03, -04, -05, -08 and -10 sizes, the threads of the SAE 45° flare and the SAE 37° flare are the same. However, the sealing surface angles are not the same.

Inch Size	Dash Size	Nominal Thread Size	Male O.D. (Inch) Fraction	Male O.D. (Inch) Decimal	Female I.D. (Inch) Fraction	Female I.D. (Inch) Decimal
1/8	02	5/16-24	5/16	.31	9/32	.27
3/16	03	3/8-24	3/8	.38	11/32	.34
1/4	04	7/16-20	7/16	.44	13/32	.39
5/16	05	1/2-20	1/2	.50	15/32	.45
3/8	06	9/16-18	9/16	.56	17/32	.51
1/2	08	3/4-16	3/4	.75	11/16	.69
5/8	10	7/8-14	7/8	.88	13/16	.81
3/4	12	1 1/16-12	1 1/16	1.06	1	.98
1	16	1 5/16-12	1 5/16	1.31	1 1/4	1.23
1 1/4	20	1 5/8-12	1 5/8	1.63	1 9/16	1.54
1 1/2	24	1 7/8-12	1 7/8	1.88	1 13/16	1.79
2	32	2 1/2-12	2 1/2	2.50	2 7/16	2.42

American Connections

ORS® SAE J1453
O-Ring Face Seal

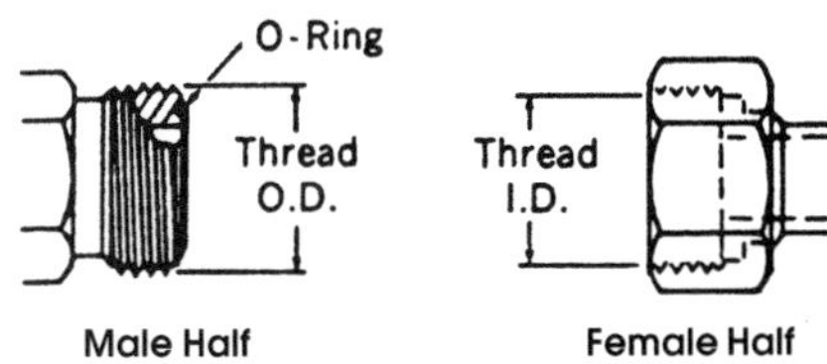

This connection offers the very best leakage control available today. The male connector has a straight thread and an o-ring in the face. The female has a straight thread and a machined flat face. The seal takes place by compressing the o-ring onto the flat face of the female, similar to the split flange type fitting. The threads hold the connection mechanically.

Inch Size	Dash Size	Nominal Thread Size	Male O.D. (Inch)		Female I.D. (Inch)	
			Fraction	Decimal	Fraction	Decimal
1/4	04	9/16-18	9/16	.56	17/32	.51
3/8	06	11/16-16	11/16	.69	5/8	.63
1/2	08	13/16-16	13/16	.82	3/4	.75
5/8	10	1-14	1	1.00	15/16	.93
3/4	12	1 3/16-12	1 3/16	1.19	1 1/8	1.11
1	16	1 7/16-12	1 7/16	1.44	1 3/8	1.36
1 1/4	20	1 11/16-12	1 11/16	1.69	1 5/8	1.61
1 1/2	24	2-12	2	2.00	1 15/16	1.92

SAE J512 Inverted

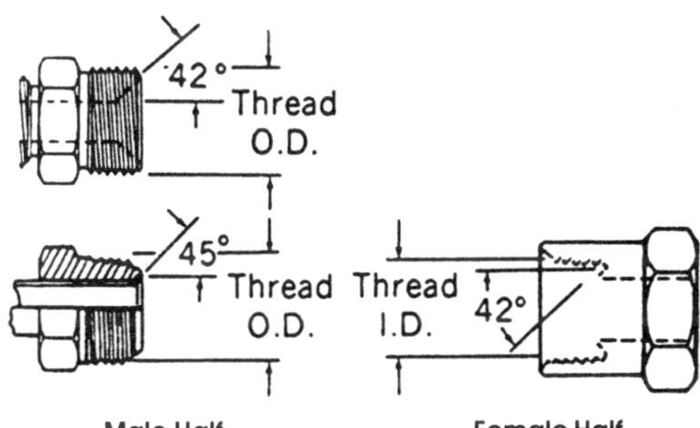

This connection is frequently used in automotive systems. The male connector can either be a 45° flare in the tube fitting form or a 42° seat in the machined adapter form. The female has a straight thread with a 42° inverted flare. The seal takes place on the flared surfaces. The threads hold the connection mechanically.

Inch Size	Dash Size	Nominal Thread Size	Male O.D. (Inch)		Female I.D. (Inch)	
			Fraction	Decimal	Fraction	Decimal
1/8	02	5/16-28	5/16	.32	9/32	.28
3/16	03	3/8-24	3/8	.38	11/32	.34
1/4	04	7/16-24	7/16	.44	13/32	.40
5/16	05	1/2-20	1/2	.50	15/32	.45
3/8	06	5/8-18	5/8	.63	9/16	.57
7/16	07	11/16-18	11/16	.69	5/8	.63
1/2	08	3/4-18	3/4	.75	23/32	.70
5/8	10	7/8-18	7/8	.88	13/16	.82
3/4	12	1 1/16-16	1 1/16	1.06	1	1.00

SAE J512 45°

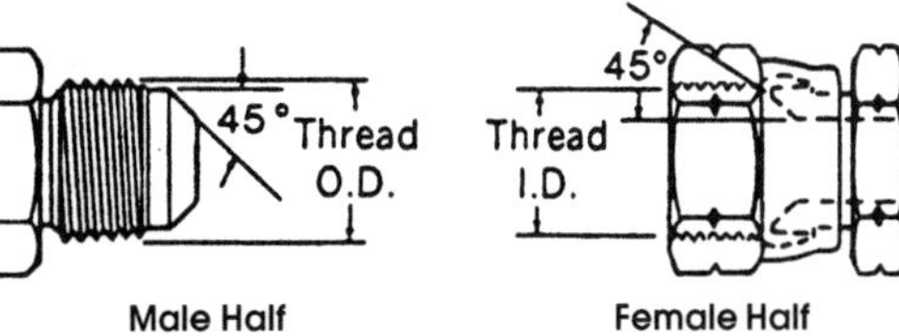

This connection is commonly used in refrigeration, automotive and truck piping systems. The connector is frequently made of brass. Both the male and the female connectors have 45° seats. The seal takes place between the male and the female cone seat. The threads hold the connection mechanically.

CAUTION: In the -02, -03, -04, -05, -08 and -10 sizes, the threads of the SAE 45° flare and the SAE 37° flare are the same. However, the sealing surface angles are not the same.

Inch Size	Dash Size	Nominal Thread Size	Male O.D. (Inch)		Female I.D. (Inch)	
			Fraction	Decimal	Fraction	Decimal
1/8	02	5/16-24	5/16	.31	9/32	.27
3/16	03	3/8-24	3/8	.38	11/32	.34
1/4	04	7/16-20	7/16	.44	13/32	.39
5/16	05	1/2-20	1/2	.50	15/32	.45
3/8	06	5/8-18	5/8	.63	9/16	.57
1/2	08	3/4-16	3/4	.75	11/16	.69
5/8	10	7/8-14	7/8	.88	13/16	.81
3/4	12	1 1/16-14	1 1/16	1.06	1	.99
7/8	14	1 1/4-12	1 1/4	1.25	1 5/32	1.16
1	16	1 3/8-12	1 3/8	1.38	1 9/32	1.29

Staplok® (SAE J1467)

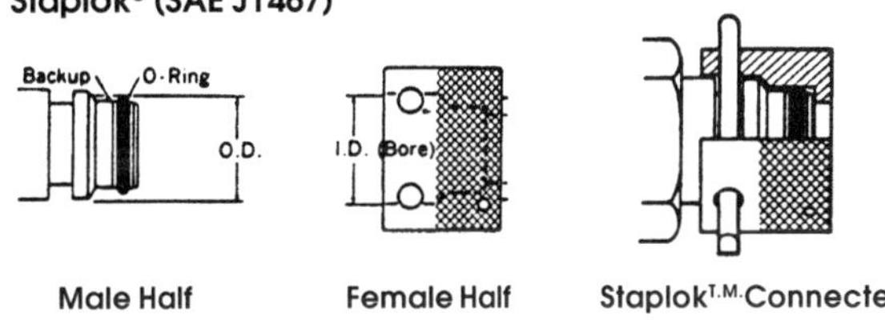

This is a radial o-ring seal connection developed in Germany and commonly used for hydraulic applications in underground mines. The male contains an exterior o-ring and backup ring, plus a groove to accept the "staple". The female has a smooth bore with two holes for the staple. A "U" shaped retaining clip or staple is inserted through the two holes, passing through the groove in the male to lock the connection together. The seal takes place by contact between the o-ring in the male and the smooth bore of the female.

Inch Size	Dash Size	Male O.D. (Inch)		Female I.D. (Inch)	
		Fraction[1]	Decimal	Fraction[1]	Decimal
1/4	04	19/32	.586	19/32	.597
3/8	06	25/32	.783	51/64	.794
1/2	08	15/16	.940	61/64	.951
3/4	12	1 9/64	1.137	1 9/64	1.148
1	16	1 17/32	1.529	1 35/64	1.540
1 1/4	20	1 13/16	1.806	1 13/16	1.817
1 1/2	24	2 5/32	2.163	2 11/64	2.174
2	32	2 33/64	2.517	2 17/32	2.528

[1] Measure to the closest 1/64-inch.

American Connections

SAE J518
4-Bolt Flange[1]

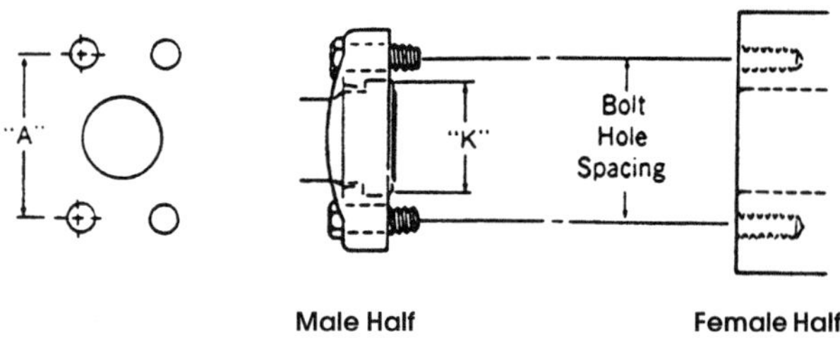

Male Half **Female Half**

This connection is commonly used in fluid power systems. There are two pressure ratings. Code 61 is referred to as the "standard" series and Code 62 is the "6000 psi" series. The design concept for both series is the same, but the bolt hole spacing and flanged head diameters are larger for the higher pressure, Code 62 connection.

The female (port) is an unthreaded hole with four bolt holes in a rectangular pattern around the port. The male consists of a flanged head, grooved for an o-ring, and either a captive flange or split flange halves with bolt holes to match the port. The seal takes place on the o-ring, which is compressed between the flange head and the flat surface surrounding the port. The threaded bolts hold the connection together.

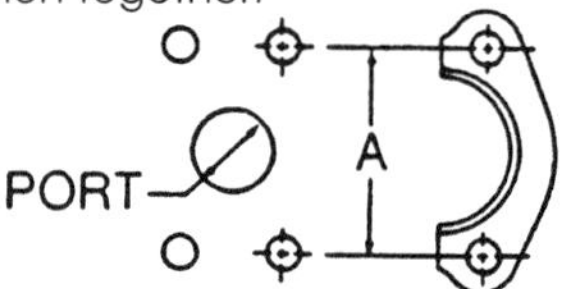

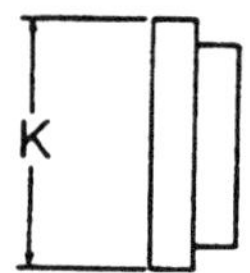

Inch size (Dash Size)	Port Hole I.D. Inch Fraction (Dec.)	Bolt Dimensions Inch Cd. 61	Bolt Dimensions Inch Cd. 62	Bolt Hole Spacing "A" Inch (Dec.) Cd. 61	Bolt Hole Spacing "A" Inch (Dec.) Cd. 62	Flanged Head Dia. "K" Inch (Dec.) Cd. 61	Flanged Head Dia. "K" Inch (Dec.) Cd. 62
1/2 (08)	1/2 (.50)	5/16-18 x 1 1/4	5/16-18 x 1 1/4	1 1/2 (1.50)	1 19/32 (1.59)	1 3/16 (1.19)	1 1/4 (1.25)
3/4 (12)	3/4 (.75)	3/8-16 x 1 1/4	3/8-16 x 1 1/2	1 7/8 (1.88)	2 (2.00)	1 1/2 (1.50)	1 5/8 (1.63)
1 (16)	1 (1.00)	3/8-16 x 1 1/4	7/16-14 x 1 3/4	2 1/16 (2.06)	2 1/4 (2.25)	1 3/4 (1.75)	1 7/8 (1.88)
1 1/4 (20)	1 1/4 (1.25)	7/16-14 x 1 1/2	1/2-13 x 1 3/4	2 5/16 (2.31)	2 5/8 (2.63)	2 (2.00)	2 1/8 (2.13)
1 1/2 (24)	1 1/2 (1.50)	1/2-13 x 1 1/2	5/8-11 x 2 1/4	2 3/4 (2.75)	3 1/8 (3.13)	2 3/8 (2.38)	2 1/2 (2.50)
2 (32)	2 (2.00)	1/2-13 x 1 1/2	3/4-10 x 2 3/4	3 1/16 (3.06)	3 13/16 (3.81)	2 13/32 (2.81)	3 1/8 (3.13)

How to Measure

Four Bolt Flange — First measure the port hole diameter using the caliper. Next, measure the longest bolt hole spacing from center-to-center (Dimension "A") or measure the flanged head diameter.

[1] SAE J518, JIS B 8363, ISO/DIS 6162 and DIN 20066 are interchangeable, except for bolt sizes.

ISO Connections

ISO/DIS 6162
4-Bolt Flange[2]

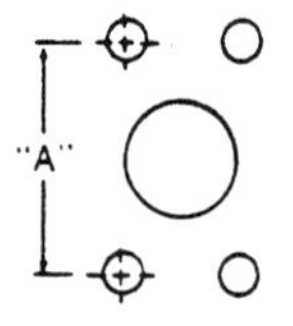

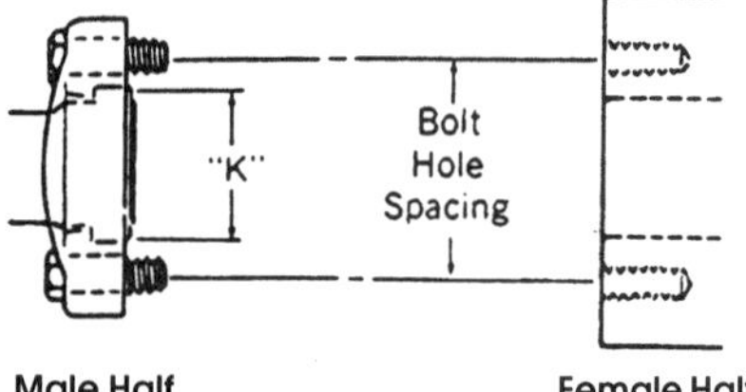

Male Half **Female Half**

This connection is commonly used in fluid power systems. There are two pressure ratings. PN 35/350 bar (Code 61) is the "standard" series and PN 415 bar (Code 62) is the high pressure series. The design concept for both series is the same, but the bolt hole spacing and flanged head diameters are larger for the higher pressure, PN 415 bar connection. Both metric and inches bolts are used. The port will have an "M" stamped on it if metric bolts are required.

The female (port) is an unthreaded hole with four bolt holes in a rectangular pattern around the port. The male consists of a flanged head, grooved for an o-ring, and either a captive flange or split flange halves with bolt holes to match the port. The seal takes place on the o-ring, which is compressed between the flanged head and flat surface surrounding the port. The threaded bolts hold the connection together.

Size mm (Inch) (Dash)	Port Hole mm (Inch)	Bolt Dimensions mm and Inch PN 35/350 Bar (Cd. 61)	Bolt Dimensions mm and Inch PN 415 Bar (Cd. 62)	Bolt Hole Spacing "A" mm (Inch) PN 35/350 Bar (Cd. 61)	Bolt Hole Spacing "A" mm (Inch) PN 415 Bar (Cd. 62)
13 (1/2) (08)	12.7 (.50)	M8 x 1.25 x 30 5/16-18 x 1 1/4	M8 x 1.25 x 30 5/16-18 x 1 1/4	38.10 (1.50)	40.49 (1.57)
19 (3/4) (12)	19.1 (.75)	M10 x 1.5 x 35 3/8-16 x 1 1/4	M10 x 1.5 x 40 3/8-16 x 1/2	47.63 (1.88)	50.80 (2.00)
25 (1) (16)	25.4 1.00	M10 x 1.5 x 35 3/8-16 x 1 1/4	M12 x 1.75 x 45 7/16-14 x 1 3/4	52.37 (2.06)	57.15 (2.25)
32 (1 1/4) (20)	31.8 (1.25)	M12 x 1.75 x 40 7/16-16 x 1 1/2	M14 x 2 x 50 1/2-13 x 1 3/4	58.72 (2.31)	66.68 (2.63)
38 (1 1/2) (24)	38.1 (1.50)	M14 x 2 x 40 1/2-13 x 1 1/2	M16 x 2 x 55 5/8-11 x 2 1/4	69.85 (2.75)	79.38 (3.13)
51 (2) (32)	50.8 (2.00)	M14 x 2 x 40 1/2-13 x 1 1/2	M20 x 2.5 x 70 3/4-10 x 2 3/4	77.77 (3.06)	96.82 (3.81)

Inch Size	Flanged Head Diameter "K" mm (Inch) PN 35/350 Bar (Cd. 61)	Flanged Head Diameter "K" mm (Inch) PN 415 Bar (Cd. 62)
1/2	30.18 (1.19)	31.75 (1.25)
3/4	38.10 (1.50)	41.28 (1.63)
1	44.45 (1.75)	47.63 (1.88)
1 1/4	50.80 (2.00)	53.98 (2.13)
1 1/2	60.33 (2.38)	63.50 (2.50)
2	71.42 (2.81)	79.38 (3.13)

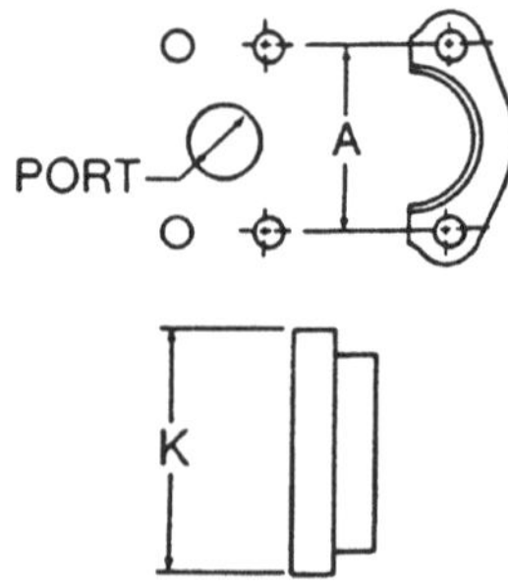

[2] DIN 20066, ISO/DIS 6162, JIS B 8363 and SAE J518 are interchangeable, except for both bolt sizes.

ISO Connections

ISO 6149

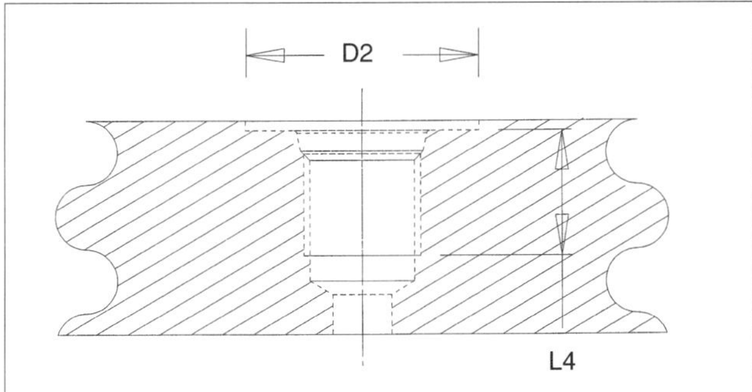

Thread Size ISO 6149-2	L4 Dia. Thread Depth ISO 6149-2	D2 Dia. Spot Face Dia. ISO 6149
M8-1	10 mm	17 mm
M10-1	10	20
M12-1.5	11.5	23
M14-1.5	11.5	23
M16-1.5	13	28
M18-1.5	14.5	30
M22-1.5	15.5	34
M27-2	19	40
M33-2	19	49
M42-2	19.5	60
M48-2	22	66
M60-2	24.5	76

German Connections

DIN 7631 Series

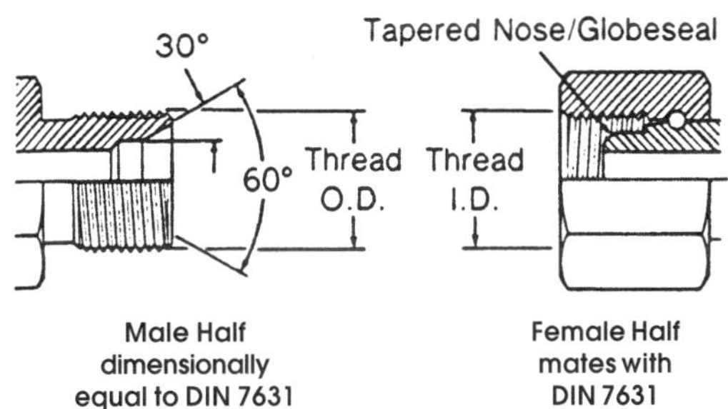

Male Half dimensionally equal to DIN 7631 **Female Half mates with DIN 7631**

This connection is frequently used in hydraulic systems. The male has a straight metric thread and a 60⁰ (included angle) recessed cone. The female has a straight thread and a tapered nose/Globeseal™ seat. The seal takes place by contact between the cone of the male and the nose of the tapered nose/Globeseal flareless swivel. The threads hold the connection mechanically.

Use with Pipe/Tube O.D.		Metric Thread Size	Male Thread O.D.		Female Thread I.D.	
mm	Inch		mm	Inch	mm	Inch
6	.24	M12 x 1.5	12	.47	10.5	.41
8	.32	M14 x 1.5	14	.55	12.5	.49
10	.39	M16 x 1.5	16	.63	14.5	.57
12	.47	M18 x 1.5	18	.71	16.5	.65
15	.59	M22 x 1.5	22	.87	20.5	.81
18	.71	M26 x 1.5	26	1.02	24.5	.96
22	.87	M30 x 1.5	30	1.18	28.5	1.12
28	1.10	M38 x 1.5	38	1.50	36.5	1.44
35	1.38	M45 x 1.5	45	1.77	43.5	1.71
42	1.65	M52 x 1.5	52	2.04	50.5	1.99

German Connections

DIN 3902 Series

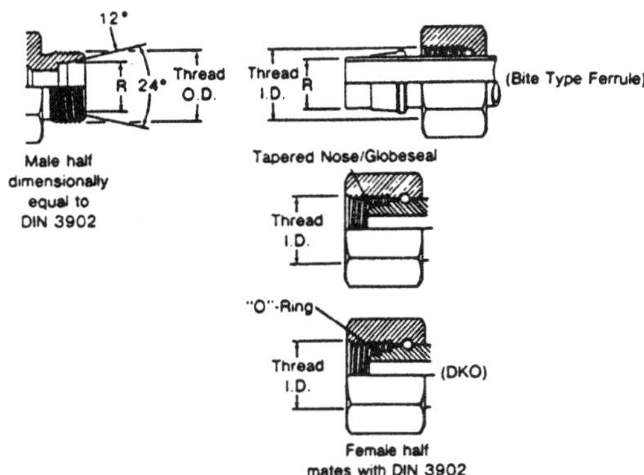

This connection style consists of a common male and three different female halves.

The male has a straight metric thread, a 24⁰ included angle and a recessed counterbore that matches the tube O.D. used with it. The female may be a tube, nut and ferrule, a tapered nose/Globeseal flareless swivel or a tapered nose/ Globeseal flareless swivel with an o-ring in the nose (DKO type).

Tube O.D. "R" Dim. l.Rh.[1] mm (Inch)	Tube O.D. "R" Dim. s.Rh[2] mm (Inch)	Metric Thread Size	Male Thread O.D.		Female Thread I.D.	
			mm	Inch	mm	Inch
6 (.24)		M12 X 1.5	12	.47	10.5	.41
8 (.32)	6 (.24)	M14 X 1.5	14	.55	12.5	.49
10 (.39)	8 (.32)	M16 X 1.5	16	.63	14.5	.57
12 (.47)	10 (.39)	M18 X 1.5	18	.71	16.5	.65
	12 (.47)	M20 X 1.5	20	.78	18.5	.73
15 (.59)	14 (.55)	M22 X 1.5	22	.87	20.5	.81
	16 (.63)	M24 X 1.5	24	.94	22.5	.89
18 (.71)		M26 X 1.5	26	1.02	24.5	.96
22 (.87)	20 (.78)	M30 X 2.0	30	1.18	28	1.11
28 (1.10)	25 (.98)	M36 X 2.0	36	1.41	34	1.34
	30 (1.18)	M42 X 2.0	42	1.65	40	1.57
35 (1.38)		M45 X 2.0	45	1.77	43	1.70
42 (1.65)	38 (1.50)	M52 X 2.0	52	2.04	50	1.97

[1] l.Rh. is a light duty system.
[2] s.Rh. is a heavy duty system.

German Connections

DIN 20066
4-Bolt Flange[1]

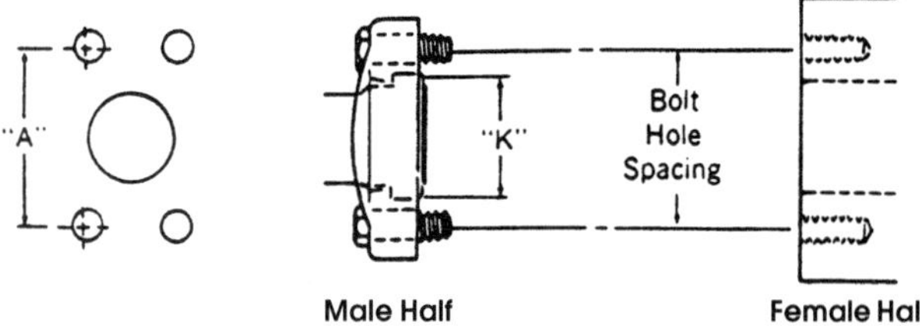

This connection is commonly used in fluid power systems. There are two pressure ratings. Form R (Code 61) is referred to as the "standard duty" series and Form S (Code 62) is the "heavy duty" series. The design concept for both series is the same, but the bolt hole spacing and flanged head diameters are larger for the higher pressure, Form S connection. Both metric and inch bolts are used.

The female (port) is an unthreaded hole with four bolt holes in a rectangular pattern around the port. The male consists of a flanged head, grooved for an o-ring, and either a captive flange or split flange halves with bolt holes to match the port. The seal takes place on the o-ring, which is compressed between the flanged head and the flat surface surrounding the port. The threaded bolts hold the connection together.

Size mm (Inch) (Dash)	Port Hole mm (Inch)	Bolt Dimensions mm and Inch		Bolt Hole Spacing "A" mm (Inch)	
		Form R (Cd. 61)	Form S (Cd. 62)	Form R (Cd. 61)	Form S (Cd. 62)
12 (1/2) (08)	12.7 (.50)	M8 X 1.25 X 30 5/16-18 x 1 1/4	M8 X 1.25 X 30 5/16-18 x 1 1/4	38.10 (1.50)	40.49 (1.57)
20 (3/4) (12)	19.1 (.75)	M10 x 1.5 x 30 3/8-16 X 1 1/4	M10 X 1.5 X 40 3/8-16 X 1 1/2	47.63 (1.88)	50.80 (2.00)
25 (1) (16)	25.4 (1.00)	M10 X 1.5 X 30 3/8-16 X 1 1/4	M12 X 1.75 X 45 7/16-14 X 1 3/4	52.37 (2.06)	57.15 (2.25)
32 (1 1/4) (20)	31.7 (1.25)	M10 X 1.5 X 40 7/16-14 X 1 1/2	M14 X 2 X 45 1/2-13 X 1 3/4	58.72 (2.31)	66.68 (2.63)
40 (1 1/2) (24)	38.0 (1.50)	M12 X 1.75 X 40 1/2-13 X 1 1/2	M16 X 2 X 55 5/8-11 X 2 1/4	69.85 (2.75)	79.38 (3.13)
50 (2) (32)	50.8 (2.00)	M12 X 1.75 X 40 1/2-13 X 1 1/2	M20 X 2.5 X 70 3/4-10 X 2 3/4	77.77 (3.06)	96.82 (3.81)

Inch Size	Flanged Head Diameter "K" mm (Inch) Form R Bar (Cd. 61)	Form S Bar (Cd. 62)
1/2	30.18 (1.19)	31.75 (1.25)
3/4	38.10 (1.50)	41.28 (1.63)
1	44.45 (1.75)	47.63 (1.88)
1 1/4	50.80 (2.00)	53.98 (2.13)
1 1/2	60.33 (2.38)	63.50 (2.50)
2	71.42 (2.81)	79.38 (3.13)

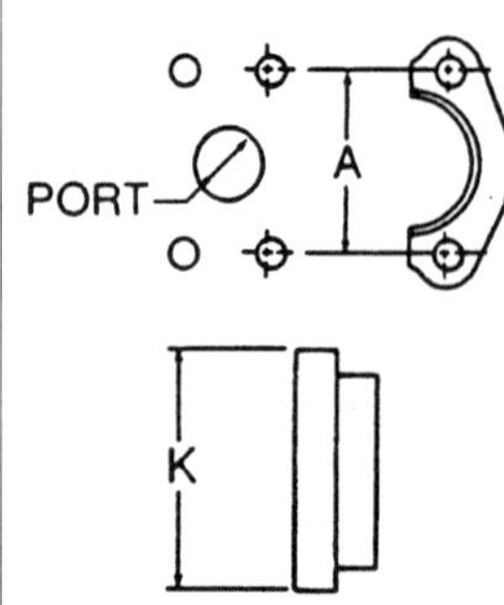

[1] ISO/DIS 6162, DIN 20066, JIS B 8363 and SAE J518 are interchangeable, except for bolt sizes.

DIN 3852
Male Connectors and Female Ports

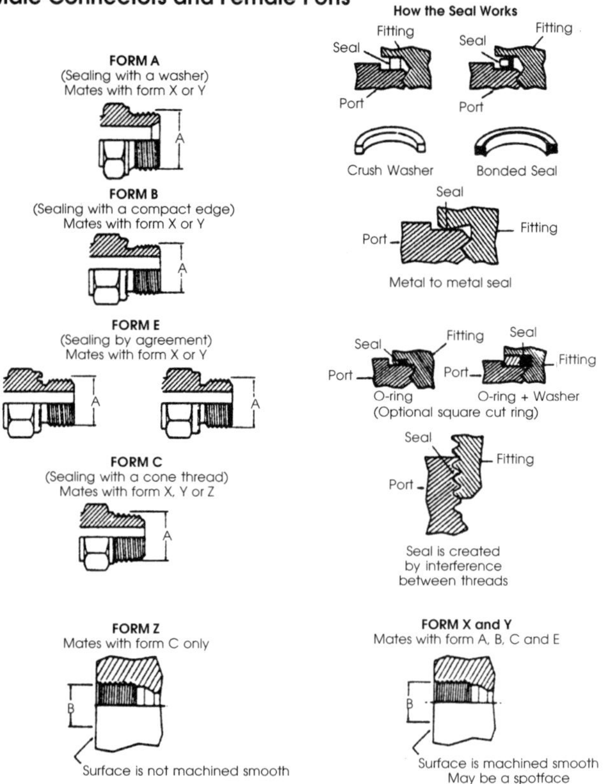

This DIN is controlled by Germany, but other countries may use it as a reference for their connector and port designs. The chart above illustrates the various forms and how they seal. For DIN 3852 Whitworth pipe thread dimensions, see BSPT/BSPP dimensions. They are the same.

Metric Thread	Male Thread O.D. "A"		Female Thread I.D. "B"	
	mm	Inch	mm	Inch
M12 x 1.5	12	.47	10.5	.41
M14 x 1.5	14	.55	12.5	.49
M16 x 1.5	16	.63	14.5	.57
M18 x 1.5	18	.71	16.5	.65
M20 x 1.5	20	.78	18.5	.73
M22 x 1.5	22	.87	20.5	.81
M24 x 1.5	24	.94	22.5	.89
M26 x 1.5	26	1.02	24.5	.96
M27 x 2	27	1.06	25	.98
M30 x 1.5	30	1.18	28.5	1.12
M30 x 2	30	1.18	28	1.10
M33 x 2	33	1.30	31	1.22
M36 x 1.5	36	1.41	34.5	1.36
M36 x 2	36	1.41	34	1.33
M38 x 1.5	38	1.49	36.5	1.43
M38 x 2	38	1.49	36	1.41
M42 x 1.5	42	1.65	40.5	1.60
M42 x 2	42	1.65	40	1.57
M45 x 1.5	45	1.77	43.5	1.71
M45 x 2	45	1.77	43	1.69
M48 x 1.5	48	1.89	46.5	1.83
M48 x 2	48	1.89	46	1.81
M52 x 1.5	52	2.04	50.5	1.89
M52 x 2	52	2.04	50	1.97

French Connections
Millimetrique and GAZ Series

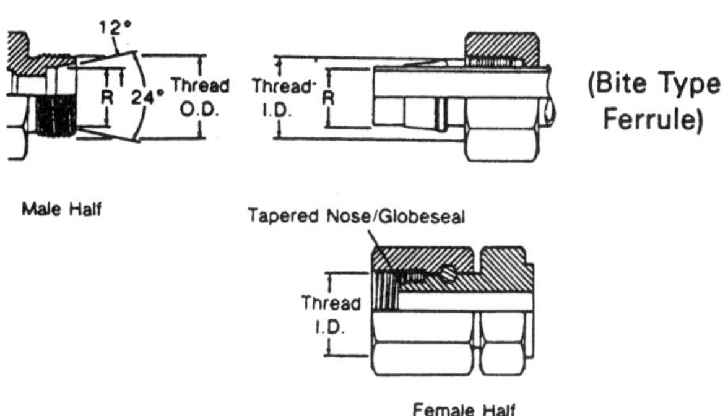

This connection consists of a common male and two different females. The Millimetrique Series is used with whole number metric O.D. tubing and the GAZ Series is used with fractional number metric O.D. pipe size tubing.

Millimetrique and GAZ Threads

Tubing O.D. "R" Dim. mm (Inch)	"GAZ" Pipe O.D. "R" Dim. mm (Inch)	Metric Thread Size	Male Thread O.D.		Female Thread I.D.	
			mm	Inch	mm	Inch
6 (.24)		M12 x 1.0	12	.47	11	.43
8 (.32)		M14 x 1.5	14	.55	12.5	.49
10 (.39)		M16 x 1.5	16	.63	14.5	.57
12 (.47)		M18 x 1.5	18	.71	16.5	.65
14 (.55)	13.25 (.52)	M20 x 1.5	20	.78	18.5	.73
15 (.59)		M22 x 1.5	22	.87	20.5	.81
16 (.63)	16.75 (.66)	M24 x 1.5	24	.94	22.5	.89
18 (.71)		M27 x 1.5	27	1.06	25.5	1.00
22 (.87)	21.25 (.83)	M30 x 1.5	30	1.18	28.5	1.12
25 (.98)		M33 x 1.5	33	1.30	31.5	1.24
28 (1.10)	26.75 (1.05)	M36 x 1.5	36	1.41	34.5	1.36
30 (1.18)		M39 x 1.5	39	1.54	37.5	1.48
32 (1.25)		M42 x 1.5	42	1.65	40.5	1.60
35 (1.38)	33.50 (1.32)	M45 x 1.5	45	1.77	43.5	1.71
38 (1.50)		M48 x 1.5	48	1.89	46.5	1.83
40 (1.57)	42.5 (1.66)	M52 x 1.5	52	2.04	50.5	1.99
45 (1.77)		M54 x 2.0	54	2.12	52	2.05
	48.25 (1.90)	M58 x 2.0	58	2.28	55	2.16

British Connections
British Standard Pipe (BSP)

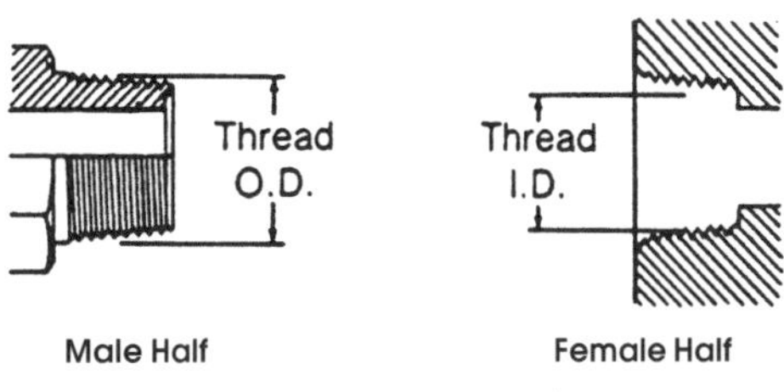

The BSPT (tapered) connection is similar to the NPT, except that the thread pitches are different in most sizes, and the thread form and O.D.'s are close but not the same. Sealing is accomplished by thread distortation. A thread sealant is recommended.

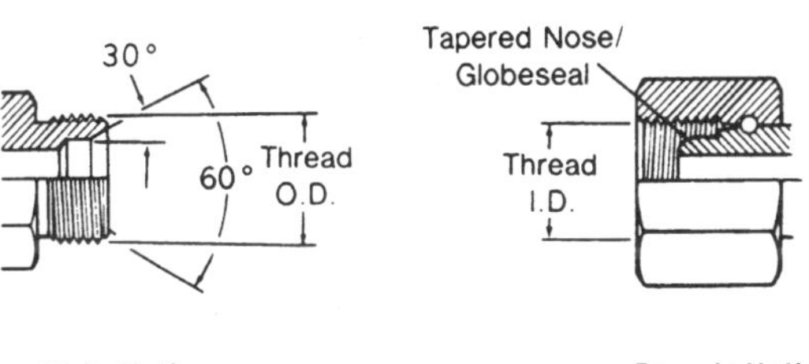

The BSPP (parallel) male is similar to the NPSM male except the thread pitches are different in most sizes. The female swivel BSPP has a tapered nose/Globeseal flareless swivel which seals on the cone seat of the male.

BSPT/BSPP Threads

Inch Size	Dash Size	Nominal Thread Size[1]	Male Thread O.D. (Inch)		Female Thread I.D. (Inch)	
			Fraction	Decimal	Fraction	Decimal
1/8	02	1/8-28	3/8	.38	11/32	.35
1/4	04	1/4-19	33/64	.52	15/32	.47
3/8	06	3/8-19	21/32	.65	19/32	.60
1/2	08	1/2-14	13/16	.82	3/4	.75
5/8	10	5/8-14	7/8	.88	13/16	.80
3/4	12	3/4-14	1 1/32	1.04	31/32	.97
1	16	1-11	1 5/16	1.30	1 7/32	1.22
1 1/4	20	1 1/4-11	1 21/32	1.65	1 9/16	1.56
1 1/2	24	1 1/2-11	1 7/8	1.88	1 25/32	1.79
2	32	2-11	2 11/32	2.35	2 1/4	2.26

[1] Frequently, the thread size is expressed as a fractional dimension preceded by the letter "G" or the letter "R". The "G" represents a parallel thread and the "R" indicates a tapered thread. For example, BSPP 3/8-19 may be expressed as G3/8, and BSPT 3/8-19 may be expressed as R3/8.

Japanese Connections

JIS 30° Male Inverted Seat, Parallel Pipe Threads (Threads Per JIS B 0202)

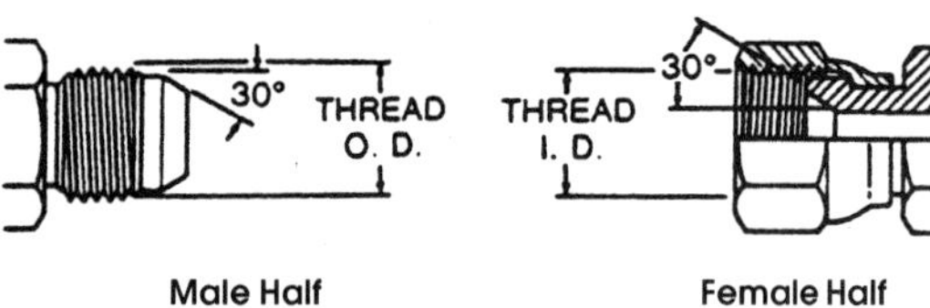

Male Half   Female Half

The JIS parallel is similar to the BSPP connection. The JIS parallel thread and the BSPP connection are interchangeable.

JIS Tapered Pipe (PT) (Threads Per JIS B 0203)

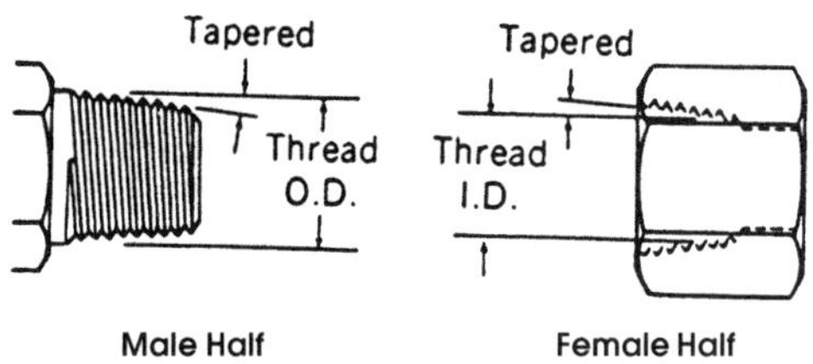

Male Half   Female Half

The JIS tapered thread is similar to the BSPT connection in design, appearance and dimensions. The JIS tapered thread and the BSPT connection are interchangeable.

Inch Size	Size mm	Nominal Thread Size (Similar to BSPP[1] & BSPT[2])	Male Thread O.D. Frac.	Male Thread O.D. mm	Female Thread I.D. Frac.	Female Thread I.D. mm
1/4	6 (04)	1/4-19	33/64	13.2	15/32	11.9
3/8	9 (06)	3/8-19	21/32	16.7	19/32	15.3
1/2	12 (08)	1/2-14	13/16	21.0	3/4	19.2
3/4	19 (12)	3/4-14	1 1/32	26.4	31/32	24.6
1	25 (16)	1-11	1 5/16	33.3	1 7/32	30.9
1 1/4	32 (20)	1 1/4-11	1 21/32	41.9	1 9/16	39.6
1 1/2	38 (24)	1 1/2-11	1 7/8	47.8	1 25/32	45.5
2	50 (32)	2-11	2 11/32	59.7	2 1/4	57.4

[1] Applies to JIS 30° male inverted seat, parallel pipe threads.
[2] Applies to JIS tapered pipe.

JIS 30° Male (Inverted) Seat, Metric Threads (Threads per JIS B 0207)

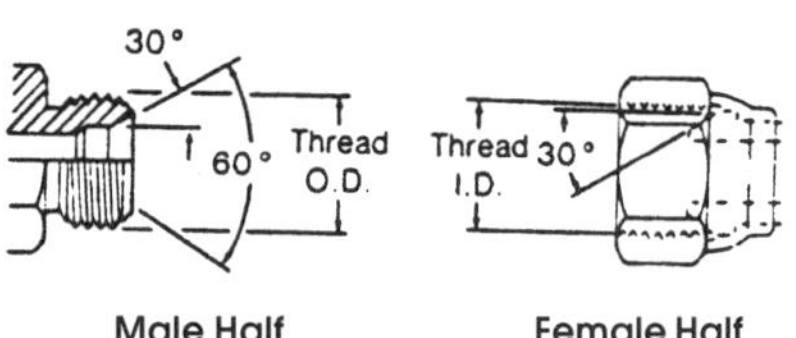

Male Half   Female Half

The JIS parallel (metric) is the same as the JIS parallel (PF), except for the thread difference.

Size mm	Dash Size Equivalent	Thread Size	Male Thread O.D. mm	Male Thread O.D. Inch	Female Thread I.D. mm	Female Thread I.D. Inch
6	04	M14 x 1.5	14	.55	12.5	.49
9	06	M18 x 1.5	18	.71	16.5	.65
12	08	M22 x 1.5	22	.87	20.5	.81
19	12	M30 x 1.5	30	1.18	28.5	1.12
25	16	M33 x 1.5	33	1.30	31.5	1.24
32	20	M42 x 1.5	42	1.65	40.5	1.60

JIS 30° Female (Cone) Seat, Parallel Pipe Threads (Threads per JIS B 0202)

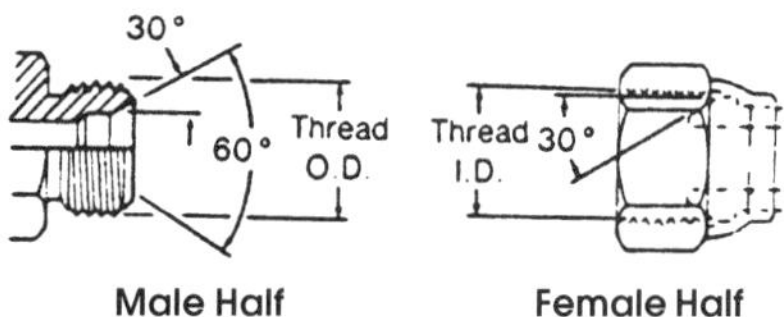

Male Half   Female Half

The Japanese JIS 30° flare is similar to the American SAE 37° flare connection in application as well as sealing principles. However, the flare angle and dimensions are different. The threads are similar to BSPP.

Inch Size	Size mm	Thread Size (Similar to BSPP)	O.D. Frac.	O.D. mm	I.D. Frac.	I.D. mm
1/4	6 (04)	1/4-19	33/64	13.2	15/32	11.9
3/8	9 (06)	3/8-19	21/32	16.7	19/32	15.3
1/2	12 (08)	1/2-14	13/16	21.0	3/4	19.2
3/4	19 (12)	3/4-14	1 1/32	26.4	31/32	24.6
1	25 (16)	1-11	1 5/16	33.3	1 7/32	30.9
1 1/4	32 (20)	1 1/4-11	1 1/32	41.9	1 9/16	39.6
1 1/2	38 (24)	1 1/2-11	1 7/8	47.8	1 25/32	45.5
2	50 (32)	2-11	2 11/32	59.7	2 1/4	57.4

Japanese Connections

JIS B 8363 4-Bolt Flange[1]

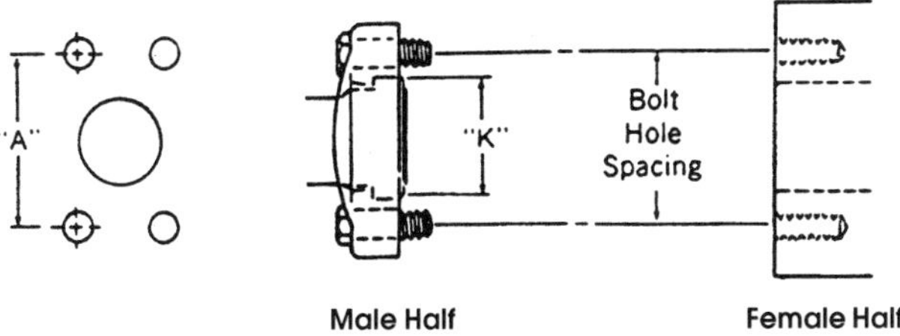

This connection is commonly used in fluid power systems. There are two pressure ratings. Type I (Code 61) is referred to as the "standard" series and Type II (Code 62) is the "6000 psi" series. The design concept for both series is the same, but the bolt hole spacing and flanged head diameters are larger for the higher pressure, Type II connection. Both metric and inch bolts are used.

The female (port) is an unthreaded hole with four bolt holes in a rectangular pattern around the port. The male consists of a flanged head, grooved for an o-ring, and either a captive flange or split flange halves with bolt holes to match the port. The seal takes place on the o-ring, which is compressed between the flanged head and the flat surface surrounding the port. The threaded bolts hold the connection together.

Size mm (Inch) (Dash)	Port Inch mm (Inch)	Bolt Dimensions mm and Inch		Bolt Hole Spacing "A" mm (Inch)	
		Type I (Cd. 61)	Type II (Cd. 62)	Type I (Cd. 61)	Type II (Cd. 62)
12 (1/2) (08)	12.7 (.50)	M8 x 1.25 x 30 5/16-18 x 1 1/4	M8 x 1.25 x 30 5/16-18 x 1 1/4	38.10 (1.50)	40.49 (1.57)
19 (3/4) (12)	19.1 (.75)	M10 x 1.5 x 30 3/8-16 x 1 1/4	M10 x 1.5 x 40 3/8-16 x 1 1/2	47.63 (1.88)	50.80 (2.00)
25 (1) (16)	25.4 (1.00)	M10 x 1.5 x 30 3/8-16 x 1 1/4	M12 x 1.75 x 45 7/16-14 x 1 3/4	52.37 (2.06)	57.15 (2.25)
32 (1 1/4) (20)	31.7 (1.25)	M10 x 1.5 x 40 7/16-14 x 1 1/2	M14 x 2 x 45 1/2-13 x 1 3/4	58.72 (2.31)	66.68 (2.63)
38 (1 1/2) (24)	38.0 (1.50)	M12 x 1.75 x 40 1/2-13 x 1 1/2	M16 x 2 x 55 5/8-11 x 2 1/4	69.85 (2.75)	79.38 (3.13)
50 (2) (32)	50.8 (2.00)	M12 x 1.75 x 40 1/2-13 x 1 1/2	M20 x 2.5 x 70 3/4-10 x 2 3/4	77.77 (3.06)	96.82 (3.81)

Inch Size	Flanged Head Diameter "K" mm (Inch)	
	Type I (Cd. 61)	Type II (Cd. 62)
1/2	30.18 (1.19)	31.75 (1.25)
3/4	38.10 (1.50)	41.28 (1.63)
1	44.45 (1.75)	47.63 (1.88)
1 1/4	50.80 (2.00)	53.98 (2.13)
1 1/2	60.33 (2.38)	63.50 (2.50)
2	71.42 (2.81)	79.38 (3.13)

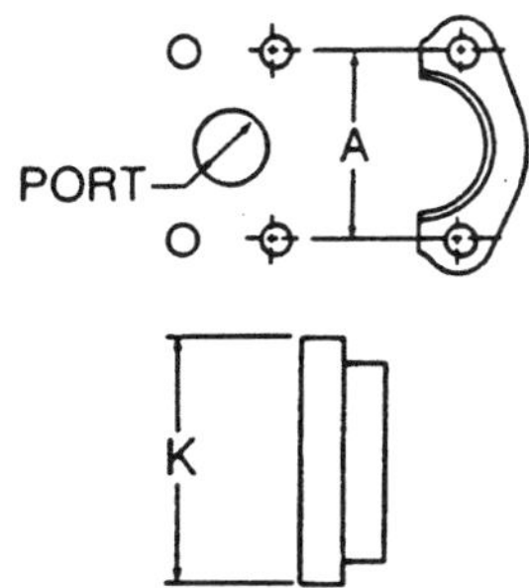

[1] JIS B 8363, ISO/DIS 6162, DIN 20066 and SAE J518 are interchangeable, except for bolt sizes.

JIS 210 Kgf/cm² 4-Bolt Square Flange

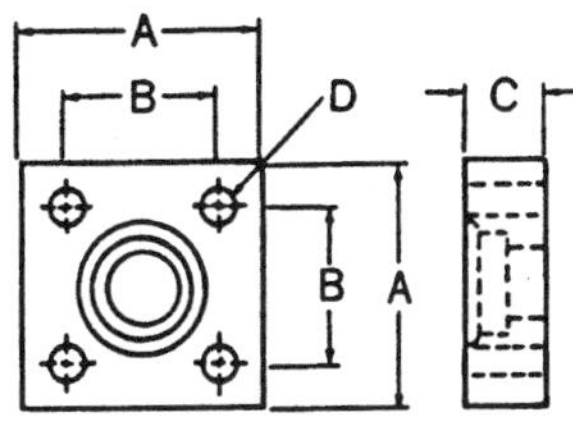

The JIS 4-Bolt square flange connection is similar in concept to the SAE 4-Bolt flange connection, except that the JIS bolt pattern is square and the flange itself is different.

Size mm	Approx. Inch Size	Bolt Size mm (Bolt length for long Design)	Dim. "A" mm (Inch)	Dim. "B" mm (Inch)	Dim. "C" mm (Inch)	Bolt Hole "D" mm (Inch)
12	1/2	M10 x 1.5 x 55 (80)	63 (2.48)	40 (1.57)	22 (.87)	11 (.43)
19	3/4	M10 x 1.5 x 55 (80)	68 (2.67)	45 (1.77)	22 (.87)	11 (.43)
25	1	M12 x 1.75 x 70 (100)	80 (3.15)	53 (2.09)	28 (1.10)	13 (.51)
32	1 1/4	M12 x 1.75 x 70 (100)	90 (3.54)	63 (2.48)	28 (1.10)	13 (.51)
38	1 1/2	M16 x 2.0 x 90 (130)	100 (3.94)	70 (2.76)	36 (1.42)	18 (.71)
50	2	M16 x 2.0 x 90 (130)	112 (4.41)	80 (3.15)	36 (1.42)	18 (.71)

JIS 210 Kgf/cm² O-Ring

Nominal Size mm	Dim. "D" mm	Dim. "W" mm
12	24.4 ± 0.15	3.1 ± 0.1
19	29.4 ± 0.15	3.1 ± 0.1
25	34.4 ± 0.15	3.1 ± 0.1
32	39.4 ± 0.15	3.1 ± 0.1
38	49.4 ± 0.15	3.1 ± 0.1
50	59.4 ± 0.15	3.1 ± 0.1

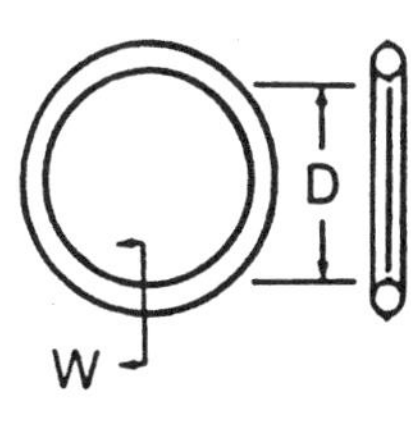

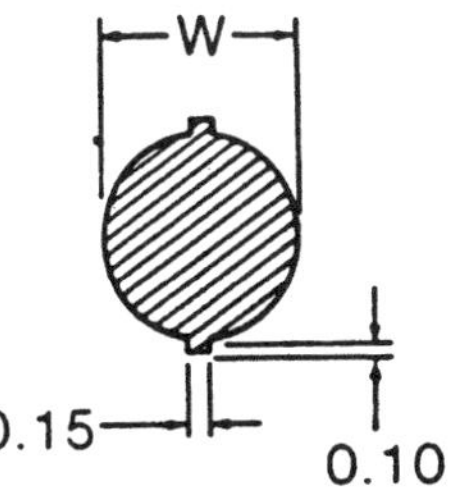

Solenoid Valve Interfaces

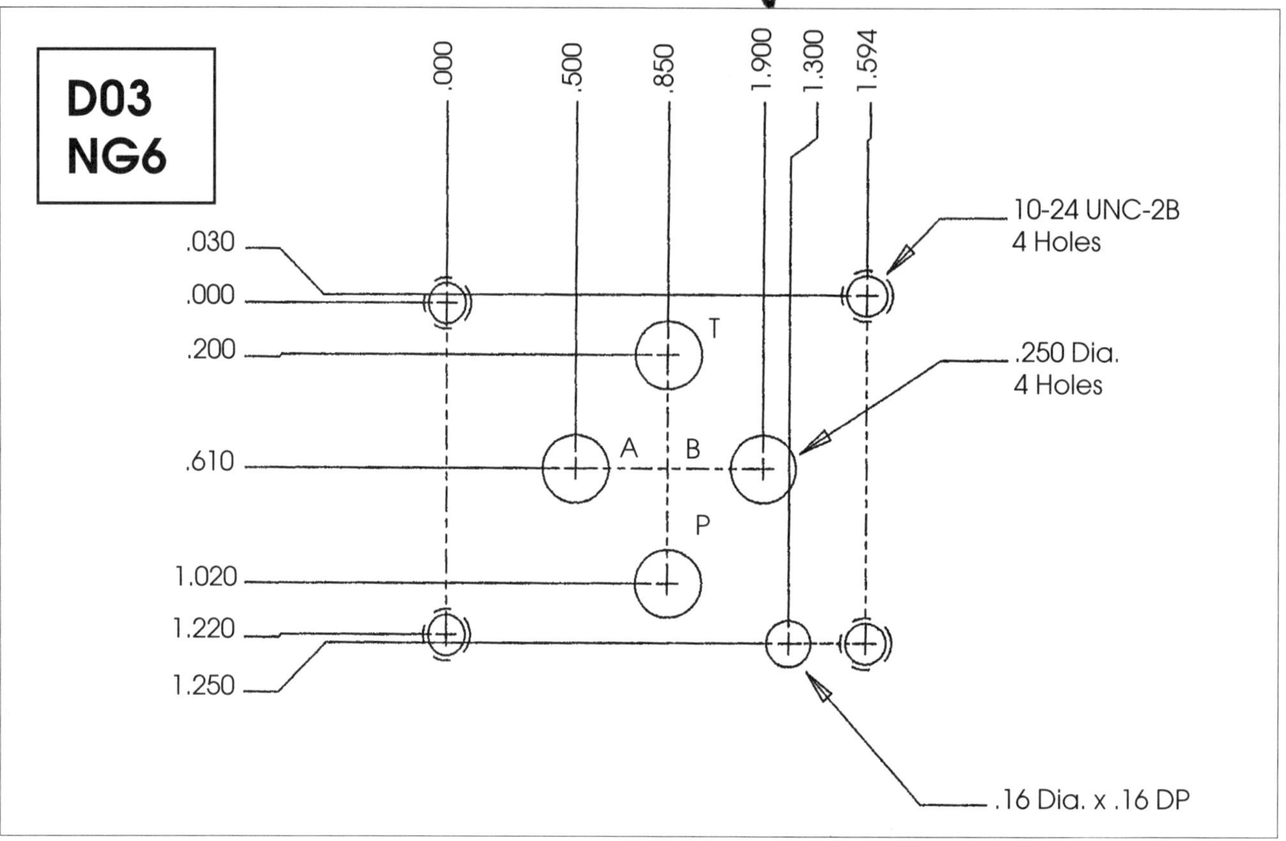

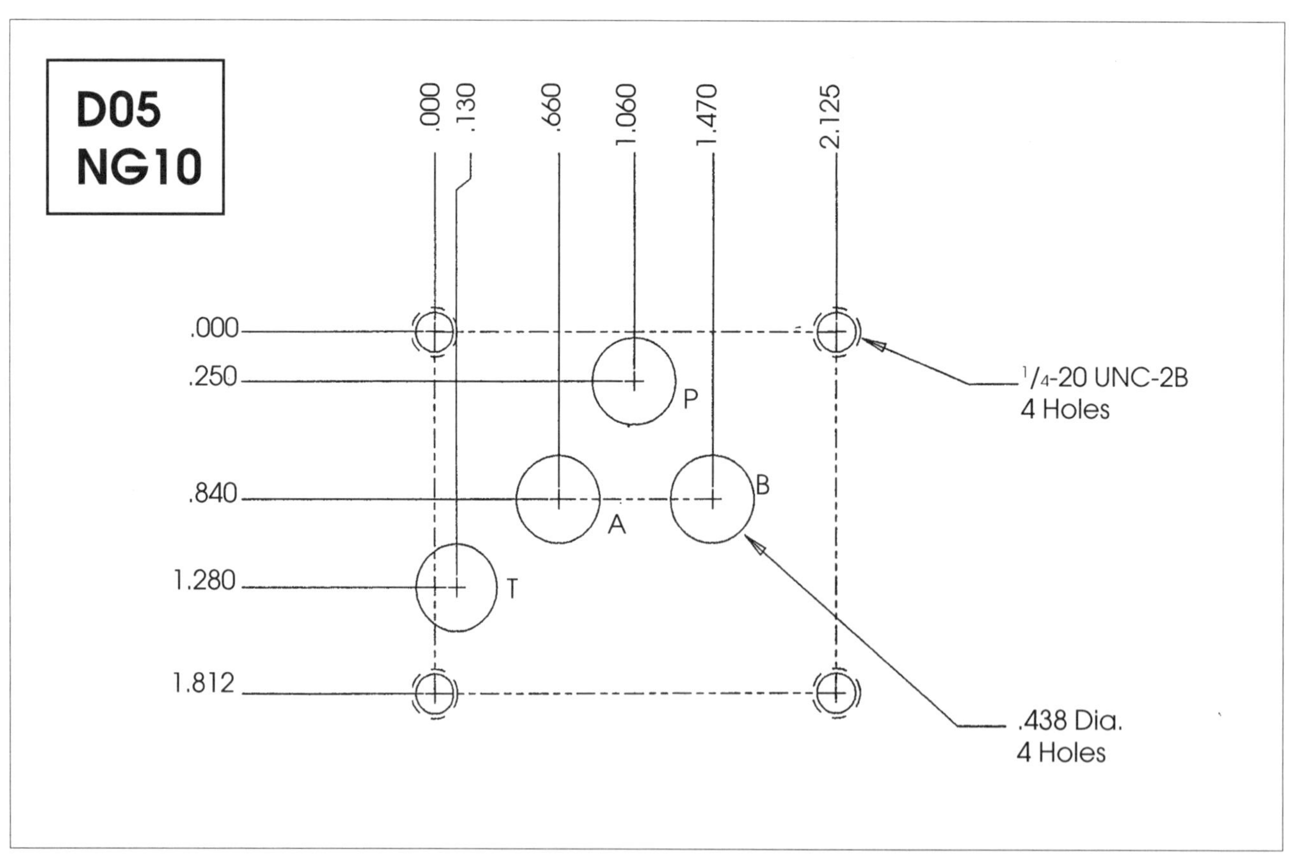

Solenoid Valve Interfaces

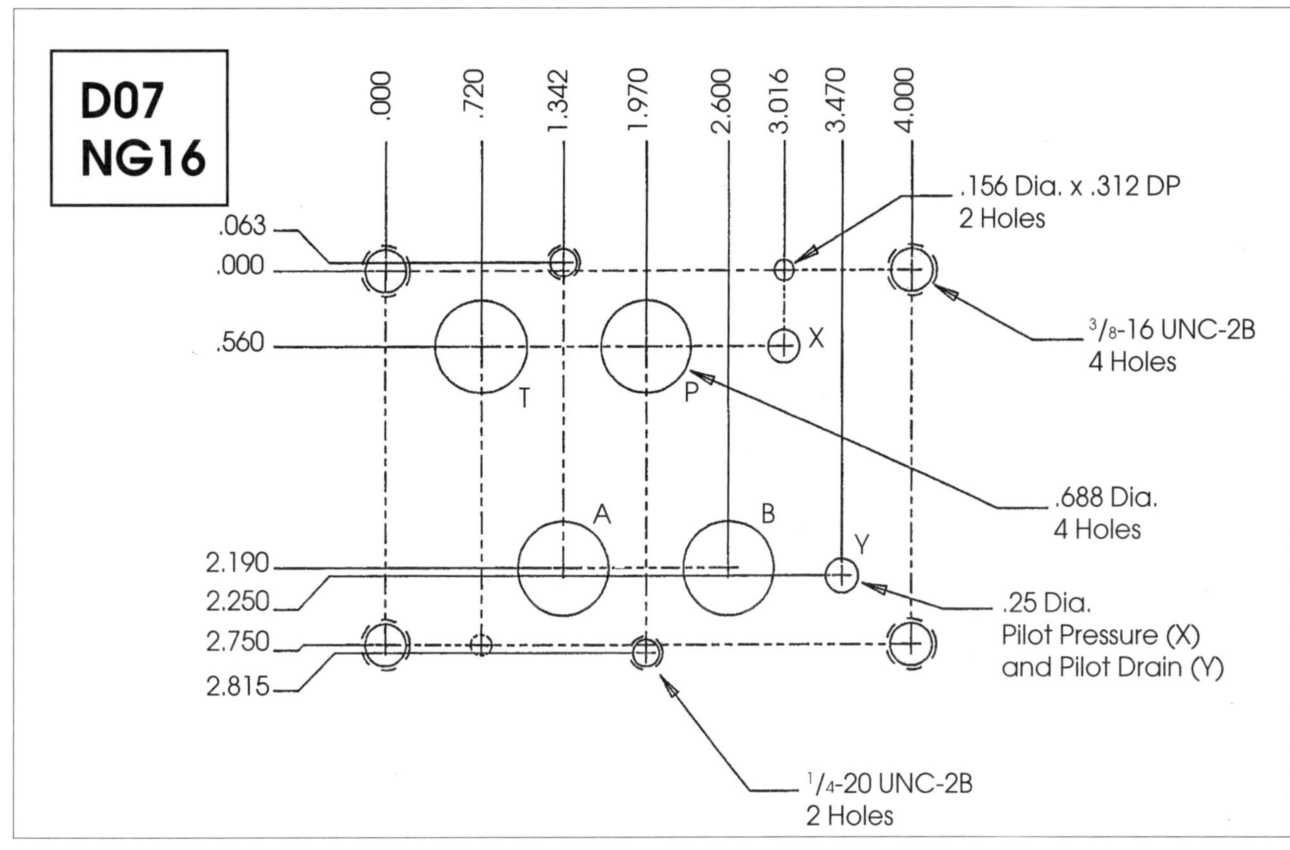

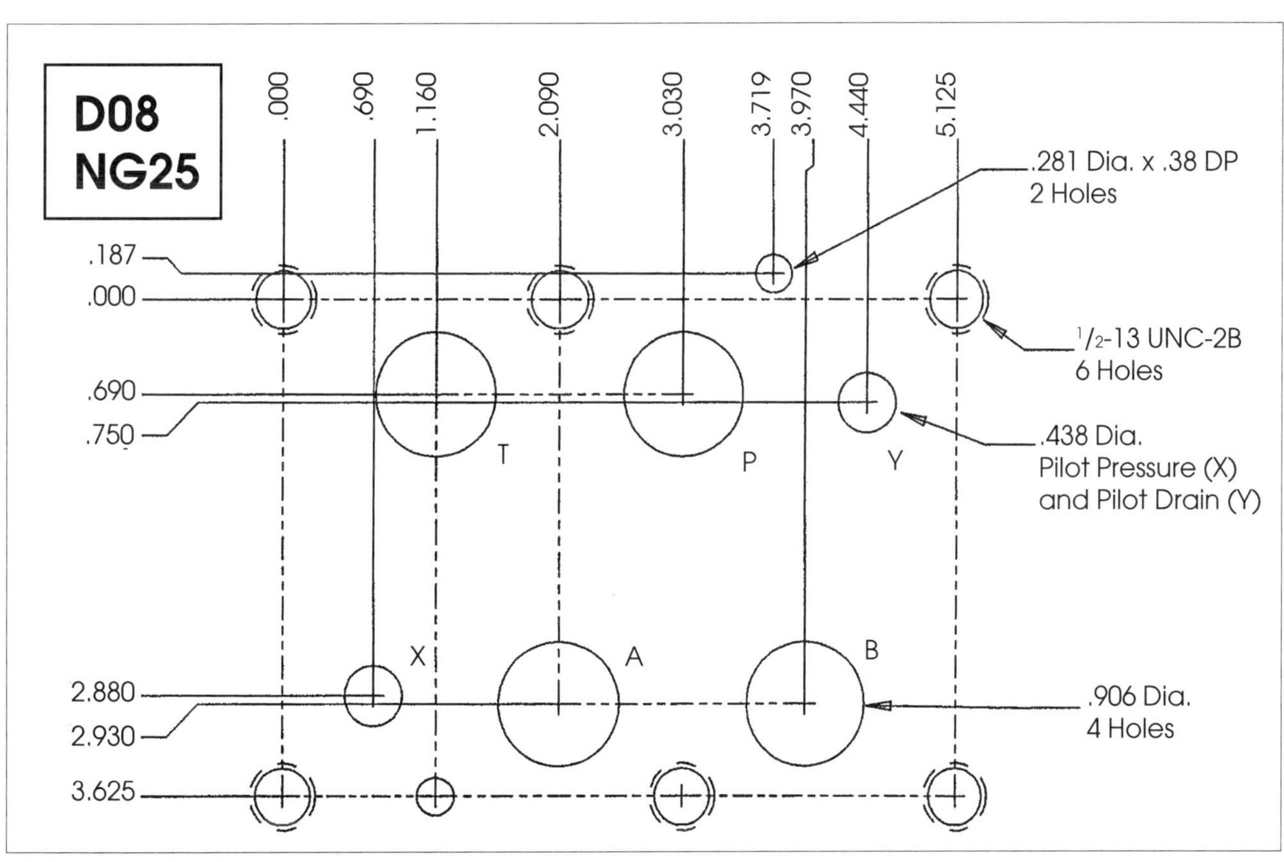

Solenoid Valve Interfaces

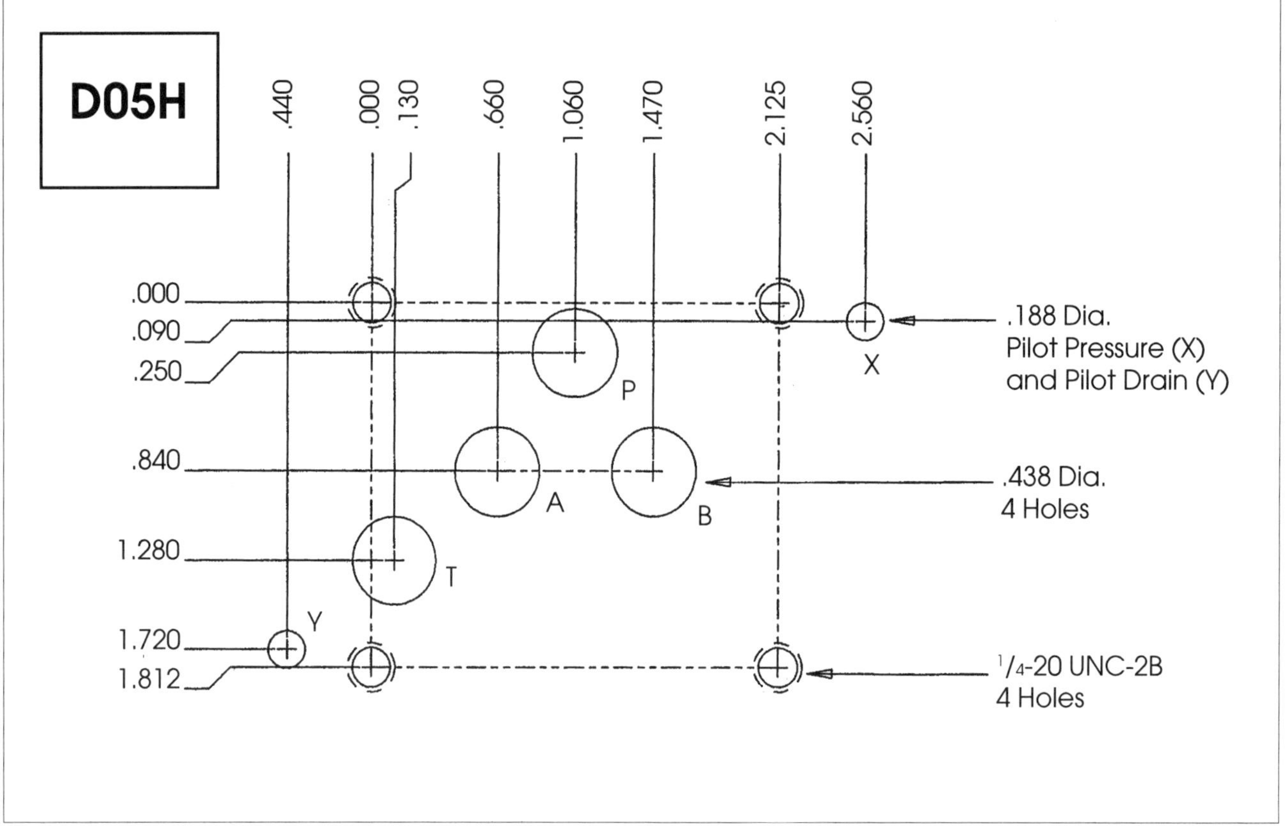

O-Ring Materials and Durometers

To determine material, base your selection on:

- How the unit will operate. Will it be static or dynamic sealing?
- The media to be sealed (fluid, gas, specific chemicals, etc.) Or will the o-ring have some other use?
- Temperature extremes, length of exposure
- Pressure range (higher pressure requires higher durometer)

Material Designation **BN** Durometers 40 thru 80°

Buna-N (Nitrile) (NBR)
The standard for most general applications: petroleum base lubricants, hydraulic oils, gasoline, fuels, alcohol, L.P. gases, water and many other media.

Temperature range -40 to +275°F (-40 to +135°C)

Material Designation **FS** Durometers 60 thru 80°

Fluorosilicone (FS)
Recommended where extreme temperature properties are required. For petroleum oils, fuels and other lubricants; synthetic ester and diester lubricants.

Temperature range -80 to +350°F (-64 to +177°C)

Material Designation **EP** Durometers 40 thru 80°

Ethylene-Propylene (EPDM)
For hot water, steam, acids, alcohols, alkalis, keytones, phosphate esters, brake fluids, drive belts and exposure to oxygen, ozone and weathering.

Temperature range -67 to +302°F (-55 to +150°C)

Material Designation **CR** Durometers 40 thru 80°

Neoprene (Chloroprene) (CR)
Recommended for refrigeration freon gases, carbon dioxide gases, chlorine, ozone, sunlight exposure and for use as drive belts.

Temperature range -60 to +300°F (-51 to +149°C)

Material Designation **SL** Durometers 25 thru 80°

Silicone (SL)
Compatible with a variety of fluids, air, oxygen, ozone and other media for extremes of low and high temperature applications.

Temperature range -85 to +482°F (-65 to +250°C)

Material Designation **VT** Durometers 60 thru 90°

Fluorocarbon (Viton) (Fluorel) (VT)
High temperature toughness, stability and compatibility with a wide range of fluid and chemical types including acids, oils, fuels, solvents and gases.

Temperature range -40 to +500°F (-40 to +260°C)

Material Designation **CP** Durometers 70 thru 90°

Cast Polyurethane (CP)
Extremely high tensile abrasion resistance with elongation that allows easy installation. Highly recommended for cylinder design for a wide variety of hydraulic fluids.

Temperature range -65 to +225°F (-54 to +107°C)

O-Ring Sizes — ARP, AN, MS

ARP 568 Uniform Dash#	AN6227B AN6230B	MS 29513 Dash#	I.D.	Nom. O.D.	W.
-001			1/32	3/32	1/32
-002			3/64	9/64	3/64
-003			1/16	3/16	1/16
-004		-4	5/64	13/64	1/16
-005		-5	3/32	7/32	1/16
-006	AN6227B-96	-6	1/8	1/4	1/16
-007	AN6227B-2	-7	5/32	9/32	1/16
-008	AN6227B-3	-8	3/16	5/16	1/16
-009	AN6227B-4	-9	7/32	11/32	1/16
-010	AN6227B-5	-10	1/4	3/8	1/16
-011	AN6227B-6	-11	5/16	7/16	1/16
-012	AN6227B-7	-12	3/8	1/2	1/16
-013		-13	7/16	9/16	1/16
-014		-14	1/2	5/8	1/16
-015		-15	9/16	11/16	1/16
-016		-16	5/8	3/4	1/16
-017		-17	1 1/16	13/16	1/16
-018		-18	3/4	7/8	1/16
-019		-19	13/16	15/16	1/16
-020		-20	7/8	1	1/16
-021		-21	15/16	1 1/16	1/16
-022		-22	1	1 1/8	1/16
-023		-23	1 1/16	1 3/16	1/16
-024		-24	1 1/8	1 1/4	1/16
-025		-25	13/16	15/16	1/16
-026		-26	1 1/4	1 3/8	1/16
-027		-27	15/16	1 7/16	1/16
-028		-28	1 3/8	1 1/2	1/16
-029			1 1/2	1 5/8	1/16
-030			1 5/8	1 3/4	1/16
-031			1 3/4	1 7/8	1/16
-032			1 7/8	2	1/16
-033			2	2 1/6	1/16
-034			2 1/8	2 1/4	1/16
-035			2 1/4	2 3/8	1/16
-036			2 3/8	2 1/2	1/16
-037			2 1/2	2 5/8	1/16
-038			2 5/8	2 3/4	1/16
-039			2 3/4	2 7/8	1/16
-040			2 7/8	3	1/16
-041			3	3 1/8	1/16
-042			3 1/4	3 3/8	1/16
-043			3 1/2	3 5/8	1/16
-044			3 3/4	3 7/8	1/16
-045			4	4 1/8	1/16
-046			4 1/4	4 3/8	1/16
-047			4 1/2	4 5/8	1/16
-048			4 3/4	4 7/8	1/16
-049			5	5 1/8	1/16
-050			5 1/4	5 3/8	1/16
-110	AN6227B-8	-110	3/8	9/16	3/32
-111	AN6227B-9	-111	7/16	5/8	3/32
-112	AN6227B-10	-112	1/2	11/16	3/32
-113	AN6227B-11	-113	9/16	3/4	3/32
-114	AN6227B-12	-114	5/8	13/16	3/32
-115	AN6227B-13	-115	11/16	7/8	3/32
-116	AN6227B-14	-116	3/4	15/16	3/32
-117		-117	13/16	1	3/32
-118		-118	7/8	1 1/16	3/32
-119		-119	15/16	1 1/8	3/32
-120		-120	1	13/16	3/32
-121		-121	1 1/16	1 1/4	3/32

ARP 568 Uniform Dash#	AN6227B AN6230B	MS 29513 Dash#	I.D.	Nom. O.D.	W.
-122		-122	1 1/8	1 5/16	3/32
-123		-123	1 3/16	1 3/8	3/32
-124		-124	1 1/4	1 7/16	3/32
-125		-125	1 5/16	1 1/2	3/32
-126		-126	1 3/8	1 9/16	3/32
-127		-127	1 7/16	1 5/8	3/32
-128		-128	1 1/2	1 11/16	3/32
-129		-129	1 9/16	1 3/4	3/32
-130		-130	1 5/8	1 13/16	3/32
-131		-131	1 11/16	1 7/8	3/32
-132		-132	1 3/4	1 15/16	3/32
-133		-133	1 13/16	2	3/32
-134		-134	1 7/8	2 1/16	3/32
-135		-135	1 15/16	2 1/8	3/32
-136		-136	2	2 3/16	3/32
-137		-137	2 1/16	2 1/4	3/32
-138		-138	2 1/8	2 5/16	3/32
-139		-139	2 3/16	2 3/8	3/32
-140		-140	2 1/4	2 7/16	3/32
-141		-141	2 5/16	2 1/2	3/32
-142		-142	2 3/8	2 9/16	3/32
-143		-143	2 7/16	2 5/16	3/32
-144		-144	2 1/2	2 11/16	3/32
-145		-145	2 9/16	2 3/4	3/32
-146		-146	2 5/8	2 13/16	3/32
-147		-147	2 11/16	2 7/8	3/32
-148		-148	2 3/4	2 15/16	3/32
-149		-149	2 13/16	3	3/32
-150			2 7/8	3 1/16	3/32
-151			3	3 3/16	3/32
-152			3 1/4	3 7/16	3/32
-153			3 1/2	3 11/16	3/32
-154			3 3/4	3 15/16	3/32
-155			4	4 3/16	3/32
-156			4 1/4	4 7/16	3/32
-157			4 1/2	4 11/16	3/32
-158			4 3/4	4 15/16	3/32
-159			5	5 3/16	3/32
-160			5 1/4	5 7/16	3/32
-161			5 1/2	5 11/16	3/32
-162			5 3/4	5 15/16	3/32
-163			6	6 3/16	3/32
-164			6 1/4	6 7/16	3/32
-165			6 1/2	6 11/16	3/32
-166			6 3/4	6 15/16	3/32
-167			7	7 3/16	3/32
-168			7 1/4	7 7/16	3/32
-169			7 1/2	7 11/16	3/32
-170			7 3/4	7 15/16	6/32
-171			8	8 3/16	3/32
-172			8 1/4	8 7/16	3/32
-173			8 1/2	8 11/16	3/32
-174			8 3/4	8 15/16	3/32
-175			9	9 3/16	3/32
-176			9 1/4	9 7/16	3/32
-177			9 1/2	9 11/16	3/32
-178			9 3/4	9 15/16	3/32
-210	AN6227B-15	-210	3/4	1	1/8
-211	AN6227B-16	-211	13/16	1 1/16	1/8
-212	AN6227B-17	-212	7/8	1 1/8	1/8
-213	AN6227B-18	-213	15/16	1 3/16	1/8
-214	AN6227B-19	-214	1	1 1/4	1/8

ARP 568 Uniform Dash#	AN6227B AN6230B	MS 29513 Dash#	I.D.	Nom. O.D.	W.
-215	AN6227B-20	-215	1 1/16	1 5/16	1/8
-216	AN6227B-21	-216	1 1/8	1 3/8	1/8
-217	AN6227B-22	-217	1 3/16	1 7/16	1/8
-218	AN6227B-23	-218	1 1/4	1 1/2	1/8
-219	AN6227B-24	-219	1 5/16	1 9/16	1/8
-220	AN6227B-25	-220	1 3/8	1 5/8	1/8
-221	AN6227B-26	-221	1 7/16	1 11/16	1/8
-222	AN6227B-27	-222	1 1/2	1 3/4	1/8
-223	AN6227B-28	-223	1 5/8	1 7/8	1/8
-224	AN6230B-2	-224	1 3/4	2	1/8
-225	AN6230B-3	-225	1 7/8	2 1/8	1/8
-226	AN6230B-4	-226	2	2 1/4	1/8
-227	AN6230B-5	-227	2 1/8	2 3/8	1/8
-228	AN6230B-6	-228	2 1/4	2 1/2	1/8
-229	AN6230B-7	-229	2 3/8	2 5/8	1/8
-230	AN6230B-8	-230	2 1/2	2 3/4	1/8
-231	AN6230B-9	-231	2 5/8	2 7/8	1/8
-232	AN6230B-1	-232	2 3/4	3	1/8
-233	AN6230B-11	-233	2 7/8	3 1/8	1/8
-234	AN6230B-12	-234	3	3 1/4	1/8
-235	AN6230B-13	-235	3 1/8	3 3/8	1/8
-236	AN6230B-14	-236	3 1/4	3 1/2	1/8
-237	AN6230B-15	-237	3 3/8	3 5/8	1/8
-238	AN6230B-16	-238	3 1/2	3 3/4	1/8
-239	AN6230B-17	-239	3 5/8	3 7/8	1/8
-240	AN6230B-18	-240	3 3/4	4	1/8
-241	AN6230B-19	-241	3 7/8	4 1/8	1/8
-242	AN6230B-20	-242	4	4 1/4	1/8
-243	AN6230B-21	-243	4 1/8	4 3/8	1/8
-244	AN6230B-22	-244	4 1/2	4 1/2	1/8
-245	AN6230B-23	-245	4 3/8	4 5/8	1/8
-246	AN6230B-24	-246	4 1/2	4 3/4	1/8
-247	AN6230B-25	-247	4 5/8	4 7/8	1/8
-248	AN6230B-26	-248	4 3/4	5	1/8
-249	AN6230B-27	-249	4 7/8	5 1/8	1/8
-250	AN6230B-28	-250	5	5 1/4	1/8
-251	AN6230B-29	-251	5 1/8	5 3/8	1/8
-252	AN6230B-30	-252	5 1/4	5 1/2	1/8
-253	AN6230B-31	-253	5 3/8	5 5/8	1/8
-254	AN6230B-32	-254	5 1/2	5 3/4	1/8
-255	AN6230B-33	-255	5 5/8	5 7/8	1/8
-256	AN6230B-34	-256	5 3/4	6	1/8
-257	AN6230B-35	-257	5 7/8	6 1/8	1/8
-258	AN6230B-36	-258	6	6 1/4	1/8
-259	AN6230B-37	-259	6 1/4	6 1/2	1/8
-260	AN6230B-38	-260	6 1/2	6 3/4	1/8
-261	AN6230B-39	-261	6 3/4	7	1/8
-262	AN6230B-40	-262	7	7 1/4	1/8
-263	AN6230B-41	-263	7 1/4	7 1/2	1/8
-264	AN6230B-42	-264	7 1/2	7 3/4	1/8
-265	AN6230B-43	-265	7 3/4	8	1/8
-266	AN6230B-44	-266	8	8 1/4	1/8
-267	AN6230B-45	-267	8 1/4	8 1/2	1/8
-268	AN6230B-46	-268	8 1/2	8 3/4	1/8
-269	AN6230B-47	-269	8 3/4	9	1/8
-270	AN6230B-48	-270	9	9 1/4	1/8
-271	AN6230B-49	-271	9 1/4	9 1/2	1/8
-272	AN6230B-50	-272	9 1/2	9 3/4	1/8
-273	AN6230B-51	-273	9 3/4	10	1/8
-274	AN6230B-52	-274	10	10 3/4	1/8
-275			10 1/2	10 3/4	1/8
-276			11	11 1/4	1/8

O-Ring Sizes — ARP, AN, MS

ARP 568 Uniform Dash#	AN6227B AN6230B	MS 29513 Dash#	I.D.	Nom. O.D.	W.
-277			11 1/2	11 3/4	1/8
-278			12	12 1/4	1/8
-279			13	13 1/4	1/8
-280			14	14 1/4	1/8
-281			15	15 1/4	1/8
-282			16	16 1/4	1/8
-283			17	17 1/4	1/8
-284			18	18 1/4	1/8
-325	AN6227B-28	-325	1 1/2	1 7/8	3/16
-326	AN6227B-29	-326	1 5/6	2	3/16
-327	AN6227B-30	-327	1 3/4	2 1/8	3/16
-328	AN6227B-31	-328	1 7/8	2 1/4	3/16
-329	AN6227B-32	-329	2	2 3/8	3/16
-330	AN6227B-33	-330	2 1/8	2 1/2	3/16
-331	AN6227B-34	-331	2 1/4	2 5/8	3/16
-332	AN6227B-35	-332	2 3/8	2 3/4	3/16
-333	AN6227B-36	-333	2 1/2	2 7/8	3/16
-334	AN6227B-37	-334	2 5/8	3	3/16
-335	AN6227B-38	-335	2 3/4	3 1/8	3/16
-336	AN6227B-39	-336	2 7/8	3 1/4	3/16
-337	AN6227B-40	-337	3	3 3/8	3/16
-338	AN6227B-41	-338	3 1/8	3 1/2	3/16
-339	AN6227B-42	-339	3 1/4	3 5/8	3/16
-340	AN6227B-43	-340	3 3/8	3 3/4	3/16
-341	AN6227B-44	-341	3 1/2	3 7/8	3/16
-342	AN6227B-45	-342	3 5/8	4	3/16
-343	AN6227B-46	-343	3 3/4	4 1/8	3/16
-344	AN6227B-47	-344	3 7/8	4 1/4	3/16
-345	AN6227B-48	-345	4	4 3/8	3/16
-346	AN6227B-49	-346	4 1/8	4 1/2	3/16
-347	AN6227B-50	-347	4 1/4	4 5/8	3/16
-348	AN6227B-51	-348	4 3/8	4 3/4	3/16
-349	AN6227B-52	-349	4 1/2	4 7/8	3/16
-350			4 5/8	5	3/16
-351			4 3/4	5 1/8	3/16
-352			4 7/8	5 1/4	3/16
-353			5	5 3/8	3/16
-354			5 1/8	5 1/2	3/16
-355			5 1/4	5 5/8	3/16
-356			5 3/8	5 3/4	3/16
-357			5 1/2	5 7/8	3/16
-358			5 5/8	6	3/16
-359			5 3/4	6 1/8	3/16
-360			5 7/8	6 1/4	3/16
-361			6	6 3/8	3/16
-362			6 1/4	6 5/8	3/16
-363			6 1/2	6 7/8	3/16
-364			6 3/4	7 1/8	3/16
-365			7	7 3/8	3/16
-366			7 1/4	7 5/8	3/16
-367			7 1/2	7 7/8	3/16
-368			7 3/4	8 1/8	3/16
-369			8	8 3/8	3/16
-370			8 1/4	8 5/8	3/16
-371			8 1/2	8 7/8	3/16
-372			8 3/4	9 1/8	3/16
-373			9	9 3/8	3/16
-374			9 1/4	9 5/8	3/16
-375			9 1/2	9 7/8	3/16
-376			9 3/4	10 1/8	3/16
-377			10	10 3/8	3/16
-378			10 1/2	10 7/8	3/16
-379			11	11 3/8	3/16
-380			11 1/2	11 7/8	3/16
-381			12	12 3/8	3/16
-382			13	13 3/8	3/16
-383			14	14 3/8	3/16
-384			15	15 3/8	3/16
-385			16	16 3/4	3/16
-386			17	17 3/8	3/16
-387			18	18 3/8	3/16
-388			19	19 3/8	3/16
-389			20	20 3/8	3/16
-390			21	21 3/8	3/16
-391			22	22 3/8	3/16
-392			23	23 8/8	3/16
-393			24	24 3/8	3/16
-394			25	25 3/8	3/16
-395			26	26 3/8	3/16
-425	AN6227B-88	-425	4 1/2	5	1/4
-426	AN6227B-53	-426	4 5/8	5 1/8	1/4
-427	AN6227B-54	-427	4 3/4	5 1/4	1/4
-428	AN6227B-55	-428	4 7/8	5 3/8	1/4
-429	AN6227B-56	-429	5	5 1/2	1/4
-430	AN6227B-57	-430	5 1/8	5 5/8	1/4
-431	AN6227B-58	-431	5 1/4	5 3/4	1/4
-432	AN6227B-59	-432	5 3/8	5 7/8	1/4
-433	AN6227B-60	-433	5 1/2	6	1/4
-434	AN6227B-61	-434	5 5/8	6 1/8	1/4
-435	AN6227B-62	-435	5 3/4	6 1/4	1/4
-436	AN6227B-63	-436	5 7/8	6 3/8	1/4
-437	AN6227B-64	-437	6	6 1/2	1/4
-438	AN6227B-65	-438	6 1/4	6 3/4	1/4
-439	AN6227B-66	-439	6 1/2	7	1/4
-440	AN6227B-67	-440	6 3/4	7 1/4	1/4
-441	AN6227B-68	-441	7	7 1/2	1/4
-442	AN6227B-69	-442	7 1/4	7 3/4	1/4
-443	AN6227B-70	-443	7 1/2	8	1/4
-444	AN6227B-71	-444	7 3/4	8 1/4	1/4
-445	AN6227B-72	-445	8	8 1/2	1/4
-446	AN6227B-73	-446	8 1/2	9	1/4
-447	AN6227B-74	-447	9	9 1/2	1/4
-448	AN6227B-75	-448	9 1/2	10	1/4
-449	AN6227B-76	-449	10	10 1/2	1/4
-450	AN6227B-77	-450	10 1/2	11	1/4
-451	AN6227B-78	-451	11	11 1/2	1/4
-452	AN6227B-79	-452	11 1/2	12	1/4
-453	AN6227B-80	-453	12	12 1/2	1/4
-454	AN6227B-81	-454	12 1/2	13	1/4
-455	AN6227B-82	-455	13	13 1/2	1/4
-456	AN6227B-83	-456	13 1/2	14	1/4
-457	AN6227B-84	-457	14	14 1/2	1/4
-458	AN6227B-85	-458	14 1/2	15	1/4
-459	AN6227B-86	-459	15	15 1/2	1/4
-460	AN6227B-87	-460	15 1/2	16	1/4
-461			16	16 1/2	1/4
-462			16 1/2	17	1/4
-463			17	17 1/2	1/4
-464			17 1/2	18	1/4
-465			18	18 1/2	1/4
-466			18 1/2	19	1/4
-467			19	19 1/2	1/4
-468			19 1/2	20	1/4
-469			20	20 1/2	1/4
-470			21	21 1/2	1/4
-471			22	22 1/2	1/4
-472			23	23 1/2	1/4
-473			24	24 1/2	1/4
-475			26	26 1/2	1/4

O-Ring Sizes for Bosses and Tube Fittings

The following o-rings are intended for use with internal straight thread bosses and tube fittings, and 10049 and 10050, MS33656, MS3657, SAE straight thread o-ring boss and mating fittings.

ARP 568 Uniform Dash#	AN-6290	Tube Size	Actual Size I.D.	Actual Size W.	Nominal Size I.D.	Nominal Size O.D.	Nominal Size W.
-901		3/32	.185	.056	3/16	19/64	1/16
-902	-2	1/8	.239	.064	15/64	3/8	1/16
-903	-3	3/16	.301	.064	19/64	27/64	1/16
-904	-4	1/4	.351	.072	11/32	1/2	5/64
-905	-5	5/16	.414	.072	13/32	9/16	5/64
-906	-6	3/8	.468	.078	15/32	5/8	5/64
-907		7/16	.530	.082	17/32	43/64	5/64
-908	-8	1/2	.644	.087	41/64	13/16	3/32
-909		9/16	.706	.097	45/64	29/32	3/32
-910	-10	5/8	.755	.097	3/4	61/64	3/32
-911		11/16	.863	.116	55/64	1 3/32	7/64
-912	-12	3/4	.924	.116	59/64	1 5/32	7/64
-913		13/16	.986	.116	63/64	1 7/32	7/64
-914		7/8	1.047	.116	13/64	1 9/32	7/64
-916	-16	1	1.171	.116	1 11/64	1 13/32	7/64
-918		1 1/8	1.355	.116	1 23/64	1 19/32	7/64
-920	-20	1 1/4	1.475	.118	1 15/32	1 23/32	1/8
-924	-24	1 1/2	1.720	.118	1 23/32	1 61/64	1/8
-928	-28	1 3/4	2.090	.118	2 3/32	2 21/64	1/8
-932	-32	2	2.337	.118	2 11/32	2 37/64	1/8

O-Ring Sizes for SAE Flanges

The following o-rings are intended for use with SAE 4-bolt o-ring flanges.

ARP 568 Uniform Dash#	AN6227B AN6230B	SAE Flange Size	O-ring Size Nominal I.D.	O-ring Size Nominal O.D.	O-ring Size Nominal W.
-210	AN6227B-15	1/2	3/4	1	1/8
-214	AN6227B-19	3/4	1	1 1/4	1/8
-219	AN6227B-24	1	1 5/16	1 9/16	1/8
-222	AN6227B-27	1	1 1/2	1 3/4	1/8
-225	AN6230B-3	1 1/2	1 7/8	2 1/8	1/8
-228	AN6230B-6	2	2 1/4	2 1/2	1/8
-232	AN6230B-10	2 1/2	2 3/4	3	1/8
-233	AN6230B-15	3	3 3/8	3 5/8	1/8
-241	AN6230B-19	3 1/2	3 7/8	4 1/8	1/8
-245	AN6230B-23	4	4 3/8	4 5/8	1/8
-253	AN6230B-31	5	5 3/8	5 5/8	1/8

Elastomer Compounds For Hydraulic Fluids

High-Water-Base Fluids (95-5 fluids)

Temperature range: +40 to +120ºF
O-ring compounds: N674-70, nitrile
E540-80, ethylene propylene
E893-80, ethylene propylene (purple Chromassure)

Hydrocarbon Base Hydraulic Fluids (including petroleum base)

Temperature range: -65 to +300ºF
O-ring compounds: -30 to +250ºF, N741-75, nitrile
-20 to +275ºF, N741-75, nitrile
-65 to +275ºF, N756-75, nitrile
-15 to +400ºF, V747-75, fluorocarbon
-15 to +400ºF, V884-75, fluorcarbon (brown Chromassure)

Phosphate Esters

Aircraft types (alkyl phosohate esters)
Temperature range: -65 to +300ºF
O-ring compounds: E515-80, ethylene propylene (NAS1613)
E893-80, ethylene propylene (purple Chromassure)
E692-75, ethylene propylene

Industrial types (aryl phosphate esters))
Temperature range: -30 to +200ºF
O-ring compounds: E540-80, ethylene propylene
E893-80, ethylene propylene, (purple Chromassure)
V747-75, fluorcarbon
V884-75, fluorcarbon (brown Chromassure)

Phosphate Ester-Petroleum Oil Blends

Temperature range: +30 to +212ºF
O-ring compounds: V747-75, fluorcarbon
V884-75, fluorocarbon (brown Chromassure)

Silicate Esters

Temperature range: -65 to +550ºF
O-ring compounds: -15 to +400ºF, V747-75, fluorocarbon
-15 to +400ºF, V884-75, fluorocarbon (brown Chromassure)
-65 to +300ºF, C873-70, neoprene

Silicone Hydraulic Fluids

Temperature range: -100 to +550ºF
O-ring compounds: -100 to +350ºF, L677-70, fluorosilicone (static only) (blue Chromassure)
-65 to +300ºF, E540-80, ethylene propylene
-65 to +300ºF, E893-80, ethylene propylene (purple Chromassure)
-15 to +400ºF, V747-75, fluorocarbon
-15 to +400ºF, V884-75, fluorocarbon (brown Chromassure)

Water-Glycol

Temperature range: 0 to +140ºF (but wider range as a coolant)
O-ring compounds: E540-80, ethylene propylene
E893-80, ethylene propylene (purple Chromassure)
N674-70, nitrile (limited life as dynamic seal anticipated above 110ºF)
N741-75, nitrile (for higher temperature coolant use)

Water-in-Oil Emulsions ("Invert" emulsions)

Temperature range: +10 to +120ºF
O-ring compounds: N674-70, nitrile

Due to variations in each type of fluid, and the many variables possible in the application of o-rings, these compound listings are intended only as general guides. Users must test under their own operating conditions to determine the suitability of any compound in a particular application.

Mil-Spec Hydraulic Fluids and Major Lubricants

	ELASTOMER SEALING COMPOUNDS											
Lubricant	Butyl	Butadiene	Ethyl-Propylene	Fluoro-Carbon	Fluoro-Silicone	Hypalon	Isoprene	Nitrile	Neoprene	Poly-Acrylate	Polysulfide	Silicone
MIL-L-2105	U	U	U	1	1	1	U	1	1	1	1	3
MIL-L-3150	U	U	U	1	1	2	U	1	2	2	1	U
MIL-C-4339	U	1	U	1	1	U	U	1	U	1	1	3
MIL-G-4343	3	U	3	1	1	1	U	2	2	1	1	U
MIL-H-5606	U	U	U	1	1	2	U	1	2	2	1	U
MIL-L-6081	U	U	U	1	1	2	U	1	2	1	1	U
MIL-L-6083	U	U	U	1	1	2	U	1	2	1	1	U
MIL-L-6085	U	U	U	1	2	U	U	2	U	2	2	U
MIL-C-6529	X	X	X	1	1	3	X	2	2	2	3	U
MIL-C-7024	U	U	U	1	1	U	U	1	2	2	1	U
MIL-L-7808	U	U	U	1	2	U	U	2	U	2	1	U
MIL-L-7870	U	U	U	1	2	U	U	1	2	1	1	U
MIL-L-9000	U	U	U	1	2	2	U	1	2	1	1	U
MIL-G-10924	U	U	U	1	1	2	U	1	2	2	1	U
MIL-L-14107	U	U	U	1	1	U	U	3	X	X	U	1
MIL-F-17111	1	U	U	1	2	2	U	1	2	1	1	U
MIL-L-17331	U	U	U	1	1	2	U	1	2	1	1	U
MIL-G-18709	U	U	U	1	1	1	U	1	1	1	1	3
MIL-L-21260	U	U	U	1	1	2	U	1	2	1	1	3
MIL-L-23699	U	U	U	1	2	3	U	2	3	3	2	U
MIL-G-23827	3	U	U	1	1	3	U	1	3	3	3	3
MIL-L-25681	1	2	1	1	2	2	2	2	2	2	2	U
MIL-G-26760	U	U	U	1	2	2	U	2	2	2	2	U
MIL-H-27601	U	U	U	1	2	3	U	1	2	1	3	U
MIL-G-27617	1	2	1	1	1	X	X	U	X	X	X	U
MIL-L-48000	3	U	U	1	1	3	U	1	3	3	2	3
MIL-H-46004	U	U	U	1	1	2	U	1	2	1	1	3
MIL-H-81019	U	U	U	1	1	2	U	1	2	1	1	3
MIL-H-83282	U	U	U	1	1	2	U	1	2	2	1	U

RATING 1 = Satisfactory 2 = Fair 3 = Doubtful U = Unsatisfactory X = Insufficient Data

This chart is based on published secondary material, and should be used for information purposes only. Further evaluation should be conducted to determine suitable elastomer material for all applications. Please consult your distributor for elastomer recommendations.

Gland Dimensions — Face Seal Glands

Design Table For O-Ring Face Seal Glands

These dimensions are intended primarily for face type seals and low temperature applications

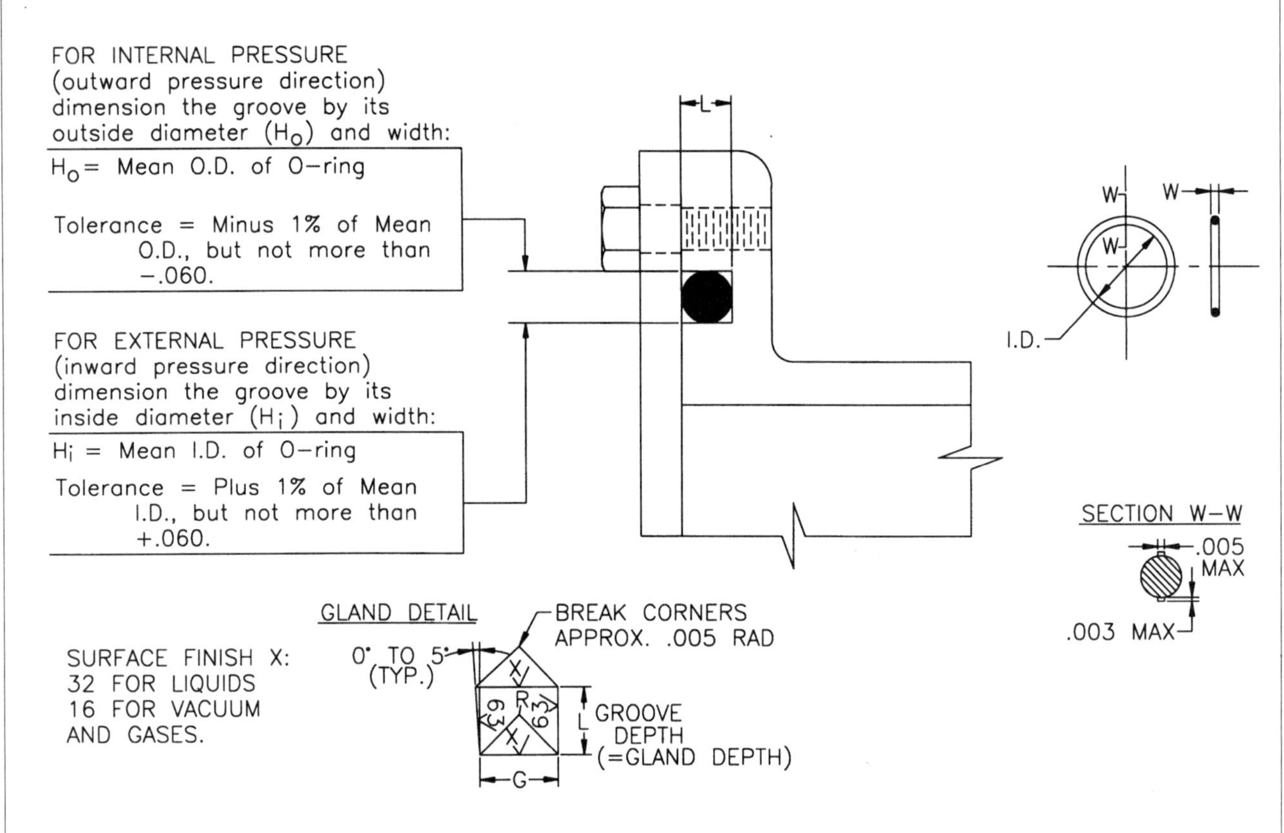

W Cross Section		L	Squeeze		G Groove Width		R
Nominal	Actual	Gland Depth	Actual	%	Liquids	Vacuum & Gases	Groove Radius
1/16	.070 ±.003	.050 to .054	.013 to .023	19 to 32	.101 to .107	.083 to .088	.005 to .015
3/32	.103 ±.003	.074 to .080	.020 to .032	20 to 30	.136 to .142	.118 to .123	.005 to .015
1/8	.139 ±.004	.101 to .107	.028 to .042	20 to 30	.177 to .187	.157 to .163	.010 to .025
3/16	.210 ±.005	.152 to .162	.043 to .063	21 to 30	.270 to .290	.236 to .241	.020 to .035
1/4	.275 ±.006	.201 to .211	.058 to .080	21 to 29	.342 to .362	.305 to .310	.020 to .035
3/8	.375 ±.007	.276 to .286	.082 to .108	22 to 28	.475 to .485	.419 to .424	.030 to .045
1/2	.500 ±.008	.370 to .380	.112 to .138	22 to 27	.638 to .645	.560 to .565	.030 to .045

Dynamic Seals
Industrial Reciprocating

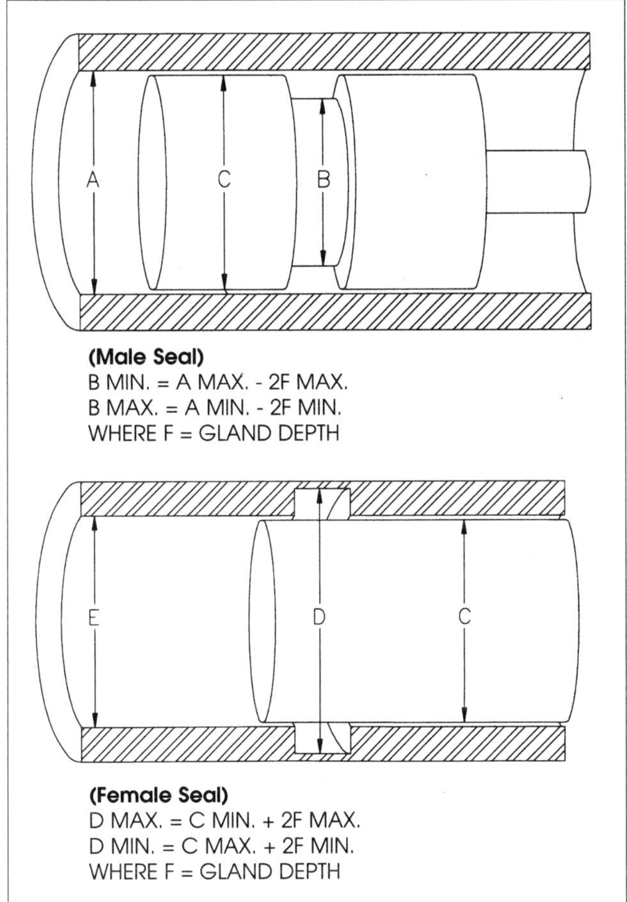

(Male Seal)
B MIN. = A MAX. - 2F MAX.
B MAX. = A MIN. - 2F MIN.
WHERE F = GLAND DEPTH

(Female Seal)
D MAX. = C MIN. + 2F MAX.
D MIN. = C MAX. + 2F MIN.
WHERE F = GLAND DEPTH

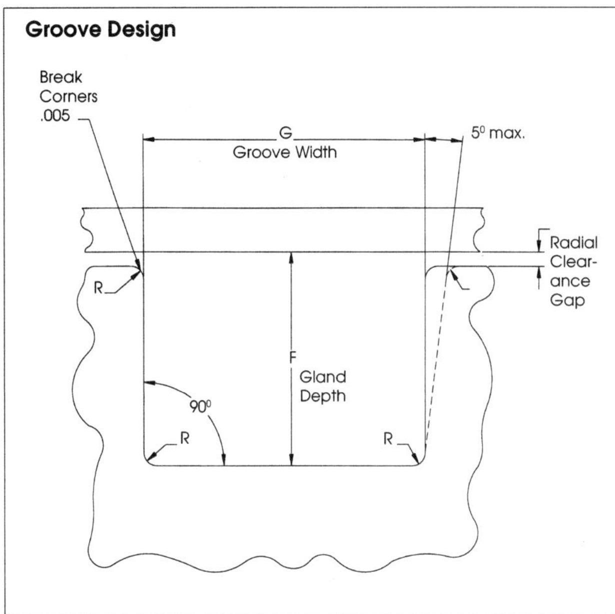

GROOVE SURFACE FINISHES:
Groove sides, no backups — 32
Groove sides with backups — 63
Groove bottom — 32
Rubbing surfaces — 8 to 16

O-ring Cross Section	Gland Depth F Radial Static In.	Static Squeeze for Radial Seals Actual In.	Static Squeeze for Radial Seals %	G Standard Groove Width	Compact Groove No Backup Washer (2) In.	Compact Groove One Backup Washer (2) In.	Compact Groove Two Backup Washer (2) In.	Max Allowable Eccentricity (1) In.	Clearance (Diametral)[3]
.040	.031/.032	.005/.009	13.5 to 22.5	.063	.053			.002	.002* min
.050	.040/.041	.006/.013	12.8 to 24.5	.073	.064			.002	.002* min
.060	.048/.049	.08/.015	14 to 23.8	.084	.075			.002	.002* min
.070	.055/.057	.010/.018	15 to 25	.093	.083	.138	.205	.002	.002* min
.103	.087/.090	.010/.018	10 to 18	.140	.120	.171	.238	.002	.002* min
.139	.119/.123	.012/.024	9 to 17	.185	.160	.208	.275	.003	.003* min
.210	.183/.188	.017/.032	8 to 15	.285	.235	.311	.410	.004	.003* min
.275	.234/.240	.029/.047	11 to 16	.375	.310	.408	.538	.005	.004* min

(1) TIR between groove and adjacent bearing surface.
(2) These groove widths are for compounds that free swell less than 15%. Suitable allowances should be made for higher swell compounds.
(3) For maximum allowable clearance, refer to the O-ring Extrusion table to determine value based upon pressure requirement and compound durometer.
* The piston dimension for male glands must be calculated by using the maximum gap derived from the extrusion tables on page 83, and the minimum gap listed above.
* The bore diameter for female glands must be calculated by using the maximum gap derived from the extrusion tables on page 83, and the minimum gap listed above.

Radial Static Seals (Industrial)

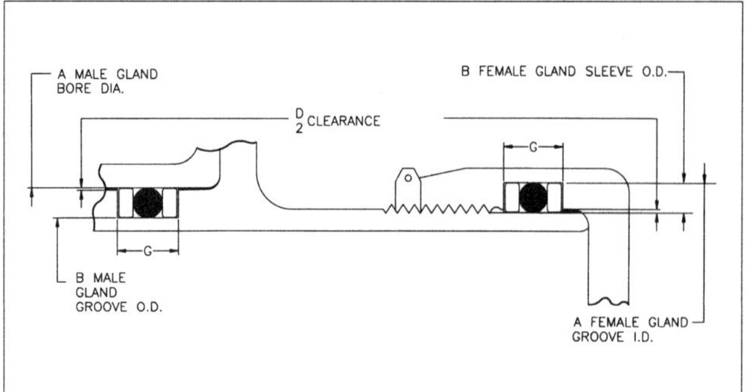

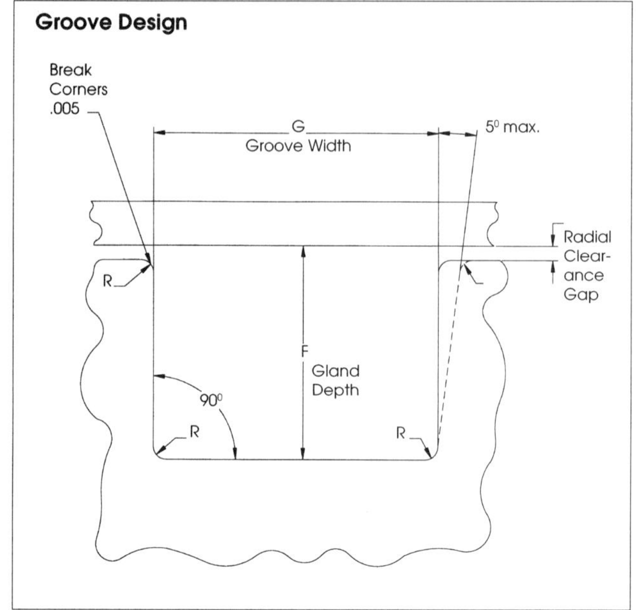

Gland Detail:

X: 63 Surface Finish
32 For Vacuum or Gas

Y: 32 Without Backups
63 With Backups

O-ring Cross Section	Gland Depth F Radial Static In.	Static Squeeze for Radial Seals Actual In.	%	G Standard Groove Width	Compact Groove No Backup Washer (2) In.	One Backup Washer (2) In.	Two Backup Washer (2) In.	Max Allowable Eccentricity (1) In.	Clearance (Diametral)[3]
.040	.029/.030	.007 .014	16 38	.063	.053			.002	.002* min.
.050	.036/.037	.010 .017	19 32	.073	.064			.002	.002* min.
.060	.044/.045	.012 .019	21 30	.084	.075			.002	.002* imn.
.070	.050/.052	.015 .023	22 32	.093	.083	.138	.205	.002	.002* imn.
.103	.081/.083	.017 .025	17 24	.140	.120	.171	.238	.002	.002* min.
.139	.111/.115	.020 .032	15 22	.185	.160	.208	.275	.003	.003* min.
.210	.175/.180	.025 .040	12 19	.285	.235	.311	.410	.004	.003* min.
.275	.228/.234	.035 .053	13 19	.375	.310	.408	.538	.005	.004* min.

(1) TIR between groove and adjacent bearing surface.
(2) These groove widths are for compounds that free swell less than 15%. Suitable allowances should be made for higher swell compounds.
(3) For maximum allowable clearance, refer to the O-ring Extrusion table to determine value based upon pressure requirement and compound durometer.
* Maximum clearance should be reduced by 1/2 for compounds exhibiting poor strength such as silicone and fluorosilicone.

Male plug dimensions and female throat (bore) dimensions must be calculated based upon maximum and minimum clearance gaps.

O-Ring Extrusion Limits

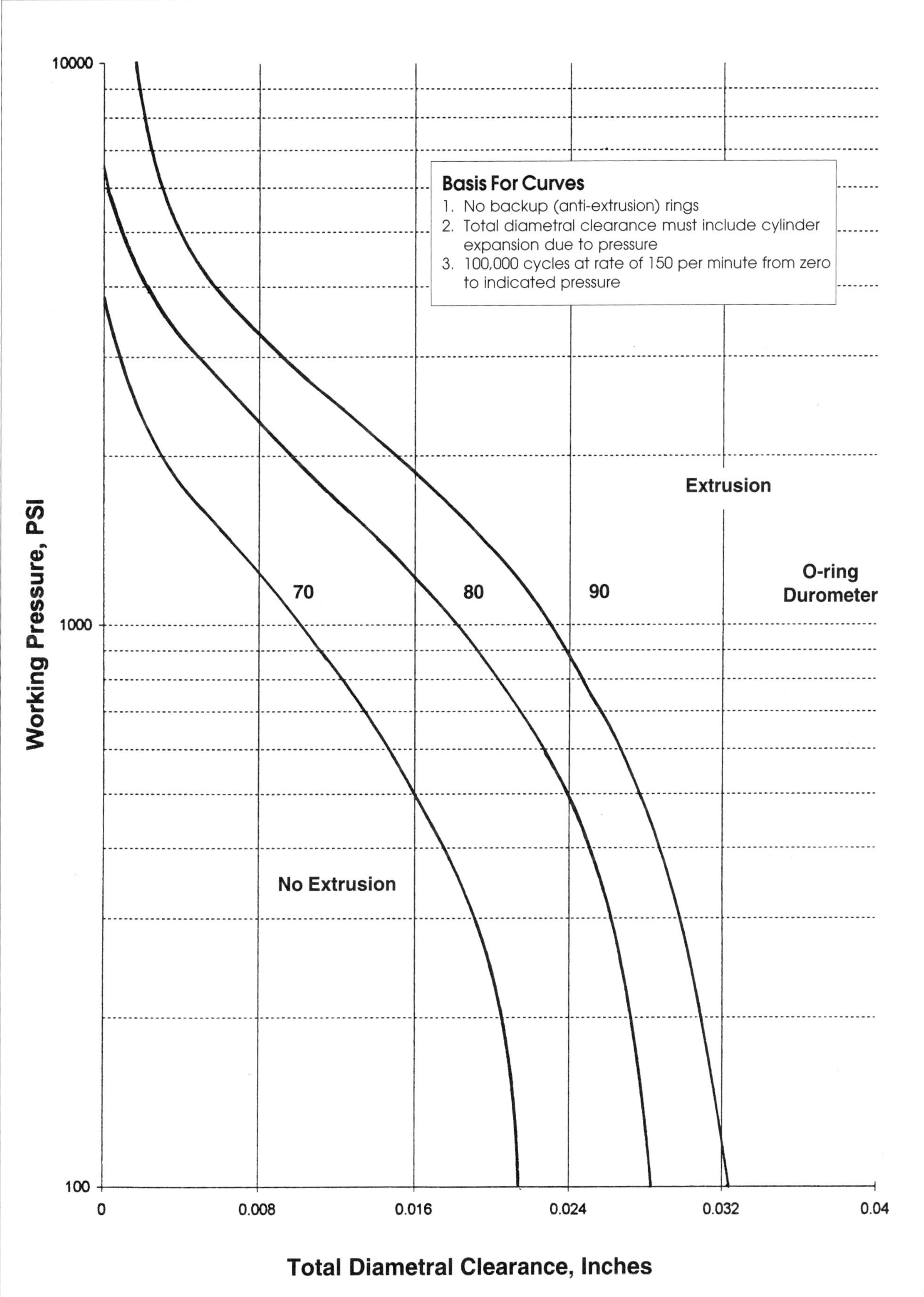

Reference Formulas For Fluids

Fluid

COMPRESSIBILITY of a Fluid

$$C(b) = \frac{1}{BM}$$

C(b) = Compressibility of a fluid (in^2/lb)
BM = Bulk Modulus of Elasticity (lb/in^2)
BM ≈ 280,000 psi for normal hydraulic oils
For normal hydraulic oil compressibility is ≈ 1/2% per 1000 psi

SPECIFIC GRAVITY of a Fluid

$$SG = \frac{\text{Weight of 1 cu. ft of fluid}}{\text{Weight of 1 cu. ft of water}}$$

SG = Specific Gravity of a fluid (no units)
Weight of 1 cu. ft of a fluid (lb/cu. ft)
Weight of 1 cu. ft of water = 62.4283 (lb/cu. ft)

BULK MODULUS of Elasticity (Isothermal Secant)

$$BM = BM_{STD} \times C_A \times C_{PR} \times C_T$$

BM = Bulk Modulus of Elasticity, corrected (lb/in^2)
BM_{STD} = Bulk Modulus of Elasticity, STD (lb/in^2)
C_A = Correction for Air Entrapment (no units)
C_{PR} = Correction for Pressure (no units)
C_T = Correction for Temperature (no units)
See charts for Correction Factors

$$BM_{ISO} = (SECANT^{-1} \{TAN(BM_{AT})\}) \times C_V/C_P$$

BM_{ISO} = Bulk Modulus of Elasticity, Isothermal Secant (lb/in^2)
BM_{AT} = Bulk Modulus of Elasticity Adiabitic Tangent (lb/in^2)
C_V = Specific heat at constant volume (btu/lbm - ^0R)
C_P = Specific heat at constant pressure (btu/lbm - ^0R)

$$BM_{STD} = BM_{ISO}$$

(With any correction for temperature, pressure, etc. from the test conditions of BM_{AT} to the standard conditions for BM_{STD})

BULK MODULUS of Elasticiy

Gasoline	=	137,000 (psi)
Methyl Alcohol	=	161,000 (psi)
Kerosene	=	178,000 (psi)
Hydraulic oil	=	280,000 (psi) (range of 150,000 to 350,000 (psi))
Water	=	320,000 (psi)

Standard at 100^{0}F, 1 ATM. The values given in the chart are average values of the Isothermal Secant Bulk Modulus of Elasticity and are satisfactory for most engineering calculations. If the particular application is critical, the appropriate Modulus of Elasticity should be obtained from the manufacturer of the particular fluid used

SPECIFIC GRAVITY Conversion

$$Sp.\ Gr. = \frac{141.5}{131.5 + {}^0A.P.I.}$$

Bulk Modulus of Elasticity

The Bulk Modulus of Elasticity is a measure of the resistance of a fluid to compression (reciprocal of compressibility). This property is a function of pressure, temperature and entrapped & dissolved air. Bulk Modulus is important in hydraulics in the areas of: (1) Hydraulic and servo system efficiencies and reaction time, and (2) resonance and water-hammer effects in pressurized systems. The two normal methods of determining the bulk modulus of a fluid are: (1) to measure the density change for a pressure change or (2) velocity of sound in the fluid at a given pressure (and temperature, etc.). The first method, measuring density change with a pressure change, is an average measure of the bulk modulus over the pressure range used in determining the bulk modulus (Isothermal Secant) and is the normal value used in engineering calculations. The bulk modulus determined by the sound measurement is the bulk modulus for the specific pressure that it was measured at and is the Adiabatic Tangent Bulk Modulus of Elasticity. The bulk modulus is a physical property of the fluid and is not significantly effected by additives. Entrained and dissolved air in hydraulic fluids will reduce the bulk modulus of the fluid (i.e. increased compressibility). Air saturation at atmospheric pressure does not significantly effect the bulk modulus for normal engineering calculations as can airation from a poorly designed system.

BULK MODULUS
Pressure & Temperature Correction

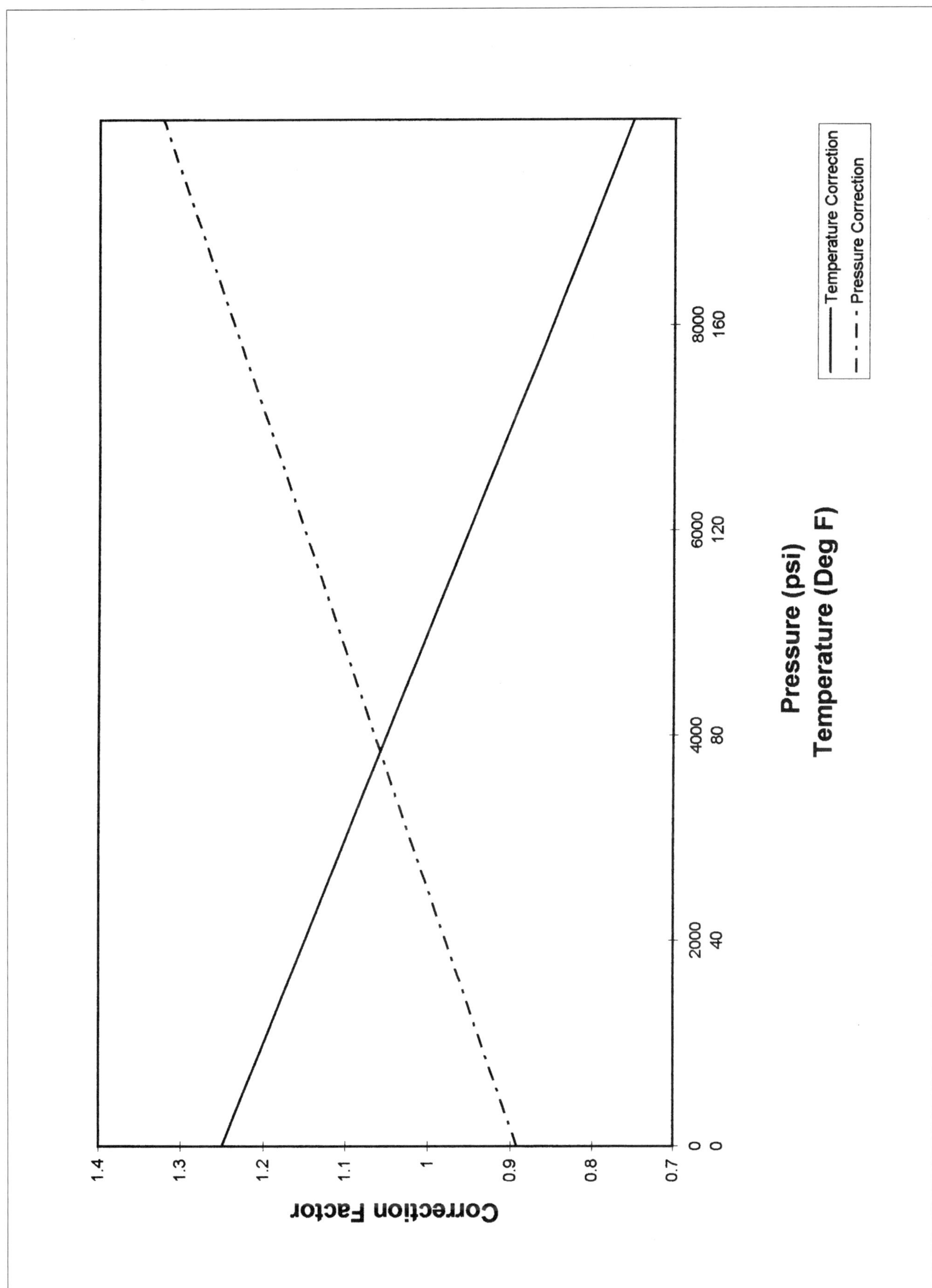

ENTRAPPED AIR CORRECTION
Pressure & Temperature Correction

2000 PSI
1000 PSI
500 PSI
200 PSI
100 PSI
50 PSI

ENTRAPPED AIR CORRECTION
1
0.9
0.8
0.7
0.6
0.5
0.4
0.3
0.2
0.1
0

AIR ENTRAPPED AT 1 ATM
0
1
2
3
4
5

Hydraulic Oil

Hydraulic oil must perform three basic functions:

1. Transmit power effectively
2. Lubricate the hydraulic components
3. Transmit heat away from hydraulic components

There are several characteristics used to describe the performance of oil:

- **Viscosity Index** — Expresses change in viscosity with change in temperature. The higher the viscosity index, usually abbreviated "V.I.", the smaller the change in viscosity with a given change in temperature.
- **Specific Gravity** — The ratio of the weight of any volume of a substance to the weight of an equal volume of water. See formula on page 84.
- **Pour Point** — An indication of the relative ability of an oil to flow at low temperature.
- **Flash Point** — The temperature, expressed in degrees Fahrenheit, to which an oil must be heated in order that its vapor will form a flammable mixture in air.
- **Viscosity** — The measure of an oil's resistance to flow (its internal resistance to shear) and is considered to be hydraulic oil's most important characteristic. If a hydraulic oil does not have the right viscosity, it cannot perform satisfactorily, no matter how superior it may be in other respects.

Use of an oil that is too light usually results in:

- Excessive leakage
- Lower volumetric pump efficiency
- Increased wear
- Loss of pressure
- Lack of positive hydraulic control
- Lower overall efficiency

Use of an oil that is too heavy usually results in:

- Higher oil temperatures
- Sluggish operation
- Lower mechanical efficiency
- Higher power consumption

Changes in viscosity:

- Viscosity decreases as the temperature of the oil increases
- Viscosity increases as pressure on the oil increases
- The viscosity of hydrocarbon oils increases when exposed to high nuclear radiation

Newtonian Fluids

A fluid for which the shear stress is directly proportional to the rate of strain is called a Newtonian fluid. Hydraulic fluid falls into this category. A fluid that is not directly proportional to the rate of strain is called a non-Newtonian fluid, an example of this is grease. The information provided in this section will only address Newtonian fluids.

Absolute Viscosity

Based on the assumption that a fluid would exert an internal frictional force similar to friction in a solid material, Newton developed an apparatus to measure this friction now known as viscosity. The apparatus consisted of two concentric cylinders, one placed inside of the other with space for a fluid in the annular space in between the two cylinders. The two cylinders were immersed in still water and a rotational force was applied to one of the cylinders increasing in rate until the second cylinder also started to rotate at a fixed ratio to the first. The force required to bring the two cylinders into rotation is a measure of the internal friction of the fluid. Newton applied this apparatus to several fluids with the same general outcome:

- The force is proportional to the cylinder's area
- The force is proportional to the cylinder's rotational speed
- The force is inversely proportional to the film thickness

Stated mathematically, the force is acted upon by a resistance to shear that is constant for a given fluid at a specific temperature. This constant is the Absolute Viscosity of the fluid tested at the test temperature. The formula for this relationship is as follows:

$$F = \mu \times A\,(\upsilon/h)$$

F = Force (lbs)
μ = Absolute Viscosity (lb - in - sec)/(in^3)
A = Area (in^2)
υ = Velocity (in/sec)
h = Film thickness (in)

μ = Absolute Viscosity = (lb x in x sec)/(in^3) = Reyn
μ = Absolute Viscosity = (dyn x cm x sec)/(cm^3) = Posie

A second method of determining the viscosity of a fluid has been developed, that of a fluid flow in a capillary tube and is defined by Poiseuilles equation:

$$\mu = \frac{P \times \pi \times R^4}{8 \times \upsilon \times L}$$

μ = Absolute Viscosity (lb - in - sec)/(in^3)
P = Pressure (psi)
π = 3.1415926
R = Tube radius (in)
υ = Velocity (in/sec)
L = Length (in)

The equation can be rewritten using the fact that pressure = gh, where g is gravity and h is head and in consideration of the effect of density (ρ) on the flow rate to:

$$\frac{\mu}{\rho} = \frac{g \times h \times \pi \times R^4}{8 \times \upsilon \times L}$$

Kinematic Viscosity

The ratio of Absolute Viscosity of a fluid to its Density is Kinematic Viscosity:

$$\upsilon = \frac{\mu}{\rho} = \frac{\text{poise}}{\text{Density}} = \frac{(\text{dyn - cm - sec})/(\text{cm}^3)}{(\text{gram/cm}^3)}$$

The more common usage in units is a Centistoke:

$$\text{Centistoke} = \frac{\text{Stoke}}{100}$$

The American equivalent is the Saybolt Universal Seconds (SUS); which is the time, in sec, for 60 cm^3 to flow through a tube section in a viscometer at the reported temperature. The test is valid for fluids that exceed 32 SUS only. For fluids that have high viscosity a second instrument is used, a Saybolt Furol viscometer which has a tube diameter almost twice the diameter of the Universal instrument. The values from the Furol instrument have units of Saybolt Furol Seconds (SFS). The approximate conversion from SUS to SFS is to multiply SUS by 10. The test temperatures for SUS are normally at 100^{0}F and 210^{0}F. The test temperature for SFS is normally 122^{0}F or 210^{0}F.

SUS is sometimes noted as SSU (Saybolt Seconds Universal).

Conversions

Specific gravity at T = (Specific Gravity at 60^{0}F) - .0004 x (T - 60)

Rule of thumb for conversion from Saybolt Universal Seconds to ISO is to divide SUS by 5 and choose the nearest ISO grade.

Viscosity Conversion Factors and Formulas

Multiply	By	To Obtain
Poises	100	Centipoises
Lbs-seconds sq/ft	47,880.1	Centipoises
Reyns	6.89473×10^{6}	Centipoises
Centipoises	2.08855×10^{-5}	Lbs-seconds/sq ft
Centipoises	1.45038×10^{-7}	Reyns
Stokes	100	Centistokes
Sq ft/seconds	92,903.4	Centistokes
Centistokes	1.07639 x 10-5	Sq ft/second
Sq ft/second	$1488.16 \times \gamma \text{ in } \frac{\text{lbs}}{\text{cu ft}}$	Centipoises
Centipoises	$\frac{6.71970 \times 10^{-4}}{\gamma \text{ in lbs/cu ft}}$	Sq ft/second

Viscosity Conversion Chart

Conversion of centistokes (cSt) to saybolt universal seconds (SUS).

Viscosity Centistokes	Saybolt Viscosity @ 100°F	@ 210°F	Viscosity Centistokes	Saybolt Viscosity @ 100°F	@ 210°F
2	32.6	32.9	41	190.8	192.1
3	36.0	36.3	42	195.3	196.7
4	39.1	39.4	43	199.8	201.2
5	42.4	42.7	44	204.4	205.9
6	45.6	45.9	45	209.1	210.5
7	48.8	49.1	46	213.7	215.2
8	52.1	52.5	47	218.3	219.8
9	55.5	55.9	48	222.9	224.5
10	58.9	59.3	49	227.5	229.1
11	62.4	62.9	50	232.1	233.8
12	66.0	66.5	51	236.7	238.4
13	69.8	70.3	52	241.4	243.0
14	73.6	74.1	53	246.0	247.7
15	77.4	77.9	54	250.6	252.3
16	81.3	81.9	55	255.2	257.0
17	85.3	85.9	56	259.8	261.6
18	89.4	90.1	57	264.4	266.3
19	93.6	94.2	58	269.1	270.9
20	97.8	98.5	59	273.7	275.6
21	102.0	102.8	60	278.3	280.2
22	106.4	107.1	61	282.9	284.9
23	110.7	111.4	62	287.5	289.5
24	115.0	115.8	63	292.1	294.5
25	119.3	120.1	64	296.7	298.8
26	123.7	124.5	65	301.4	303.5
27	128.1	129.0	66	306.0	308.1
28	132.5	133.4	67	310.6	312.8
29	136.9	137.9	68	315.2	317.4
30	141.3	142.3	69	319.8	322.1
31	145.7	146.8	70	324.4	326.7
32	150.2	151.2			
33	154.7	155.8			
34	159.2	160.3			
35	163.7	164.9	Over 70 Centistokes @ 100°F	Centistokes x 4.6673 = SUS	
36	168.2	169.4			
37	172.7	173.9			
38	177.3	178.5			
39	181.8	183.0			
40	186.3	187.6			

Physical Properties of Several Fluids

Fluid	cSt Viscosity at 40°C	Viscosity Index	Density (g/cm³ at 15°C)	Pour Point (°C)	Flash Point (°C)
Mineral Oils					
Naphtenic	17	70	0.876	-43	177
Paraffinic	17	100	0.861	-18	190
Superrefined	14	92	0.842	-60	204
Polymer-thickened	15	226	0.827	<-62	107
Polyolefin	16	121	0.816	<-62	204
Esters					
Dibasic acid	13	153	0.915	<-54	227
Polyol	15	124	0.960	<-54	252
Tricresyl phosphate	38	-67	1.180	-23	235
Tributyl phosphate	2.7	90	0.970	<-46	171
Compounded phosphate	11	239	1.070	<-60	182
Polyglycol ether	35	148	0.980	-46	210
Silicate	6.8	151	0.910	<-60	188
Silicones					
Dimethyl	79	400	0.970	<-60	260
Phenl methyl	74	360	0.990	<-60	260
Chlorophenyl methyl	56	164	1.030	<-60	288
Alkyl methyl	80	200	0.910	<-60	260
Perfluoro methyl	44	158	1.210	<-60	
Phenyl ether					
4P-3E	75	-20	1.180	9.5	263
5P-4E	355	-74	1.210	4.4	343
Perfluoro polyether					
Low viscosity	36	90	1.880	-48	
High viscosity	287	115	1.870	-26	

Approximate Operating Temperature Limits For Oils

Lubricant	Long Term[1] °C	Long Term[1] °F	Short Term[2] °C	Short Term[2] °F
Petroleum oils	95-120	200-250	135-150	275-300
Superrefined petroleum oils	175-230	350-450	315-345	600-650
Synthetic hydrocarbons	175-230	350-450	315-345	600-650
Organic esters	175-190	350-375	220-230	425-450
Phosphate esters	95-175	200-350	135-230	275-450
Polyglycols	160-175	325-350	205-220	400-425
Polyphenyl ethers	315-370	600-700	425-480	800-900
Silicate esters	190-220	375-425	260-290	500-550
Silicones	220-275	425-525	315-345	600-650
EP Oils	60-80	140-175	65-85	150-185

[1] Long term = hundreds of hours
[2] Short term = few hours

Lubricating Oil Additives

Most modern lubricants require more than one additive to meet all performance demands. In some cases, individual additives are blended directly into the base oil. In other cases, a group of additives are blended into an additive "package", which is subsequently blended into the base oil. Since most additives are active chemicals, they can interact in the package or in the lubricant to form new compounds. These interactions can decrease additive effectiveness and lead to insoluble or otherwise undesirable by-products.

Wear Inhibitors

Wear inhibitors and lubricity agents are polar materials that deposit onto the surface of a metal and provide a film that reduces metal-to-metal contact. Extreme pressure (EP) additives are a special class of boundary lubrication additive which react with the metal surface to form compounds with a lower shear strength than the base metal. This low-shear compound provides the lubrication. Friction modifiers can either absorb or react with the surface to reduce friction by forming a very low shear-strength film.

Corrosion Inhibitors

Rust, the formation of hydrated iron oxide, is a particularly important form of corrosion. Corrosion inhibitors can react with metal surfaces to form a protective coating or deactivate corrosive contaminants in the lubricant.

Oxidation Inhibitors

Oxidation, the most common form of lubricant deterioration, proceeds through free-radical reactions which are accelerated by heat and catalyzed by metals. Oxidation inhibitors generally function by one or more of three mechanisms: free radical inhibition, peroxide decomposition, or metal deactivation. Each mechanism inhibits oxidation at a different link in the chain reaction.

Detergent and Dispersants

Both detergent and dispersant additives are polar materials which provide a cleaning function. Detergency is a surface phenomenon of cleaning surface deposits. Dispersancy is a bulk lubricant phenomenon of keeping contaminants suspended in the lubricant.

Viscosity Modifiers

Viscosity Index (VI) improvers are polymers that cause minimal increase in lubricant viscosity at low temperature, but considerable increase at high temperature. Many of the same chemicals are also used as thickeners to increase the viscosity of products for special applications such as gear oils.

Pour Point Depressants

Petroleum oils contain paraffinic wax which crystallizes in a lattice-like structure as the lubricant cools and prevents the lubricant from flowing. The lowest temperature at which the lubricant flows is called the pour point.

Emulsion Modifiers

Emulsifiers give stable emulsions of water-in-oil or oil-in-water. They are used where high amounts of water improve cooling due to the high specific heat and thermal conductivity of water. Lubricants which contain water are increasingly being used to conserve petroleum base stocks. Demulsifiers make emulsions unstable, which permits separation of water and lubricant. Emulsion modifiers change the interfacial tension of oil and water.

Foam Decomposers

Excessive lubricant foaming can cause an overflow of the lubricating system, displace lubricant in pumps, increase response time of hydraulic systems, and disrupt the lubricant supply. Two theories predominate on the function of foam decomposers. The first is that they increase gas-lubricant interfacial tension to the point where the bubbles collapse. The second is that these partially soluble compounds with low-surface tension cause openings in the bubbles which allow the gas to escape.

Tackiness Agents

Tackiness agents help the lubricant adhere to machine surfaces, or to itself, rather than flow, splatter, leak, mist, or otherwise contaminate the surrounding area.

Seal Swell Agents

Shrinkage of seals results in leaks while excessive swelling and softening cause wear or extrusion from the seal seat — again resulting in leaks. Most lubricants are intended to cause a minor amount of seal swell to ensure sealing without excessive softening.

Dyes

Dyes must be soluble in the lubricant, have high-coloring power and not be detrimental to other lubricant properties.

Viscosity Temperature Correction

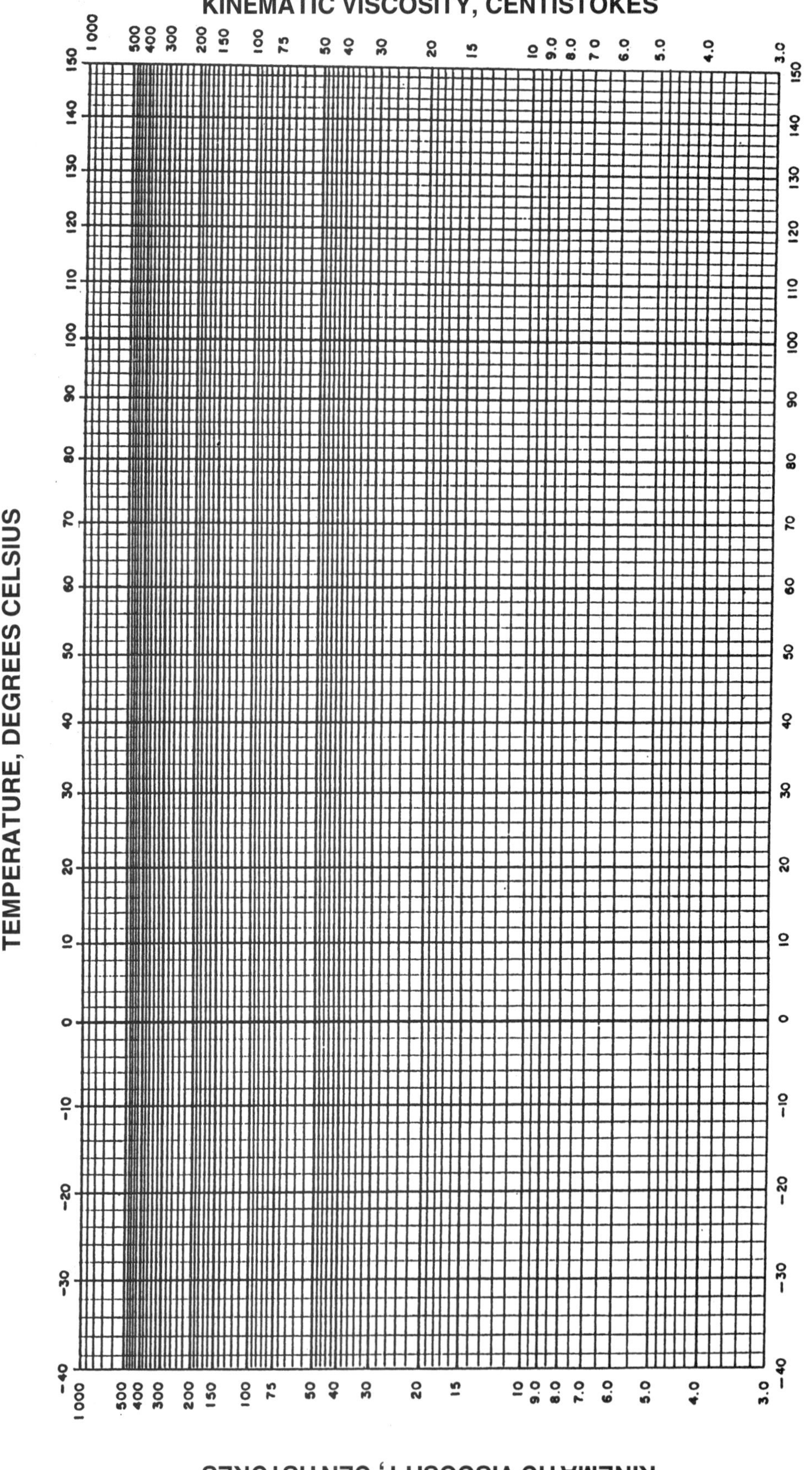

Common Liquid Viscosities

Liquid	*SP Gr at 60°F	SSU	Centistokes	At °F
Mineral Oils **Automobile Crankcase Oils** **(Average Midcontinent Paraffin Base)**				
SAE 10	** .880 to .935	165 to 240	35.4 to 51.9	100
		90 to 120	18.2 to 25.3	130
SAE 20	** .880 to .935	240 to 400	51.9 to 86.6	100
		120 to 185	25.3 to 39.9	130
SAE 30	** .880 to .935	400 to 580	86.6 to 125.5	100
		185 to 255	39.9 to 55.1	130
SAE 40	** .880 to .935	580 to 950	125.5 to 205.6	100
		255 to	55.1 to	130
		80	15.6	210
SAE 50	** .880 to .935	950 to 1,600	205.6 to 352	100
		80 to 105	15.6 to 21.6	210
SAE 60	** .880 to .935	1,600 to 2,300	352 to 507	100
		105 to 125	21.6 to 26.2	210
SAE 70	** .880 to .935	2,300 to 3,100	507 to 682	100
		125 to 150	26.2 to 31.8	210
SAE 10W	** .880 to .935	5,000 to 10,000	1,100 to 2,200	0
SAE 20W	** .880 to .935	10,000 to 40,000	2,200 to 8,800	0
Automobile Transmission Lubricants				
SAE 80	** .880 to .935	100,000 max.	22,000 max.	0
SAE 90	** .880 to .935	800 to 1500	173.2 to 324.7	100
		300 to 500	64.5 to 108.2	130
SAE 140	** .880 to .935	950 to 2300	205.6 to 507	130
		120 to 200	25.1 to 42.9	210
SAE 250	** .880 to .935	Over 2300	Over 507	130
		Over 200	Over 42.9	210
AGMA Gear Oil				
AGMA No. 1	.880 to .935	180 to 240	38.1 to 51.8	100
		45 to 48	9.9 to 10.6	210
AGMA No. 2	.880 to .935	275 to 360	59.3 to 77.5	100
		50 to 55	11.0 to 12.1	210
AGMA No. 3	.880 to .935	490 to 650	106 to 140	100
		63 to 71	13.9 to 15.6	210
AGMA No. 4	.880 to .935	650 to 1000	140 to 216	100
		71 to 87	15.6 to 19.2	210
AGMA No. 5	.880 to .935	875 to1350	189 to 291	100
		83 to 105	18.4 to 22.3	210
AGMA No. 6	.880 to .935	1,350 to 1,850	291 to 398	100
		105 to 125	22.3 to 26.5	210
AGMA No. 7	.880 to .935	1,850 to 2,500	398 to 539	100
		125 to 150	26.5 to 31.9	210
AGMA No. 8	.880 to .935	2,500 to 3,700	539 to 797	100
		150 to 190	31.9 to 40.2	210
AGMA No. 8A	.880 to .935	3,700 to 4,700	797 to 1015	100
		190 to 215	40.2 to 45.6	210
Diesel Engine Lubricating Oils (Based on Average Midcontinent Paraffin Base)				
Federal Specifications No. 9110	** .880 to .935	265 to 240	35.4 to 51.9	100
		90 to 120	18.2 to 25.3	130
Federal Specifications No. 9170	**.880 to .935	300 to 410	64.5 to 88.8	100
		140 to 180	29.8 to 38.8	130
Federal Specifications No. 9250	** .880 to .935	470 to 590	101.8 to 127.8	100
		200 to 255	43.2 to 55.1	130
Federal Specifications No. 9370	** .880 to .935	800 to 1100	173.2 to 238.1	100
		320 to 430	69.3 to 93.1	130
Federal Specifications No. 9500	** .880 to .935	490 to 600	106.1 to 129.9	130
		92 to 105	18.54 to 21.6	210
Diesel Fuel Oils[1]				
No. 2D	** .82 to .95	32.6 to 45.5	2 to 6	100
		39	1 to 3.97	130

Common Liquid Viscosities

Liquid	*SP Gr at 60° F	SSU	Centistokes	At °F
Diesel Fuel Oils, continued[1]				
No. 3D	** .82 to .95	45.5 to 65	6 to 11.75	100
		39 to 48	3.97 to 6.78	130
No. 4D	** .82 to .95	140 max.	29.8 max.	100
	** .82 to .95	70 max.	13.1 max.	130
No. 5D	** .82 to .95	400 max.	86.6 max.	122
		165 max.	35.2 max.	160
Fuel Oils[1]				
No. 1	** .82 to .95	34 to 40	2.39 to 4.28	70
		32 to 35	2.69	100
No. 2	** .82 to .95	36 to 50	3.0 to 7.4	70
		33 to 40	2.11 to 4.28	100
No. 3	** .82 to .95	35 to 45	2.69 to .584	100
		32.8 to 39	2.06 to 3.97	130
No. 5A	** .82 to .95	50 to 125	7.4 to 26.4	100
		42 to 72	4.91 to 13.73	130
No. 5B	** .82 to .95	125 to	26.4 to	100
		400	86.6	122
		72 to 130	13.63 to 67.1	130
No. 6	**. 82 to .95	450 to 3,000	97.4 to 660	122
		175 to 780	37.5 to 172	160
Fuel Oil — Navy Specification[1]	** .989 max.	110 to 225	23 to 48.6	122
		63 to 115	11.08 to 23.9	160
Fuel Oil — Navy II[1]	1.0 max.	1,500 max.	324.7 max.	122
		480 max.	104 max.	160
Gasoline[1]		.68 to .74	.46 to .88	60
			.40 to .71	100
Gasoline[1] **(Natural)**	76.5 degrees API	73	13.9	70
Gas Oil	28 degrees API	50	74	100
Insulating Oil[1]				
Transformer, switches and circuit breakers		115 max.	24.1 max.	70
		65 max.	11.75 max.	100
Kerosene	.78 to .82	35	2.69	68
		32.6	2	100
Machine Lubricating Oil (Average Pennsylvania Paraffin Base)				
Federal Specifications No. 8	**.880 to .935	112 to 160	23.4 to 34.3	100
		70 to 90	13.1 to 18.2	130
Federal Specifications No. 10	** .880 to .935	160 to 235	34.3 to 50.8	100
		90 to 120	18.2 to 25.3	130
Federal Specifications No. 20	** .880 to .935	235 to 385	50.8 to 83.4	100
		120 to 185	25.3 to 39.9	130
Federal Specifications No. 30	** .880 to .935	385 to 550	83.4 to 119	100
		185 to 255	39.9 to 55.1	130
Mineral Lard Cutting Oil				
Federal Specifications Grade 1		140 to 190	29.8 to 41	100
		86 to 110	17.22 to 23	130
Federal Specifications Grade 2		190 to 220	41 to 47.5	100
		110 to 125	23 to 26.4	130
Petrolatum	.825	100	20.6	130
		77	14.8	160
Turbine Lubricating Oil				
Federal Specifications (Penn Base)	.91 Average	400 to 440	86.6 to 95.2	100
		185 to 205	39.9 to 44.3	130
Vegetable Oils				
Castor Oil	.96 @ 68° F	1,200 to 1,500	259.8 to 324.7	100
		450 to 600	97.4 to 129.9	130

Common Liquid Viscosities

Liquid	*SP Gr at 60⁰ F	SSU	Centistokes	At ⁰F
Vegetable Oils, continued				
China Wood Oil	.943	1,425	308.5	69
		580	125.5	100
Coconut Oil	.925	140 to 148	29.8 to 31.6	100
		76 to 80	14.69 to 15.7	130
Corn Oil	.924	135	28.7	130
		54	8.59	212
Cotton Seed Oil	.88 to .925	176	37.9	100
		100	20.6	130
Linseed Oil, Raw	.925 to .939	143	30.5	100
		93	18.94	130
Olive Oil	.912 to .918	200	43.2	100
		115	24.1	130
Palm Oil	.924	221	47.8	100
		125	26.4	130
Peanut Oil	.920	195	42	100
		112	23.4	130
Rape Seed Oil	.919	250	54.1	100
		145	31	130
Rosin Oil	.980	1500	324.7	100
		600	129.9	130
Rosin (Wood)	1.09 Avg	500 to 20,000	108.2 to 4,400	200
		1,000 to 50,000	216.4 to 1,1000	190
Sesame Oil	.923	184	39.6	100
		110	23	130
Soja Bean Oil	.927 to .98	165	35.4	100
Turpentine	.86 to .87	33	2.11	60
		32.6	2.0	100

* Unless otherwise noted.
** Depends on origin or percent and type of solvent.
[1] Explosive under pressure.

Company Name	Category	Part Number	ISO (SAE) Viscosity Grade	Pour Point °F (°C)	Flash Point °F (°C)	Viscosity SUS @ 100°F/ SUS @ 210°F	Viscosity cSt @ 40°C/ cSt @ 100°C	Viscosity Index	API or Specific Gravity
Amoco	Multi-Purpose ATF	30210		-60 (-51)	365 (185)	182.8/52.4	36.2/7.8	196	31.4
	Industrial Oil	44325	32	-25	437	171/44.8	33.2/5.53	102	
		44333	46	-25	471	255/49.3	49.3/6.88	93	
		44340	68	-15	498	358/55.4	69.0/8.66	96	
		44355	100	-10	511	524/64.3	100.3/11.1	95	
		44365	150	-10	543	782/77.4	149/14.4	94	
		44375	220	15	532	1168/96.5	221/18.9	96	
		44395	320	10	554	1705/22	320/24.6	98	
		44415	460	10	568	2420/155	453/31.7	101	
		44420	680	10	547	3620/204	674/41.9	103	
	AW Oils	44760	32	-25	403	164/44.2	32.2/5.35	98	
		44762	46	-30	420	227/48.1	44.4/6.54	97	
		44764	68	-15	450	338/54.3	65.2/8.35	96	
		36675	100	-10	490	526/64.4	101/11.1	95	
		36677	150	-5	478	782/77.1	149/14.4	94	
	Rykon Oils	43458	(5W-20)	-50	417	181/52.4	35.8/7.78	197	
		43469	(5W-15)	-40	419	165/46.0	32.1/5.93	131	
		43474	(10W-20)	-40	410	235/50.9	45.8/7.37	124	
		43479	(20W-30)	-35	419	353/58.8	68.5/9.64	121	
		43487	(20W-40)	-25	424	521/70.5	101/12.7	121	
		43499	(25W-50)	-15	439	1126/110	216/21.9	123	
	Hydraulic Oil All Weather	30277	46	-40	410	235/		124	29.7
	Nonpareil Turbine Oil Medium	38862	32	+10	415	159/44.2	30.8/5.36		33.0
	Turbine Oil No. 32	45005			390	168/44.3	32.5/5.37		
	Amolite Oils	36456	11	0	360	65.8/35.6	11.1/2.77	85	34.5
		36458	22	5	414	116/40.3	22.2/4.18	93	32.5
		36463	46	5	415	168/44.4	32.6/5.40	98	30.3
		36462	46	5	435	238/48.5	46.0/6.63	94	29.7
		36465	68	5	442	354/54.3	68.0/8.35	90	28.9
		36469	100	5	490	513/62.8	98.1/10.7	91	28.2
	Dryer Bearing Oil No. 220	43254			522	1130/94.2	214/18.4		
	Food Grade Hydraulic Oil	44466	32-EL	10	410	168/44.4	32.5/5.40	100	33.1
		44467	46-EL	10	420	236/48.7	45.8/6.70	98	31.9
		44468	68-EL	15	435	340/54.4	65.5/8.38	96	30.7
		44469	100-EL	15	468	514/65.6	98.7/11.5	103	30.5
Bel Ray Company	Anti-Wear Hydraulic Fluid	1	46	-20	440	210/		100	.8681
		3	100	-15	495	500/		100	.8762
		LT-56	22	-75	220	100/		292	.8681
		LT-100	22	-40	365	100/		163	.8719
		LT-200	46	-40	365	225/		163	.8612
		LT-300	64	-40	420	330/		160	.8644
	Safety Hydraulic Fluid	S-150	32			160/			
		S-300	68			300/			
BP Oil	Industrial Fluid	Turbinol T	32	-24 (-31)	399 (204)	150.3/43.3	30.4/5.1	96	31.0
			46	-18 (-28)	419 (215)	218.0/47.9	43.7/6.5	96	30.0
			68	-13 (-25)	439 (226)	323.0/54.3	64.6/8.3	96	29.0
			100	-13 (-25)	446 (230)	500.0/63.9	96.2/11.0	96	28.0
			115	-9 (-23)	460 (238)	610/69.3	115.0/12.4	96	28.0
	Premium-Quality, Antiwear, Hydraulic Oil	Energol HLP-HD	32	-29 (-34)	400 (205)	171.6/44.7	33.3/5.5	100	31.4
			46	-26 (-32)	420 (216)	283.6/48.7	46.2/6.7	98	30.1
			68	-20 (-29)	440 (227)	323.7/53.4	62.5/8.1	97	29.6
			100	-4 (-20)	459 (237)	501.9/64.0	96.4/11.0	96	27.5
			220	0 (-18)	502 (261)	1095/93.0	207.5/18.2	96	27.2
	Transmission-Hydraulic Fluid	Tractran® UTH	(10W)	-61 (-35)	430 (221)	306/58.3	60/9.5	140	29.5
	Hydraulic Transmission Fluid	Autran C-4/TO-4	(10W)	-31 (-35)	440 (227)	248/49.3	48.0/6.9	100	28.5
			(30)	-13 (-25)	482 (250)	509/63.5	97.7/10.9	96	26.6

Company Name	Category	Part Number	ISO (SAE) Viscosity Grade	Pour Point °F (°C)	Flash Point °F (°C)	Viscosity SUS @ 100°F/ SUS @ 210°F	Viscosity cSt @ 40°C/ cSt @ 100°C	Viscosity Index	API or Specific Gravity
BP Oil, cont'd	Premium Automatic Transmission Fluid	Autran® Dexron® III/ Mercon® ATF		-40 (-40)	365 (185)		37.4/7.9	190	31.2
Castrol	Industrial & Specialty Hydraulic Fluids	776RP	22	(-29)	(210)		23.3/4.6	110	0.86
		790-150AW	32	(-29)	(204)		26.5/4.9	100	0.89
		790-300AW	68	(-29)	(236)		62/7.8	90	0.88
		831	10	(-65)	(163)		10.9/3.4	207	0.83
		852	15	(-62)	(165)		13.7/4.5	316	0.83
		853	68	(-30)	(254)		72.5/11.5	142	0.84
		872	15	(-57)	(204)		15.8/3.7	130	0.82
		873	32	(-56)	(238)		30/5.7	138	0.82
		1015	32	(-20)	(215)		30/5.1	95	0.87
		1020	46	(-17)	(229)		42.4/6.3	95	0.87
		1030	68	(-23)	(238)		62.6/8.1	95	0.87
		1050	100	(-20)	(243)		90.6/10.3	96	0.88
		1070	150	(-20)	(232)		150/13.2	91	0.89
	Turbine Oils	785		(-15)	(216)		34.9/5.7		
		98		(-57)	(243)		29.6/7.3		
		580		(-67)	(213)		24.5/5.0		
		778		(-56)	(246)		30.6/5.7		
		5000		(-57)	(260)		25.1/5.0		
Chevron U.S.A.	Heavy Duty Motor Oil	235195	(OW-30)	-65 (-54)	439 (228)		52.1/10.3	191	32.1
	Aviation Lubricant A	247707		-81 (-63)	216 (102)	74/44	13.5/5.3	300+	30.9
	F	247708		-81 (-63)	216 (102)	74/44	13.5/5.3	300+	30.9
	Automatic Transmission Fluid	226502	(10W)	-54 (-48)	396 (202)	190/51	37.5/7.5	172	32.7
	Drive Train Fluids HD	226607	(10W)	-33 (-36)	383 (195)	196/46	38/6.0	101	30.4
		226608	(30)	-11 (-24)	383 (195)	431/60	83/10.0	100	29.4
		226601	(50)	+5 (-15)	491 (255)	945/89	180/17.4	104	26.9
	Tractor Hydraulic Fluid	226606 (West)		-38 (-39)	400 (204)	267/53	52/7.9	130	30.4
		226606 (East)		-38 (-39)	400 (204)	290/57.6	56.8/9.3	145	27.6
	Lubricating Oils FM (Food Processing Lubricants)	232103	32[1]	+16 (-9)	399 (204)	157/44	30.4/5.2	100	31.5
		255110	68	+16 (-9)	421 (216)	334/55	64.6/8.5	102	30.7
		232105	100	+16 (-9)	464 (240)	494/65	95.0/11.3	105	30.4
		255106	220	+16 (-9)	464 (240)	1096/102	209/20.2	112	29.7
		232106	460	+16 (-9)	469 (243)	2308/169	437/34.7	118	28.4
	Paper Machine Oils	255050	150	+5 (-15)	475 (246)	749/78.1	143/14.6	101	29.3
		255076	200	+5 (-15)	480 (249)	1104/94.0	209/18.4	97	27.7
		255075	320	+10 (-12)	473 (248)	1678/117	315/23.5	93	26.1
	Industrial Oils EP	231732	46[1]	-38 (-39)	446 (230)	226/48	43.7/6.5	98	31.1
		231731	100	-11 (-24)	460 (238)	496/63	95.0/10.7	95	30.5
		231730	150	0 (-18)	475 (246)	752/75	143/13.9	93	28.7
		231729	220	+5 (-15)	489 (254)	1105/93	209/18.2	96	27.1
		231728	320	+10 (-12)	529 (276)	1618/116	304/23.2	95	25.6
		231727	460	+5 (-15)	518 (270)	2346/140	437/28.5	91	24.3
	Paper Machine Oils D	253019	150	+5 (-15)	505 (263)	751/76.4	143/14.2	97	29.3
		253021	220	+5 (-15)	514 (268)	1103/94.2	209/18.4	97	27.9
		253022	320	+5 (-15)	540 (282)	1617/115.7	304/23.2	95	26.2
	Cylinder Oils W	230339	220	+16 (-9)	514 (288)	1105/93	209/18.1	95	27.6
		230330	460	-18 (0)	570 (299)	2351/136	437/27.6	87	24.2
		230329	680	+10 (-12)	576 (302)	3476/191	646/39.3	99	23.5
		230338	1000	+10 (-12)	649 (343)	5136/252	950/51.9	102	22.5
	ECO Hydraulic Oils AW™	233902	32	+5 (-15)	399 (204)	165/44.4	32.0/5.4	100	31.4
		233903	46	+5 (-15)	410 (210)	237/48.8	46.0/6.7	99	31.0
		233904	68	+5 (-15)	428 (220)	353/55.5	68.0/8.7	100	30.7
	Hydraulic Oils AW	255677	10[2]	-38 (-39)	309 (154)	63/35	10.3/2.5	48	27.7
		255676	22	-33 (-36)	351 (177)	120/40	23.1/4.4	100	33.7

Company Name	Category	Part Number	ISO (SAE) Viscosity Grade	Pour Point °F (°C)	Flash Point °F (°C)	Viscosity SUS @ 100°F/ SUS @ 210°F	Viscosity cSt @ 40°C/ cSt @ 100°C	Viscosity Index	API or Specific Gravity
Chevron U.S.A., cont'd	Hydraulic Oils AW, cont'd	255675	32	-27 (-33)	399 (204)	173/45	33.6/5.5	100	32.6
		255674	46	-22 (-30)	424 (218)	238/49	46.0/6.7	98	31.8
		255673	68	-22 (-30)	455 (235)	335/54	64.6/8.4	99	31.6
	Hydraulic Oils AW MV	232953	15	-56 (-54)	302 (150)	85.3/39.7	15.8/4.0	159	27.3
		255678	32	-54 (-48)	374 (190)	155/46.6	30.4/6.1	153	32.6
		255679	46	-44 (-42)	367 (186)	222/53.0	43.7/8.0	157	31.9
		232952	68	-44 (-42)	414 (212)	329/62.3	64.6/10.6	154	30.9
		232950	100	-38 (-39)	450 (232)	487/73.4	95.0/13.5	143	30.7
	Synthetic Compressor Oils Tegra®	259129	32	-76 (-60)	466 (241)	156/46.0	30.4/5.9	142	39.1
		259130	68	-65 (-54)	475 (246)	330/61.5	64.6/10.4	149	37.6
		259128	100	-60 (-51)	489 (254)	486/75.4	95.0/14.0	151	37.2
	Turbine Oils GST	220091	EP 32	-27 (-33)	432 (222)	165/44.4	32.0/5.4	102	32.7
		253026	32	-27 (-33)	432 (222)	165/44.4	32.0/5.4	102	32.7
		253027	46	-11 (-24)	435 (224)	237/49.0	46.0/6.8	101	32.0
		253028	68	-6 (-21)	473 (245)	334/54.8	64.6/8.5	102	31.7
		253029	100	-6 (-21)	504 (262)	495/63.8	95.0/11.0	100	31.4
	Utility Oils LVI	231201	22[1]	-40 (-40)	365 (185)	122/39.8	23.1/4.0	38	27.5
		231203	32[1]	-33 (-36)	374 (190)	158/42.1	30.4/4.7	50	26.3
		231207	46[1]	-27 (-33)	376 (191)	228/45.1	43.7/5.6	43	25.1
		231209	68[1]	-27 (-33)	394 (201)	340/49.1	64.6/6.8	33	23.9
		231210	100	-22 (-30)	414 (212)	505/54.6	95.0/8.4	32	22.8
		231211	150	-24 (-11)	441 (227)	775/63.0	143/10.7	29	22.0
		231212	180	-24 (-11)	464 (240)	977/74.5	171/13.7	60	21.3
Citgo	Multipurpose ATF Dexron III\Mercon	33123		-60	367	175/51	34.5/7.5	193	31
	Tractor Hydraulic Fluid	33310		-40	400	290/57	57.2/9.1	137	28.1
	A/W Hydraulic Oils	33932	(All temp.)	-40	439	154/46	30/6.1	155	32.1
		33410	22	-45	433	113/40.5	21.5/4.2	97	31.9
		33415	32	-30	420	171/45	32/5.5	98	31.4
		33420	46	-22	453	225/49	46/6.5	95	30.2
		33430	68	0	453	352/55	68/8.5	95	29.5
		33440	100	+5	475	514/63	98.5/11	95	29
		33450	150	+5	471	782/77	149/14	95	28.2
	CP Hydraulic Fluid	33933		-50	367	130/44.2	25/5.4	155	31.2
	Glycol FR-40XD	48325		-94	None	205/	40/	200	1.09
	Glycol FR-5046HP	48346		-58	None		46/	210	1.09
	Invert FR Fluid	48401		-20		500/175	100/18.5		22
	Pacemaker T Oils	33715	T-32	-38	390	164/44	32/5.4	100	31
		37720	T-46	-27	420	238/48	46/6.7	96	30
		33730	T-68	-27	430	340/54	66/8.3	96	29.6
		33745	T-115	+5	468	593/68	114/12	94	29.3
Davis-Howland Oil Corp.	Anti-Wear Hydraulic Oil	DSL-44	32	5	405	150/44			31.5
		DSL-45	46	5	420	210/47		30.5	
		DSL-46	68	5	435	315/54			29.8
		DSL-47	100	5	465	505/64			28.7
		DSL-48	150	5	485	700/72			28.2
		DSL-49	220	10	520	985/86			27.8
	Premium grade oils for gas turbines, air compressors, & hydraulic systems not requiring anti-wear hydraulic oil	Convis OC							
		150		15	395	150/44		95	31.5
		200		15	395	200/47		95	31.0
		300		20	400	300/54		95	29.5
		500		20	410	500/64		92	29.0
		750		25	420	750/70		90	28.0
		1000		25	500	1000/87		87	27.5
Exxon Corp.	Hydraulic Oil	ESSTIC®	32	-58 (-50)	365 (185)	155/42.4	29.8/4.8	69	28.5
			68	-40 (-40)	400 (204)	329/51.4	63.0/7.5	73	27.0
			150	-13 (-25)	440 (227)	743/67.8	140/12.0	66	25.1

Company Name	Category	Part Number	ISO (SAE) Viscosity Grade	Pour Point °F (°C)	Flash Point °F (°C)	Viscosity SUS @ 100°F/ SUS @ 210°F	Viscosity cSt @ 40°C/ cSt @ 100°C	Viscosity Index	API or Specific Gravity
Exxon Corp. cont'd	Fire-resistant water-based hydraulic & circulating oil	FIREXX® 95/5 (Emulsion) FIREXX I Invert Emulsion		-35 (-37)		438/			0.928
	Anti-wear hydraulic oil	Humble® Hydraulic H	32 46 68 100	0 (-18) 0 (-18) 0 (-18) 0 (-18)	432 (222 441 (227) 464 (240) 468 (242)	155/ 224/ 330/ 511/	30.1 43.3/ 63.7/ 98./	95 95 95 95	
	Anti-wear hydraulic oil	Nuto® H	32 46 68 100 150	-35 (-37) -20 (-29) -17 (-27) -9 (-23) -6 (-21)	403 (206) 421 (216) 432 (222) 466 (241) 511 (266)	166/44.2 48.1/230 343/54.8 485/62.6 810/79.3	32.2/5.3 46.0/6.5 66.1/8.5 92.9/10.6 154.1/14.9	97 96 97 97 95	29.9 29.6 29.2 28.8 28.6
	Synthetic industrial oil	Synesstic®	32 68 100 150 EP-220	-71 (-57) -37 (77) -40 (-40) -30 (-34) -45 (-43)	510 (266) 520 (271) 485 (252) 480 (249) 345 (174)	153/44.7 339/51.4 502/60.6 740/73.0 1108/132	30.0/5.5 65.0/7.5 96.0/10.1 143/13.6 215/27	120 77 88 89 160	
	Anti-wear turbine oil	Teresstic® EP	32 68	16 (-9) 16 (-9)	410 (210) 435 (224)	157/44 336/54.4	30.3/5.2 64.8/8.4	97 97	29.9 29.4
	Paper machine oil	Teresstic® N	150 220 320 460	15 (-9) 15 (-9) 15 (-9) 15 (-9)	490 (254) 507 (264) 535 (279) 550 (288)	764/77.4 1163/96.8 1703/121 2510/146	147/14.5 230/19.0 320/24.3 427/29.2	97 97 97 97	27.4 26.6 25.9 25.3
	High viscosity-index, anti-wear hydraulic oil	Univis® N J 13 J 26		<-75 (<-59) <-75 (<-59)	219 (104) 217 (103)	43.2/75.8 134/59.0	13.9/5.1 26.9/9.8	364 382	32.5 32.1
E. F. Houghton & Co.	High Performance Synthetic Water Based Hydraulic Fluid	Houghto-Safe 250		30	—	315/			0.986
	Water Glycol Fire Resistant Hydraulic Fluid	Houghto-Safe 419-R		-33	—	200/		200	1.08
	Premium Water Glycol Fire Resistant Hydraulic Fluid For High Temps.	Houghto-Safe 620		(-50)	—	200/		200	1.08
	Water In Oil Emulsion Hydraulic Fluid	Houghto-Safe 5046		-22 (-30)	—	410/		135	0.92
Kendall Motor Oil	Hydraulic 7 Industrial Oils Kenoil R&OAW	7322 7323 7325	46 68 100	-15 (-26) -5 (-20) 0 (-18)	400 (204) 415 (213) 460 (238)	203/49.7 334/55.9 467/62.9	46.2/6.39 67.1/8.7 93.8/10.6	107 103 99	
	Hyken	7354 7357	32/46 15/22	-35 (-37) -50 (-46)	400 (204) 300 (149)	183/50.9 90/41.7	36.35/7.26 16.93/4.52	171 200	
	Four Seasons	7380 7381 7382 7384 7385	32 46 68 100 150	-25 (-32) -15 (-26) -5 (-20) 0 (-18) +10 (-12)	370 (199) 400 (204) 415 (213) 460 (238) 470 (243)	160/44.7 230/50.0 335/56.0 470/63.0 716.71/78.93	31.7/5.41 46.2/6.70 67.06/8.65 95.0/10.7 143.3/14.64	104 105 100 95 101	
	Food Machinery	7390 7391 7392	32 46 68	— — —	— — —	163.1/44.7 222.0/47.2 357.0/56.1	32.4/5.3 44.5/6.4 71.6/8.5	89 85 87	
Lubrication Engineers, Inc.	Food Processing Fluid	4010 (H1) QUINPLEX® White Oil 4030 (H1)	46 100	-40 -35	360 400	220/45.7 527/60			
	Hydraulic Oil	MONOLEC® 6105 6110 6120 6130 6520	22 46 68 100 (5W-20)	-35 -30 -20 -10 -40	400 415 440 480 400	117/40.7 238/49 348/55.2 531/66.9 175/48		95 95 95 95 135	

Company Name	Category	Part Number	ISO (SAE) Viscosity Grade	Pour Point °F (°C)	Flash Point °F (°C)	Viscosity SUS @ 100°F/ SUS @ 210°F	Viscosity cSt @ 40°C/ cSt @ 100°C	Viscosity Index	API or Specific Gravity
Lubrication Engineers, Inc., cont'd	Turbine Oil	MONOLEC®							
		6401	32	-20	400	155/43.4		95	
		6402	46	-20	420	238/48		95	
		6403	68	-20	440	353/55		95	
		6404	100	-10	460	522/65		95	
		6405	150	-10	480	788/78		95	
		6406	220	-10	500	1161/99		95	
		6407	320	0	525	1705/119		95	
		6451	32	-20	410	155/43.4		95	
	Heavy Duty R&O and AW Industrial Oil	MULTILEC®							
		6801	32	-40	415	165.3/44.4		95	
		6802	46	-35	425	244.6/49.4		95	
		6803	68	-20	425	347.6/55.5		95	
		6804	100	-15	425	516.5/65.0		95	
		6805	150	0	460	786.0/80.5		95	
		6806	220	5	460	1155.6/101.6		95	
		6807	320	5	475	1693.0/130.2		95	
Lubriplate Lubricants	Heavy Duty, Extreme Pressure, Anti-Wear Hydraulic Oils	HO-0	32	-30 (-34)	415 (213)	160/46	31/6	136	31.6
		HO-1	46	-25 (-31)	420 (216)	213/49	42/7	120	31.0
		HO-2	68	-20 (-29)	430 (221)	309/54	64/8	106	30.5
		HO-2A	100	-15 (-26)	445 (230)	518/66	99/12	104	29.5
		HO-3	150	-10 (-23)	460 (238)	690/75	135/14	102	29.4
		HO-4	220	0 (-18)	243 (525)	840/83	162/16	101	29.3
		HO-5	320	10 (-12)	480 (249)	1510/111	281/22	96	27.6
	High Performance, Anti-Wear Hydraulic Oils	HO-32	32	-30 (-34)	405 (207)	150/44	31/5	115	32.6
		HO-46	46	-25 (-32)	410 (210)	207/48	42/7	112	32.1
		HO-68	68	-20 (-29)	430 (221)	321/55	64/9	107	31.2
	Special Anti-Wear Hydraulic Oil For Extreme Low Temp. Operations	Minus 70	15	-80 slow (-62)	200 (93)	83.5/44.5	16/5.5	340	31.6
	Multi-Viscosity Hydraulic Oil	MV-HO	32	-40 (-40)	330 (166)	186/53	35/8	199	31.4
Lyondell Lubricants	Multi-Purpose ATF			(-40)	(200)		38.0/7.4	160	30.5
	Tractor Fluid			(-27)	(193)		61.0/9.7	133	24.3
	Duro AW Oils		32	(-28)	(216)	165/44	32/	95	31.0
			46	(-28)	(221)	238/49	46/	95	30.0
			68	(-28)	(227)	353/55	68/	95	29.5
			100	(-12)	(249)	523/64	100/	95	28.0
	Duro Oils		10	(-4)	(149)	61/35	10/	95	38.0
			22	(-9)	(188)	115/41	23/	95	33.0
			32	(-32)	(196)	165/44	32/	95	32.0
			46	(-32)	(207)	238/49	46/	95	31.0
			68	(-32)	(224)	353/55	68/	95	30.0
			100	(-12)	(232)	523/64	100/	95	29.0
			117	(-9)	(249)	610/69	117/	95	29.0
			150	(-9)	(252)	790/78	150/	95	29.0
			220	(-9)	(254)	1165/96	220/	95	27.5
			320	(-9)	(266)	1705/119	320/	95	27.0
	Duro FR-HD (contains 40% water)			(-26)		500/	95/		0.924
	Enviroil AW Oils		32	(-29)	(193)	165/44	32/	96	30.9
			46	(-29)	(204)	238/49	46/	96	30.5
			68	(-29)	(221)	353/55	68/	96	29.8
	Ideal Oils		32	(-7)	(188)	166/44	32/	100	33.0
			46	(-7)	(199)	238/49	46/	100	32.0
			68	(-7)	(224)	353/56	68/	100	31.0
	Ideal Plus			(-32)	(191)	158/44	-32/	111	31.8
	Polarvis Oil			(-50)	(207)	163/48	32/	170	32.0

Company Name	Category	Part Number	ISO (SAE) Viscosity Grade	Pour Point °F (°C)	Flash Point °F (°C)	Viscosity SUS @ 100°F/ SUS @ 210°F	Viscosity cSt @ 40°C/ cSt @ 100°C	Viscosity Index	API or Specific Gravity
Lyondell Lubricants, cont'd	Soluble Oils Emulsiplex			0	171	180/	42/		21.5
Mobil Oil Corp.	Hydraulic and Circulating Oil	60035-3	32	20 (-7)	380 (193)	166/45	32.0/5.5	116	33.6
	Super-Stabilized, Anti-Wear Hydraulic Oils	60260-7	22	-10 (-23)	395 (202)	115/41	22/4.5	95	32.6
		60262-3	32	-10 (-23)	395 (202)	165/44	32/5.3	95	30.4
		60263-1	46	-10 (-23)	395 (202)	238/49	46/6.7	95	28.4
		60264-9	68	-10 (-23)	400 (204)	335/55	68/8.5	95	27.5
	Seal-Control AW Hydraulic Oils	60696-2	32	20 (-7)	380 (193)	166/44	32/5.3	95	27.0
		60697-0	46	20 (-7)	380 (193)	239/49	46/6.65	95	26.0
		60698-8	68	25 (-4)	380 (193)	356/55	68/8.5	95	26.0
	Premium AW/ EP Hydraulic Oils	58301-3	32	0 (-18)	375 (191)	166/44	32/5.4	85	30.6
		58302-1	46	0 (-18)	380 (193)	240/48	46/6.4	85	30.0
		58303-9	68	0 (-18)	400 (204)	357/54	68/8.1	85	28.9
		58304-7	100	0 (-18)	400 (204)	520/61	100/10.2	85	28.5
		58305-4	150	25 (-4)	420 (216)	803/73	150/13.0	85	28.0
		58306-2	220	30 (-1)	440 (227)	1190/84	220/16.0	85	27.0
	Heavy-Duty Industrial Gear Lubricants	61085-7	68	-10 (-23)	400 (204)	356/55.1	68/8.5	95	28.0
		61092-3	100	-10 (-23)	405 (207)	527/64.3	100/11.1	95	27.0
		61086-5	150	-10 (-23)	410 (210)	796/78.1	150/14.5	95	27.0
		61087-3	220	0 (-17)	420 (216)	1175/96.1	220/18.7	95	26.5
		61088-1	320	0 (-17)	450 (232)	1700/120	320/24.0	95	26.1
		61090-1	460	20 (-6)	450 (232)	2493/150	460/30.3	95	24.1
		61091-5	680	20 (-6)	450 (232)	3726/183	680/37.3	90	22.5
	Fire-Resistant Hydraulic Fluid	60201-1		-20 (29)	—	600/	120/		21.5
	Synthetic Hydraulic Oils	60303-5	15	-60 (-51)	400 (204)	80/38	14.4/3.6	135	33.9
		60305-0	32	-55 (-48)	400 (204)	165/47	32/6.1	140	34.6
		60307-6	46	-55 (-48)	400 (204)	235/55	46/8.5	145	34.4
		60306-8	68	-55 (-48)	450 (232)	348/66	68/11.5	145	34.0
	Low-Temperature Anti-Wear Oil	58290-8	15	-40 (-40)	320 (160)	78/38	14.1/3.9	140	33.9
		58291-6	22	-40 (-40)	320 (160)	110/43	21.1/4.9	140	32.1
		58292-4	32	-40 (-40)	350 (177)	160/47	31.1/6.1	140	31.2
		58293-2	46	-40 (-40)	350 (177)	231/53	45/8.0	135	30.7
		58294-0	68	-34 (-30)	400 (204)	339/62	66/10.6	120	30.5
	Hydraulic, Non-Zinc Oils	58147-0	32	-15 (-26)	375 (191)	165/44	32/5.3	95	31.7
		58148-8	46	-15 (-26)	380 (193)	238/49	46/6.7	95	30.5
		58149-6	68	-15 (-26)	400 (204)	353/55	68/8.5	95	29.4
	Biodegradable, Virtually Nontoxic Anti-Wear Hydraulic Oil	60183-1	32/46	-25 (-32)	428 (220)	187/54	37/8.4	213	22.0
Penzoil	Premium Quality Extreme Pressure Gear Lubes	019307	46	-20 (-30)	370 (188)	6.8/49	46.0/235	100	31.0
		019308	68	-20 (-30)	370 (188)	8.7/56	68.0/335	100	29.5
		019309	100	-5 (-21)	380 (192)	11.4/65	100/500	100	29.0
		019310	150	10 (-12)	395 (201)	15/80	150/760	100	29.0
	Synthetic Extreme Pressure Industrial Gear Lubes	037941	68	-55 (-52)	445 (229)	350/54.1	68/8.9	145	30.1
		037904	100	-50 (-46)	450 (232)	500/77.1	100/14.4	147	31.9
		037374	150	-40 (-40)	450 (232)	760/101.5	150/20.0	150	31.7
		037936	220	-40 (-40)	455 (235)	1020/130.9	220/27.4	160	31.1
		037937	320	-40 (-40)	460 (238)	1484/170.3	320/36.1	160	30.9
		037380	460	-35 (-37)	475 (246)	2300/236.0	460/48.7	165	30.0
	AW Hydraulic Oils	037010	15	-30 (-34)	350 (177)	182.2/37	15/3.4	98	31.1
		037000	22	-30 (-33)	375 (190)	105/39	22/4.3	100	32.5
		037001	32	-22 (-30)	405 (208)	165/44	32/5.3	95	30.9
		037002	46	-22 (-30)	405 (208)	227/48	46/6.7	99	30.1
		037003	68	-22 (-30)	410 (210)	352/56	68/8.8	106	29.2
	Synthetic Fire Resistant Hydraulic Fluid	037329	32	-10 (-23)	495 (257)	150/42	31.5/4.6		1.16
		037330	46	-5 (-21)	485 (252)	230/44	48.9/5.1		1.13
		037331	68	5 (-15)	490 (254)	300/46	64.0/5.7		1.12

Company Name	Category	Part Number	ISO (SAE) Viscosity Grade	Pour Point °F (°C)	Flash Point °F (°C)	Viscosity SUS @ 100°F/ SUS @ 210°F	Viscosity cSt @ 40°C/ cSt @ 100°C	Viscosity Index	API or Specific Gravity
Penzoil, cont'd	Water-In-Oil Fire-Resistant Lubricant	030112		-20 (-29)	—	575/	110/	200+	0.9297
Phillips 66 Company	Tractor hydraulic fluid	HG Fluid 53130		-40 (-40)	437 (225)			145	27.9
	ATF Dexron IIE	53730		-40 (-40)	400 (204)			159	31.0
	Magnus A	81330	32	-25 (-32)	400 (204)	160/44	31.8/5.3	101	31.5
		81340	46	-25 (-33)	430 (221)	235/48	45.5/6.5	98	30.6
		81350	68	-20 (-29)	455 (235)	325/53.5	64.0/8.0	98	30.0
		81370	100	-15 (-5)	470 (243)	500/63.5	99.0/11.0	95	29.5
	Magnus "A" KV	81390	(5W-20)	-45 (-42)	390 (198)	206/52.4	40.42/7.80	167	32.0
Silogram Lubricants	Anti-Wear Hydraulic Oil		32 (AW 150)	-5	410	150/43		100	31.0
			46 (AW 200)	-5	420	200/46		100	30.0
			68 (AW 300)	-5	440	300/54		100	29.0
			32 (Light)	-20	420	150/44		105	31.8
			46 (Medium)	-20	430	205/47		105	31.0
			68 (Heavy Med.)	-20	450	325/54.6		105	30.2
Shell Oil Company	Paper Machine Fluids	Delima® 66575	150	15	460		150/14.2		
		66576	220	15	510		215/18.1		
		66577	320	15	530		320/24.2		
	Hydraulic Oils	65236	32	-20	350		31.0/4.8	60	27.0
		65237	46	-5	390		44.0/5.8	60	26.5
		65239	68	-5	400		64.0/7.2	60	25.5
		65241	100	-5	420		95.0/9.2	60	24.7
	Antiwear Hydraulic Oil	Tellus® 65206	22	-25	405		23.8/4.6		31.5
		65208	32	-25	420		32.0/5.4		31.0
		65209	46	-15	440		42.5/6.5		30.0
		65211	68	-10	455		68.0/8.3		29.5
		65214	100	-5	495		100/10.7		29.0
	High Viscosity Index Hydraulic Oils	Tellus T® 65407	15	-60	295		15.0/3.7	170	29.0
		65400	22	-50	305		22.5/5.1	172	30.0
		65401	32	54	325		31.2/6.2	154	29.5
		65402	46	-40	350		43.1/7.8	151	29.0
		65403	68	-30	390		63.6/10.0	143	28.5
		65404	100	-20	455		94.9/12.8	132	28.5
	Heavy Industrial Gas Turbine Oil	Turbo® IG 65625	32	-20	450		32/5.7		30.5
Sun Company	Tractor hydraulic fluid	453900		-45	435		58.9/9.5	145	
	Premium detergent-dispersant, anti-wear hydraulic oil (Sunvis 700 Series)	142000	22	-20/	/395	115/40.6	22.0/4.26	95	31.5/32.0[2]
		142100	32	-25/	/420	163/44.4	32.0/5.41	103	31.5
		142200	46	-25/	/440	237/49.2	46.0/6.36	104	31.0/30.5[2]
		142300	68	-25/	/465	352/56.1	68.0/8.85	103	31.0/30.0[2]
		142400	100	+5/	/485	522/64.9	100.0/11.30	98	29.5
		142500	150	0/	/495	789/77.9	150.0/14.60	95	29.0
		142800	220	+10/	/510	1164/95.7	220.0/18.70	95	28.5
	Premium, high viscosity index, anti-wear hydraulic oil with improved pump performance. Excellent water separation & filterability properties (Sunvis 800 Series)	372100	32	-25/	/425	165/44.5	32.0/5.44	105	31.5
		372200	46	-25/	/445	237/49.4	46.0/6.91	106	31.5/31.0[2]
		372300	68	-20/	/475	352/56.3	68.0/8.92	105	31.0/30.0[2]
		372400	100	+10/	/495	521/65.5	100.0/11.40	100	30.5
		372500	150	+10/	/505	788/78.4	150.0/14.61	96	29.0
	Turbine-quality R&O lubricants (Sunvis 900)	334100	15	-5/	370/	82.4/37.5	15.0/3.34	88	32.0/33.5[2]
		157200	22	+5/	375/	115.0/41.0	22.0/4.30	98	32.0/33.0[2]
		142900	32	-30/	425/	165.0/44.0	32.0/5.40	105	32.0
		143000	46	+5/-20[2]	450/	237.0/49.0	46.0/6.90	105	31.5/31.0[2]
		143100	68	+10/-15[2]	475/	352.0/56.0	68.0/8.90	104	31.0/30.5[2]
		143200	100	+10/	495/	521.0/65.7	100.0/11.50	101	30.5/30.0[2]

Company Name	Category	Part Number	ISO (SAE) Viscosity Grade	Pour Point °F (°C)	Flash Point °F (°C)	Viscosity SUS @ 100°F/ SUS @ 210°F	Viscosity cSt @ 40°C/ cSt @ 100°C	Viscosity Index	API or Specific Gravity
Sun Company, cont'd		143300	150	0/	505/	788.0/78.6	150.0/14.70	97	29.0
		143400	220	0/	515/	1165.0/95.0	220.0/18.60	94	28.0/28.5[2]
	Multi-grade hydraulic fluid used in mobile equipment operating at low or widely varying ambient temperatures	190600		-45/		240/52	47.0/7.80	138	31.5
	Sun Low Pour Hydraulic Oil for hydraulic systems operating at subzero temperatures	190900		-60/			16.6/5.52	277	30.5
	Premium quality, invert emulsion hydraulic fluid designed to provide fire resistance (Sunsafe F)	144000		-5/		410/	79.0/	—	—
Texaco Lubricants Co.	Aircraft Hydraulic Oil 15	1537		-80	180	74.3/43.0	13.5/5.0	372	30.6
	Marine Products Hydraulic Oil	Rando Oil HD							
		1629	Z-36	-40	380	183/48	35.8/6.6	141	30.2
		1540	Z-36	-40	380	185/49	36.3/6.7	143	31.0
		1693	Z-15	-50	310	77/40	15.9/4.1	170	30.1
		1657	32	-25	385	155/44	30.1/5.3	106	30.7
		1658	46	-20	425	237/49	46.2/6.9	105	29.3
		1659	68	-20	425	339/55	65.5/8.7	103	28.6
		2778	100	0	445	486/63	94/10.6	95	27.8
		1660	150	5	460	802/77	152/14.3	90	26.6
	Havoline ATF Type F	1855		-55	415	204/53	40.2/7.9	172	30.4
	Hydraulic Oil	Rando Oil HD Ashless 1636	46	-20	410	237/49	46.2/6.90	105	29.9
		Rando Oil HD 1655	22	-20	385	122/42	23.5/4.6	110	32.2
		Rando Oil Polar Ice 1527	15	below -55	295	87/42	16.3/4.6	222	30.6
		LWM Concentrate 1666	18				42.6/		27.4
	Hydraulic Safety Fluid	766		-90		205/	40/		1.09
	Transformer Oils	600		-60	305		8.9/2.4		31.0
		1515 Inhibited		-55	310		8.9/2.4		30.8
	Paper Machine Oils	797	150		485	790/76	150/14.1	90	27.2
		798	220		490	1167/93	220/18.1	90	26.7
		799	320		495	1709/116	320/23.2	90	26.4
	Food Industry Hydraulic Oils	2665	32	15	395	175/45	33.9/5.7	106	32.5
		2666	46	15	400	241/51	46.9/7.2	102	32.2
		2667	68	20	445	357/56	68.9/8.8	100	31.0
Unocal	Unax PC-AW	04632	32	-11 (-24)	417 (214)		32/5.47		
		04634	46	-11 (-24)	417 (214)		46/6.54		
		04635	68	-11 (-24)	428 (220)		68/8.25		
		04636	100	-6 (-21)	468 (242)		100/10.5		
	Unax AW	04641	32	-22 (-30)	415 (214)		32/5.5		
		04642	46	-22 (-30)	435 (224)		46/6.7		
		04643	68	-11 (-24)	455 (236)		68/8.3		
		04644	100	-6 (-21)	475 (246)		100/10.5		

Company Name	Category	Part Number	ISO (SAE) Viscosity Grade	Pour Point °F (°C)	Flash Point °F (°C)	Viscosity SUS @ 100°F/ SUS @ 210°F	Viscosity cSt @ 40°C/ cSt @ 100°C	Viscosity Index	API or Specific Gravity
Unocal, cont'd		04645	150	-6 (-21)	500 (260)		150/13.6		
		04646	220	+16 (-9)	505 (264)		220/17.3		
		04647	320	+16 (-9)	515 (268)		320/21.9		
	Unax AW-WR	4648	32	-49 (-45)	381 (194)	163.47/47.08	32.0/6.236	148	28.4
	Turbine Oil	4620	22	-27 (-33)	414 (212)	110/40	20.9/4.14	97	30.1
		4621	32	-22 (-30)	421 (216)	157/43	30.4/5.11	94	29.2
		4622	46	-24 (-11)	435 (224)	227/48	43.9/6.40	92	28.3
		4623	68	-4 (-18)	446 (230)	336/53	64.5/8.00	89	27.9
		4624	100	10 (-12)	464 (240)	495/60	94.8/9.97	82	28.3
		4625	150	16 (-9)	500 (260)	756/72	143/13.0	82	26.0
		4626	220	10 (-12)	500 (260)	1113/87	209/16.6	83	25.3
		4628	320	10 (-12)	475 (246)	1699/110	317/21.9	83	24.9
		4627	460	10 (-12)	536 (280)	2352/132	436/26.7	83	24.3
	Soluble Oil	4781	(10)				28.6/		22.1
	PM Oil	4896	150	11 (-15)	489 (254)	756/72	143/13.03	81	25.6
		4897	220	10 (-12)	506 (260)	1112/87	209/16.74	81	25.2
		4904	320	10 (-12)	511 (266)	1409/103	304/21.20	85	24.2
	Hydraulic Oil AW	4651	32	-27 (-33)	420 (216)	156.8/44.1	30.40/5.32	105	29.3
		4652	46	-11 (-24)	438 (226)	224.9/48.2	43.60/6.55	100	28.4
		4653	68	-11 (-24)	438 (226)	334.7/54.9	64.50/8.25	95	27.7
	Super ATF	3955		-54 (-48)	350 (177)	171.6/50.33	33.86/7.22	185	
	Hydraulic/ Tractor Fluid	3970		-39 (-39)	421 (216)		52.65/9.25	159	

[1] Do not use these products in high pressure systems in areas subject to fire hazard.
[2] First number, Marcus Hook/Second number , Tulsa.

Composition of Air

Component of Air	Symbol	Content - % Volume
Nitrogen	N_2	78.084 percent[1]
Oxygen	O_2	20.947 percent[1]
Argon	Ar	0.934 percent[1]
Carbon dioxide	CO_2	0.033 percent[1]
Neon	Ne	18.2 parts/million
Helium	He	5.2 parts/million
Krypton	Kr	1.1 parts/million
Sulfur dioxide	SO_2	1.0 parts/million
Methane	CH_4	2.0 parts/million
Hydrogen	H_2	0.5 parts/million
Nitrous oxide	N_2O	0.5 parts/million
Hydrogen	H_2	0.5 parts/million
Xenon	Xe	0.09 parts/million
Ozone	O_3	0.0 to 0.07 parts/milion
Ozone - Winter	O_3	0.0 to 0.02 parts/million
Nitrogen dioxide	NO_2	0.02 parts/million
Iodine	I_2	0.01 parts/million
Carbon monoxide	CO	0.0 to trace
Ammonia	NH_3	0.0 to trace

[1] 99.98%

NOTE: The above table is an average for clean, dry air at sea level.
1 part/million = 0.0001 percent

Physical Properties of Air

Density of dry air at Standard Temperature and Pressure:
1.2929 kilograms/cu meter = 0.0807 pounds/cu foot

Universal Gas Constant (R):
0.0821liter-atmosphere/°K/mole

Standard Temperature & Pressure (STP):
Standard Temperature = 0° C = 32° F = 273.15° K
Standard Pressure = 760 mm Hg =14.70 pounds/sq inch = 2116.22 pounds/sq foot =29.92 inch Hg = 1.01325 x 10^5 N/m^2

Speed of sound in dry air at STP:
331.45 meters/sec = 1087.4 ft/sec = 741.4 miles/hr

ICAO Sea Level Air Standard Values:
Atmospheric pressure = 760 mm Hg = 14.7 lbs/sq inch
Temperature = 15° C = 288.15° K = 59° F

Weight of Gases

Gas	Weight (gms/liter)	Weight (lb/cu ft)
Air	1.2928	0.08071
Air @ 59° F	1.2256	0.07651
Argon	1.7840	0.111368
Carbon dioxide	1.9770	0.123416
Carbon monoxide	1.2500	0.078033
Helium	0.1785	0.011143
Hydrogen	0.0899	0.005612
Neon	0.9002	0.056196
Nitrogen	1.2506	0.075261
Oxygen	1.4290	0.089207

Standard Atmosphere

The unit "1 Standard Atmosphere" is defined as the pressure equivalent to that exerted by a 760 mm column of mercury at 0° C (32° F), at sea level, and at standard gravity (32.174 ft/sec^2). Atmospheric pressure is simply the weight of a column of air per unit area as measured from the top of the atmosphere to the reference point being measured. Atmospheric pressure decreases as altitude increases. Equivalents to 1 atmosphere are as follows:

76 centimeters (760 mm) of mercury
29.921 inches of mercury
10.3326 meters of water
406.788 inches of water
33.899 feet of water
14.696 pounds per square inch
2116.2 pounds per square foot
1.033 kilograms per square centimeter

Horsepower & Cost For Compressing Air

Chart based on 13 cents/kwH

1-Stage Compressor			2-Stage Compressor			3-Stage Compressor		
psig	hp/ cfm	dollars/ cfm-hr	psig	hp/ cfm	dollars/ cfm-hr	psig	hp/ cfm	dollars/ cfm-hr
10	0.04	0.0039	60	0.128	0.0124	100	0.159	0.0154
20	0.067	0.0065	80	0.148	0.0143	200	0.212	0.0206
30	0.095	0.0092	100	0.164	0.0159	300	0.245	0.0238
40	0.107	0.0104	120	0.178	0.0173	400	0.269	0.0261
50	0.123	0.0119	140	0.19	0.0184	500	0.289	0.0280
60	0.136	0.0132	160	0.201	0.0195	600	0.305	0.0296
70	0.148	0.0143	180	0.211	0.0205	700	0.317	0.0307
80	0.16	0.0155	200	0.22	0.0213	800	0.329	0.0319
90	0.17	0.0165	250	0.239	0.0232	900	0.34	0.0330
100	0.179	0.0174	300	0.255	0.0247	1000	0.35	0.0339
120	0.196	0.0190	350	0.269	0.0261	1100	0.358	0.0347
140	0.211	0.0205	400	0.282	0.0273	1200	0.366	0.0355
160	0.225	0.0218	450	0.293	0.0284	1300	0.374	0.0363
180	0.239	0.0232	500	0.303	0.0294	1400	0.38	0.0368
200	0.25	0.0242				1500	0.386	0.0374

To determine the hp required to compress air for a plant application multiply the number of cfm required by the hp/ cfm value from the chart for the type of compressor used.

Example
For a 2-stage compressor supplying 50 cfm, at 160 psig the hp required is: 50 cfm x .201 hp/cfm = 10.05 hp.

To determine the cost of compressing air for a plant application, multiply the number of cfm required by the number of hours used in a day by dollars/cfm-hr from the chart for the type of compressor used.

Example
For a 2-stage compressor supplying 50 cfm at 160 psi for an 8 hour day, the cost is: 50 cfm x 8 hours x .0195 dollars/cfm-hr = 7.80 dollars per day.

Flow Coefficients

The Cv factor is the number of U.S. gallons of water that pass through a given orifice area in one minute at a pressure drop of one psi.

The flow coefficient (C_v) of a pneumatic device is a measure of the devices ability to pass air. With fixed inlet and outlet pressures the amount of air that can flow through the device is in direct proportion to the flow coefficient.

Air flow through a valve (or other device) is measured in the laboratory under carefully controlled conditions. From these measurements plus measurements of the inlet and outlet pressures, the flow coefficient can be determined. The relationship between these factors at a temperature of 68°F is expressed in the following formula:

$$Q = 0.98\, C_v \sqrt{\Delta P(P_2 + 14.7)}$$

Where Q = air flow in standard cubic feet per minute (scfm)
C_v = flow coefficient
P_1 = inlet pressure (psig)
P_2 = outlet pressure (psig)
ΔP = pressure drop across valve (psi)
= $P_1 - P_2$

Knowing the flow coefficient and the inlet and outlet pressures, the flow through a valve can be determined using this formula.

A simpler way, however, to determine air flow is to use the graph on page 105. This graph is based on the above formula when $C_v = 1$. Since air flow is proportional to C_v, this graph can be used for any C_v value; simply multiply the air flow rate given on the graph by the new C_v value.

Example: Given a supply pressure of 125 psig, a pressure drop of 7 psi, and a valve with a C_v of 4.2, find the rate of air flow through the valve.

From the 7 psi pressure drop value at the bottom of the graph, move upward to the supply pressure line for 125 psig. From this point move to the left and read the air flow rate: 30 scfm. Since this is the air flow rate for $C_v = 1$, multiply by 4.2 to obtain the air flow rate for the valve in this example.
30 x 4.2 = 126 scfm.

Determining C_v Required of a Valve Operating In a Cylinder

The charts on pages 106 and 107 can be used to determine the valve C_v required to operate cylinders of various sizes. The chart on page 106 shows the displaced volume of cylinders with various strokes and diameters. Knowing the volume of the cylinder and the number of strokes per minute the cylinder must make, the chart on page 107 can be used to determine the minimum C_v that the operating valve must have.

Parallel Connection of Pneumatic Components

If several pneumatic components are connected in parallel, the flow coefficient of the combination is simply the sum of the flow coefficients of the several units. For example:

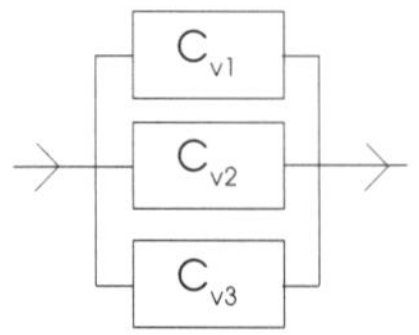

Combined Cv = $Cv^1 + Cv^2 + Cv^3$

Series Connection of Pneumatic Components

If several pneumatic components are connected in series, the flow coefficient of the combination is a little more complicated. For example:

→ C_{v1} — C_{v2} — C_{v3} →

Combined Cv: $$\frac{1}{(C_v)^2} = \frac{1}{(C_{v1})^2} + \frac{1}{(C_{v2})^2} + \frac{1}{(C_{v3})^2}$$

Using C_v values must always be tempered with good judgment. Flow coefficients are determined in the laboratory under steady flow conditions, a state that exists only part of the time in an industrial pneumatic system. Some allowance must be made for the valve response time required to establish steady flow conditions. The use of C_v values is just one area in which the experience and judgement of the pneumatic designer become important.

Standard Cubic Feet Per Minute (SCFM)

One cubic foot of gas (air) per minute at standard conditions of 68°F, 14.69 psi and a relative humidity of 36%.

Cubic Feet Per Minute (CFM)

One cubic foot of gas (air) per minute at actual conditions, i.e. at actual temperature and compressed or expanded pressure.

Free Air Flow

The volume of air at normal atmospheric conditions which enters a vacuum system due to the lower pressure caused by the pump or vacuum in a tank.

Expanded Air Flow

Air flow inside a vacuum system, same as CFM.

Conversion from SCFM to CFM:

For compressed air:

$$\text{CFM} = \text{SCFM} \times \frac{14.7}{\text{Actual System Pressure (psi)} - 14.7}$$

For vacuum:

$$\text{CFM} = \text{SCFM} \times \frac{29.9}{29.9 - \text{Actual System Pressure (in Hg)}}$$

Pressure Conversion Chart

	=psia	=kPa	=Atm	=kg/cm²	=Bar
psia x		6.9	.068	.070	.069
kPa x	.145		.00986	.0102	.01
Atm x	14.7	101.4		1.034	1.014
Kg/cm² x	14.22	98.1	.967		.981
Bar x	14.5	100	.986	1.02	

Vacuum Conversion Chart

	=H_2O	=mm Hg	=cm H_2O	=in Hg
in H_2O x		1.87	2.54	.0736
mm Hg x	.535		1.36	.0394
cm H_2O x	.394	.736		.0289
in Hg x	13.6	25.4	34.5	

1 Torr = 1 mm Hg

Flow of Air Through Orifices Under Vacuum

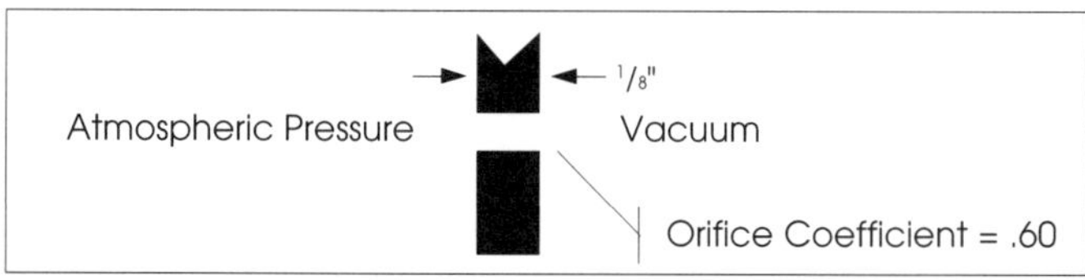

Orifice Dia.	Inches of Hg														
	1/2	1	2	3	4	5	6	7	8	9	10	11	12	13	14
1/16	.15	.21	.30	.33	.4	.45	.48	.52	.56	.58	.62	.65	.66	.70	.73
1/8	.6	.8	1.2	1.3	1.6	1.7	1.8	2	2.2	2.3	2.4	2.5	2.6	2.8	2.9
3/16	1.4	1.9	2.4	2.6	3.0	3.3	3.6	3.9	4.2	4.4	4.6	4.8	5.	5.3	5.5
1/4	2.3	3.0	4.2	5.3	6.4	6.8	7.5	8	9.3	9	9.6	10	10.5	11	11.3
5/16	3.5	4.8	7.3	8.3	10	11	11.5	13	13.5	14	15	16	16.5	17	17.7
3/8	5	7.5	10.5	11.5	14	15	16.5	18	19	20.5	22	23	23.8	24.6	25.5
7/16	7.5	10	13	16	18.5	21	23	25	26	28	29	31	32	33.5	34.8
1/2	10	13.5	17	21	24	27	30	32	34	36	39	40	42	43	45
5/8	15	21	27	33	38	43	46	50	52.5	55	59	63	65	66	71
3/4	22	28	38	41	56	62	68	72	77	82	86	90	95	100	103

Orifice Dia.	Inches of Hg														
	15	16	17	18	19	20	21	22	23	24	25	26	27	28	29
1/16	.76	.80	.83	.86	.90	.93	.95	.98	1.0	1.2	1.4	1.6	1.8	1.9	2.0
1/8	3	3.1	3.2	3.3	3.4	3.5	3.6	3.7	3.8	3.9	4	4.1	4.2	4.3	4.4
3/16	5.7	5.8	6	6.2	6.5	6.6	6.8	6.9	7	7.2	7.4	7.7	7.9	8	8.1
1/4	11.6	12	12.5	12.8	13.2	13.5	13.9	14.2	14.6	15	15.3	15.6	16	16.3	16.6
5/16	18.5	19	19.5	20	20.7	21.3	21.8	22.3	23	23.5	24	24.3	25	25.6	26
3/8	26.5	27	28	29	30	30.5	31.4	32	32.7	33.5	34	35	36	36.8	37.3
7/16	36	37	38	39.5	40.5	42	42.5	43.5	44.7	45.5	46.7	47.7	49	50	51
1/2	47.5	48.5	50	52	53	55	56	57	58	60	61	62	64	65	67
5/8	74	75	78	80	83	85	87	89	91	93	95	97	100	102	104
3/4	106	109	113	116	120	123	126	129	131	134	137	140	144	147	150

Altitude Correction For Vacuum Level

It is important to consider the relationship between atmospheric pressure and altitude as it affects vacuum pump performance. Because free air is less dense at higher altitudes (i.e. lower atmospheric pressures) operation at these higher altitudes has the effect of reducing the capacity and maximum vacuum levels attainable. Refer to the following table to correct for performance at various altitudes.

Altitude	Atmospheric Pressure	Maximum Vacuum Level Obtainable	Vacuum Level Loss At Altitude
0 Feet	14.7 psi(a)	29.921" Hg(G)	—
1,000 Feet	14.16 psi (a)	28.9" Hg(G)	3.4%
2,000 Feet	13.66 psi(a)	27.8 Hg(G)	7.1%
3,000 Feet	13.16 psi(a)	26.8" Hg(G)	10.4%
4,000 Feet	12.68 psi(a)	25.8" Hg(G)	13.8%
5,000 Feet	12.22 psi (a)	24.9" Hg(G)	16.8%
6,000 Feet	11.77 psi(a)	24.0" Hg(G)	19.8%
7,000 Feet	11.33 psi(a)	23.1" Hg(G)	22.8%
8,000 Feet	10.91 psi(a)	22.2" Hg(G)	25.9%
9,000 Feet	10.50 psi(a)	21.4" Hg(G)	28.6%
10,000 Feet	10.10 psi(a)	20.6" Hg(G)	31.3%
11,000 Feet	9.71 psi(a)	19.8" Hg(G)	33.9%
12,000 Feet	9.34 psi(a)	19.0" Hg(G)	36.5%
13,000 Feet	8.97 psi(a)	18.3" Hg(G)	39.0%
14,000 Feet	8.62 psi(a)	17.5" Hg(G)	41.4%
15,000 Feet	8.28 psi(a)	16.9" Hg(G)	43.6%

Pressure Loss Due to Air Friction In Pipes

For various flow rates, for 1/2 in., 3/4 in., 1 in., 1 1/4 in., nominal pipe diameters, at initial pressures of 60, 80, 100, 125 psi

Free Air, cfm	AIR FLOW RATES Equivalents cfm of Compressed Air				PRESSURE LOSSES FOR EACH 100 FEET OF PIPE LENGTH, PSI 1/2 in. Pipe Diameter				PRESSURE LOSSES FOR EACH 100 FEET OF PIPE LENGTH, PSI 3/4 in. Pipe Diameter			
	60 psi	**80 psi**	**100 psi**	**125 psi**	**60 psi**	**80 psi**	**100 psi**	**125 psi**	**60 psi**	**80 psi**	**100 psi**	**125 psi**
10	1.97	1.55	1.28	1.05	.59	.46	.38	.31	.14	.11	.09	.08
20	3.94	3.10	2.56	2.10	2.23	1.74	1.42	1.17	.53	.41	.34	.28
30	5.90	4.66	3.84	3.16	4.94	3.84	3.13	2.54	1.14	.90	.74	.60
40	7.87	6.21	5.13	4.21	8.90	6.93	5.55	4.53	1.99	1.55	1.28	1.05
50	9.84	7.765	6.41	5.26	14.20	10.70	8.65	7.01	3.08	2.42	2.00	1.62
60	11.81	9.31	7.69	6.31	—	—	—	—	4.45	3.47	2.84	2.33
70	13.78	10.87	8.97	7.37	—	—	—	—	6.06	4.73	3.85	3.14
80	15.74	12.42	10.25	8.42	—	—	—	—	7.96	6.14	5.01	4.08
90	17.71	13.97	11.53	9.47	—	—	—	—	10.00	7.75	6.40	5.17
100	19.68	15.50	12.82	10.52	—	—	—	—	12.60	9.62	7.80	6.33
125	24.60	19.40	16.02	13.15	—	—	—	—	21.00	15.50	12.40	9.80
150	29.51	23.30	19.22	15.78	—	—	—	—	31.50	23.00	18.10	14.40

Free Air, cfm	AIR FLOW RATES Equivalents cfm of Compressed Air				PRESSURE LOSSES FOR EACH 100 FEET OF PIPE LENGTH, PSI 1 in. Pipe Diameter				PRESSURE LOSSES FOR EACH 100 FEET OF PIPE LENGTH, PSI 1 1/4 in. Pipe Diameter			
	60 psi	**80 psi**	**100 psi**	**125 psi**	**60 psi**	**80 psi**	**100 psi**	**125 psi**	**60 psi**	**80 psi**	**100 psi**	**125 psi**
10	1.97	1.55	1.28	1.05	.05	.04	.03	.02	.011	.0086	.0071	.0058
20	3.94	3.10	2.56	2.10	.16	.13	.10	.08	.040	.032	.026	.021
30	5.90	4.66	3.84	3.16	.34	.28	.23	.19	.086	.068	.056	.046
40	7.87	6.21	5.13	4.21	.59	.46	.38	.31	.146	.116	.096	.079
50	9.84	7.76	6.41	5.26	.92	.73	.60	.49	.22	.18	.146	.120
60	11.81	9.31	7.69	6.31	1.30	1.02	.84	.69	.32	.25	.21	.17
70	13.78	10.87	8.97	7.37	1.75	1.36	.112	.92	.42	.34	.28	.23
80	15.74	12.42	10.25	8.42	2.24	1.76	1.44	1.18	.55	.44	.36	.30
90	17.71	13.97	11.53	9.47	2.88	2.23	1.85	1.49	.69	.55	.45	.37
100	19.68	15.50	12.82	10.52	3.45	2.69	2.21	1.81	.84	.66	.55	.45
125	24.60	19.40	16.02	13.15	5.38	4.18	3.41	2.79	1.31	1.03	.85	.69
150	29.51	23.30	19.22	15.78	7.81	5.75	4.91	3.99	1.87	1.47	1.20	.99
175	34.44	27.20	22.43	18.41	10.80	8.10	6.80	5.45	2.58	2.00	1.64	1.32
200	39.36	31.00	25.63	21.05	14.50	10.90	8.79	7.11	3.31	2.58	2.12	1.73
250	49.20	38.80	32.04	26.31	—	—	—	—	5.30	4.05	3.30	2.67
300	59.00	46.60	38.45	31.57	—	—	—	—	7.51	5.78	4.71	3.83
350	68.90	45.30	44.86	36.83	—	—	—	—	10.30	7.90	6.45	5.15
400	78.70	62.10	51.26	42.09	—	—	—	—	13.70	10.30	8.30	6.74

Holding Force Created By Vacuum on Suckers

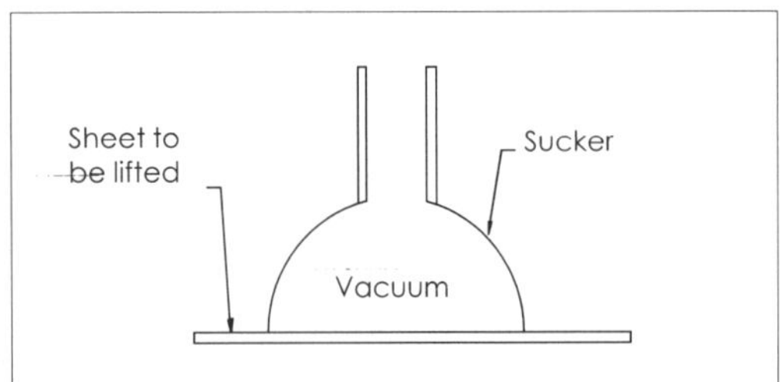

Vacuum in inches of mercury	Diameter of Sucker					
	1"	1 1/2"	2"	3"	4"	5"
5	1.9 lbs.	4.4	7.8	17.7	31.4	48 lbs.
10	3.8 lbs.	8.7	15.4	24.6	62	96 lbs.
15	5.8 lbs.	13.1	23.3	52.2	93	145 lbs.
20	7.6 lbs.	17.3	30.8	69	123	192 lbs.
25	9.6 lbs.	21.8	38.6	87	154	241 lbs.

Altitude Table

(Atmospheric Pressure at Altitudes Above Sea Level)

Altitude in feet	Atmosphere in lbs. per sq. in.	Gauge Reading in inches
Sea Level	**14.7**	**0**
1,000	14.2	1.0
2,000	13.7	2.1
3,000	13.2	3.1
4,000	12.7	4.1
5,000	12.2	5.0
6,000	11.7	6.0
7,000	11.3	6.9
8,000	10.9	7.7
9,000	10.5	8.6
10,000	10.3	9.4
15,000	8.29	13.0
20,000	6.75	16.3

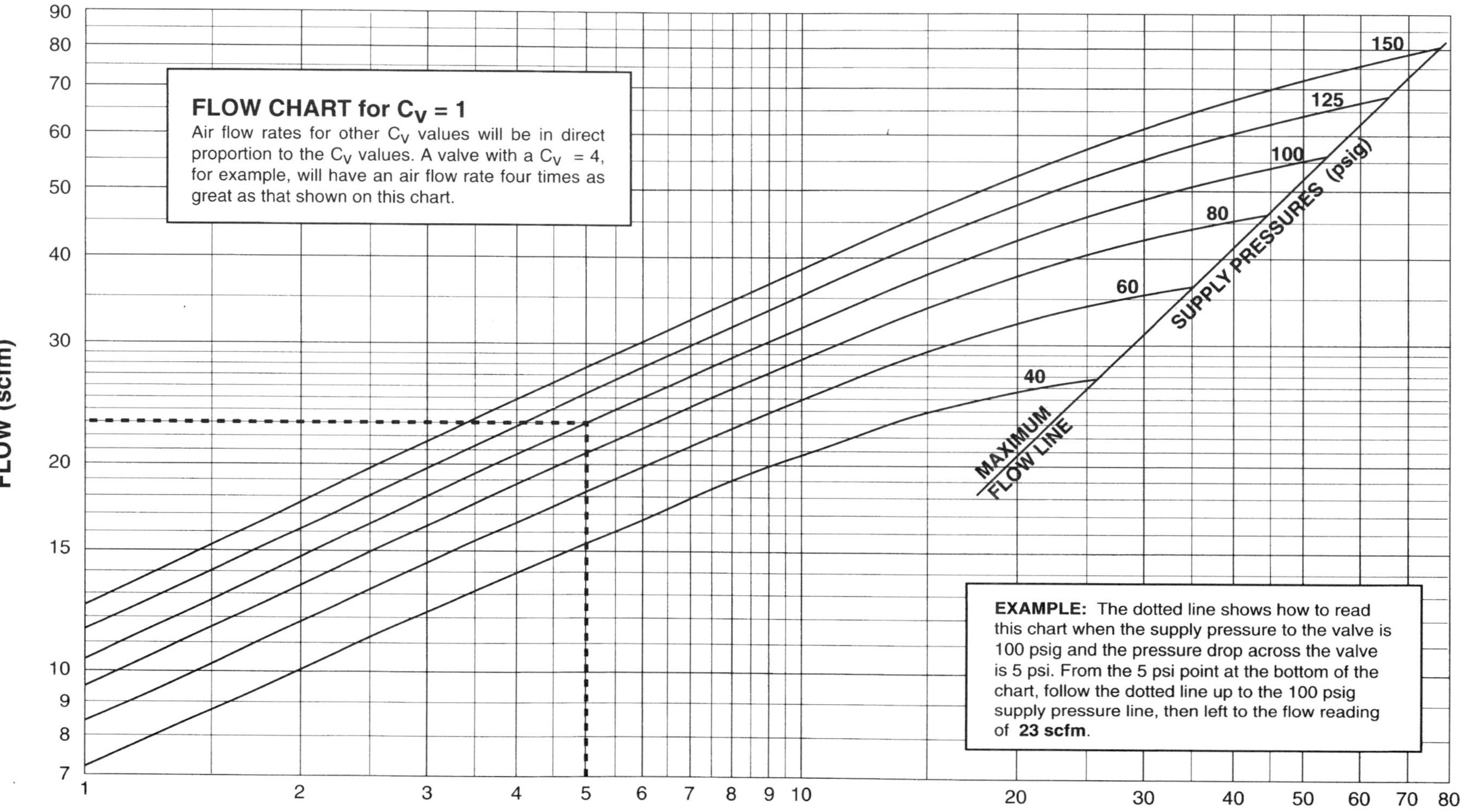
FLOW CHART for C_V = 1
Air flow rates for other C_V values will be in direct proportion to the C_V values. A valve with a C_V = 4, for example, will have an air flow rate four times as great as that shown on this chart.
EXAMPLE: The dotted line shows how to read this chart when the supply pressure to the valve is 100 psig and the pressure drop across the valve is 5 psi. From the 5 psi point at the bottom of the chart, follow the dotted line up to the 100 psig supply pressure line, then left to the flow reading of 23 scfm.
SUPPLY PRESSURES (psig)
150
125
100
80
60
40
MAXIMUM FLOW LINE
FLOW (scfm)
90
80
70
60
50
40
30
20
15
10
9
8
7
PRESSURE DROP (psi)
1
2
3
4
5
6
7
8
9
10
20
30
40
50
60
70
80

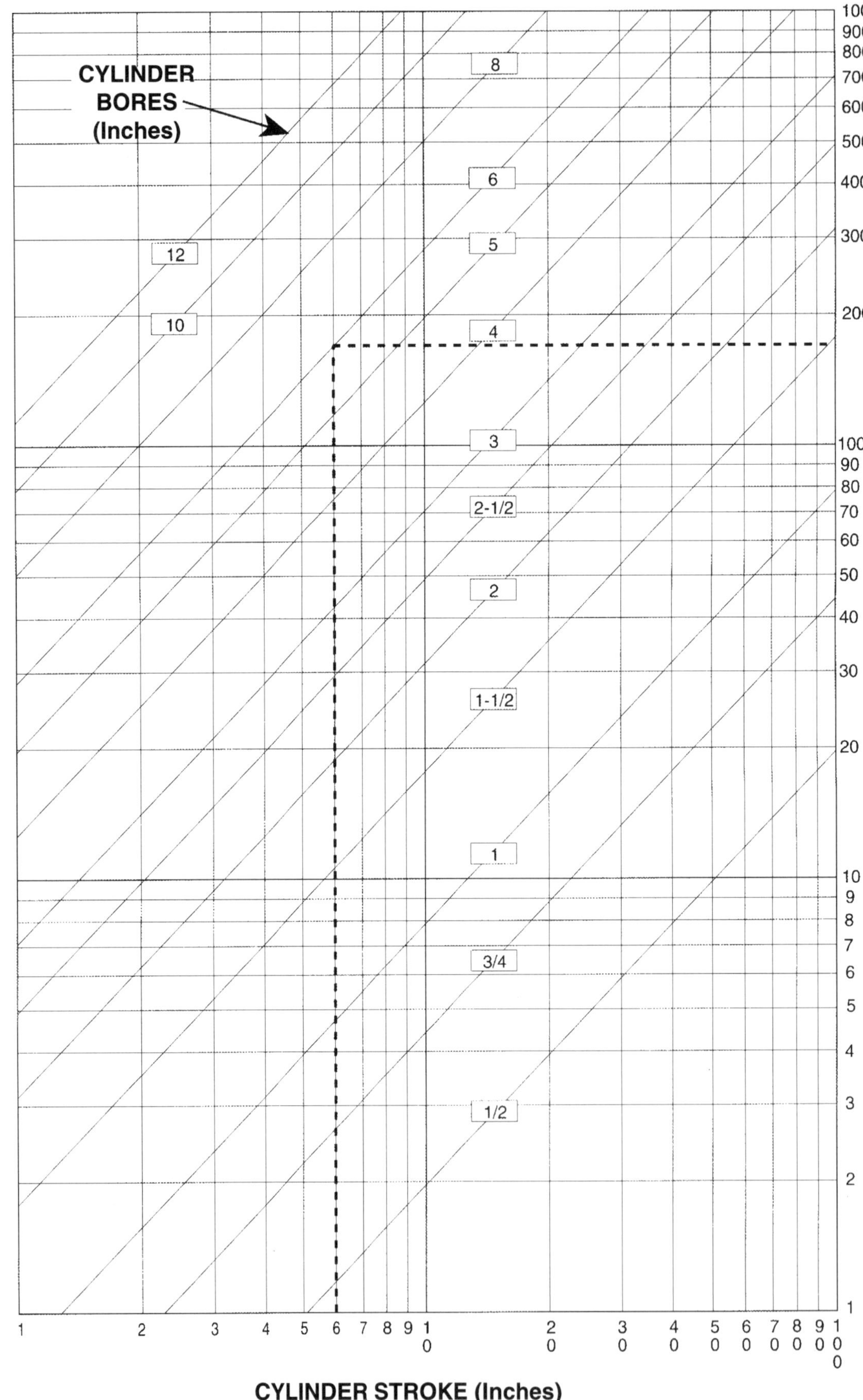

CYLINDER BORES (Inches)
12
10
8
6
5
4
3
2-1/2
2
1-1/2
1
3/4
1/2
1000
900
800
700
600
500
400
300
200
100
90
80
70
60
50
40
30
20
10
9
8
7
6
5
4
3
2
1
1
2
3
4
5
6
7
8
9
10
20
30
40
50
60
70
80
90
100
CYLINDER STROKE (Inches)
CYLINDER DISPLACED VOLUME (Cubic Inches)

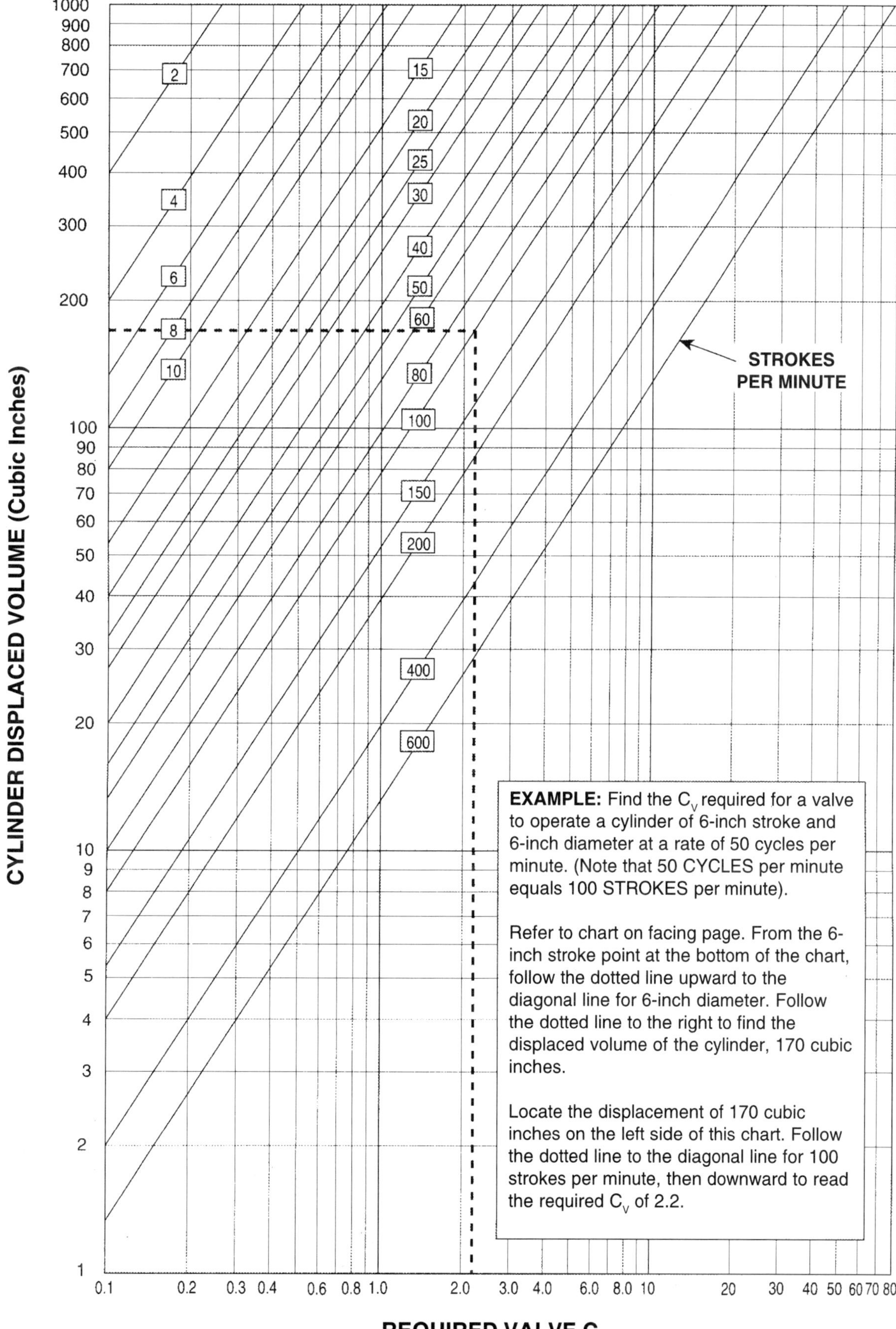

EXAMPLE: Find the C_V required for a valve to operate a cylinder of 6-inch stroke and 6-inch diameter at a rate of 50 cycles per minute. (Note that 50 CYCLES per minute equals 100 STROKES per minute).

Refer to chart on facing page. From the 6-inch stroke point at the bottom of the chart, follow the dotted line upward to the diagonal line for 6-inch diameter. Follow the dotted line to the right to find the displaced volume of the cylinder, 170 cubic inches.

Locate the displacement of 170 cubic inches on the left side of this chart. Follow the dotted line to the diagonal line for 100 strokes per minute, then downward to read the required C_V of 2.2.

Heating Requirements

Electric Heater Sizing

The following section outlines the procedure for sizing an electric resistance heater for a hydraulic power unit tank.

- Determine the required KW capacity required to bring the fluid up to operating temperature in the desired time. See the manufacturer's recommended minimum oil viscosity for all components. If that information is not available, bringing the oil to approximately 40° F is typically sufficient.

Factors that should be considered for both the start up period and the operational KW requirements are:

- Specified heat-up period
- Heat loss of the components and container
- Thermal properties of the fluid, container and any insulation

- Determine the required KW capacity required to maintain the fluid at operating temperature. In most hydraulic systems the inefficiencies in the components usually generate sufficient heat during operation to maintain an acceptable temperature level in the tank. Cooling of the oil is sometimes required (Refer to the information on Heat Exchanger Sizing).

- Select the size and number of heaters required to supply the required KW of the greater of the requirements as determined by the above two steps. With the heat generation from component inefficiencies during operation in most hydraulic systems the greater requirement will normally be the KW required to bring the fluid to operational temperature. A contingency of about 25-50% should be added to the final KW requirement. Automatic temperature control limits the KWh consumption to actual requirements.

Selection of the size, type and number of electric heaters is dependent on:

- **Watt Density Permitted** — Watt density is the heaters total wattage divided by the total contact area. The ability of a fluid to absorb energy governs the watt density that is acceptable for that fluid. Watt densities in excess of what is allowable for the fluid can damage both the fluid and the heater which could create a safety hazard. See the following chart for recommended Watt density maximum values.

Fluid	Maximum Watt Density (W/in²)
SAE 10, 90-100 *SUS @ 130° F*	23
SAE 20, 120-185 *SUS @ 130° F*	23
SAE 30, 185-225 *SUS @ 130° F*	23
SAE 40, 65-80 *SUS @ 210° F*	13
SAE 50, 80-105 *SUS @ 210° F*	13

- **Sheath Material Required** — Steel sheaths are normally used for hydraulic systems, with alloy or stainless steel in special applications. Copper sheaths can be used for systems using water as the working fluid.

After resolving the above points, choose the type of heater best suited for the particular application. For example, an oil reservoir may be heated by direct immersion heaters, or by clamping a strip, ring or tabular heater to the outside of the tank. Here the choice will involve the process, appearance, space availability, both inside and outside, economy, etc.

Initial Heating KW Requirement

$$KW = \frac{kWh}{h}$$

$$kWh = \frac{Qm}{3410} + \left(\frac{Ls}{2 \times 1000} \times h\right)$$

kWh = Kilowatt-hour
Qm = Heat input to raise the temperature of both the steel tank and the oil

$Qm = (wt_1 \times Cv_1 + wt_2 \times Cv_2) \times \Delta T$ (Btu)
wt_1 = weight of material 1 (lb)
Cv_1 = specific weight of material 1 (Btu/lb/°F)

Material	lb/ft³	Cv (Btu/lb/°F)
steel	487	.12
copper	555	.093
aluminum	168	.226
hyd. oil	54.8	.511

wt_2 = weight of material 2 (lb)
Cv_2 = specific weight of material 2 (Btu/lb/°F)
ΔT = change in temperature (F°)

Ls = A x L

Ls = Heat loss through convection (W)
A = Surface area (ft²)
L = Loss per square foot (W/ft²)

h = time to reach temperature (hours)
3410 = conversion from Btu to kWh (kWh/Btu)
2 = to find average heat loss (no units)
1000 = conversion from W to kW

Operating KW Requirement

$$KW = \frac{kWh}{h}$$

$$kWh = \frac{Qm}{3410} + \frac{Ls}{1000}$$

kWh = Kilowatt-hour
Qm = Heat input to raise the temperature of both the steel tank and the oil

$Qm = (wt_1 \times Cv_1 + wt_2 \times Cv_2) \times \Delta T$ (Btu)
wt_1 = weight of material 1 (lb)
Cv_1 = specific weight of material 1 (Btu/lb/°F)
wt_2 = weight of material 2 (lb)
Cv_2 = specific weight of material 2 (Btu/lb/°F)

Material	lb/ft³	Cv (Btu/lb/°F)
steel	487	.12
copper	555	.093
aluminum	168	.226
hyd. oil	54.8	.511

ΔT = change in temperature

Ls = A x L

Ls = Heat loss through convection (W)
A = Surface area (ft²)
L = Loss per square foot (W/ft²)
See Heat Loss Vertical Surface graph

3410 = conversion from Btu to kWh (kWh/Btu)
2 = to find average heat loss (no units)
1000 = conversion from W to kW

Heat Loss
Vertical Metal Surface

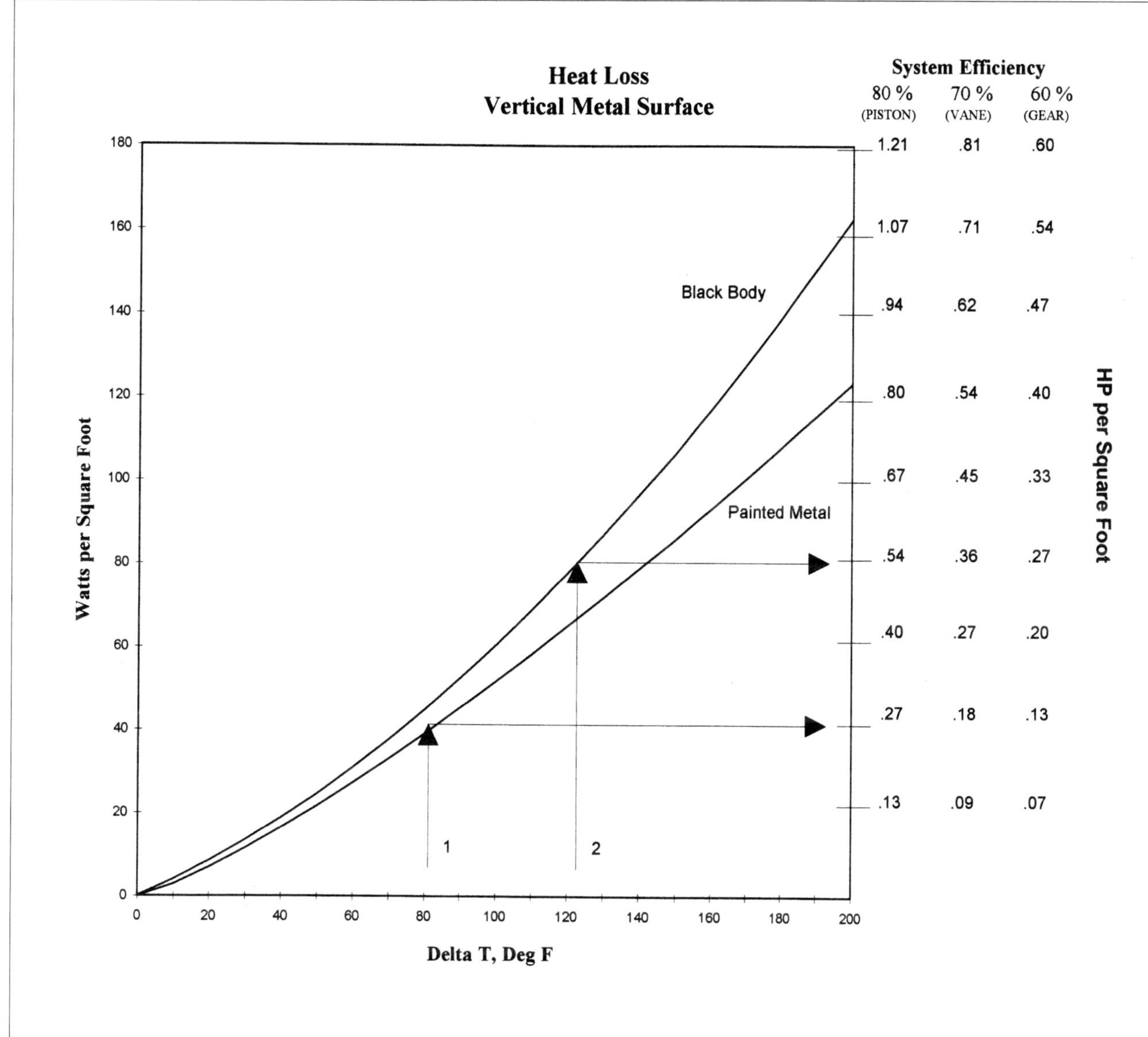

Heat Loss Examples

Example 1

Size a hydraulic reservoir with an electric motor rated at 7.5 hp, driving a vane type pump with an overall system efficiency of 70%. The summer outside air temperature is 80° F and you wish to maintain the oil temperature at 160° F.

Summer Time

Summer ΔT = 160-80 = 80° F

Project a vertical line (line 1) from 80° F to the painted metal line. Read across to 70% efficiency to find .18 hp per square foot.

hp loss per square foot = .18 (40 watts per square foot)

Divide the electric motors hp by the hp per square foot determined above, to obtain the total vertical wall area in square feet.

$$\text{Total vertical wall area} = \frac{7.5\ \text{hp}}{1.8\ \text{hp/ft}^2} = 41.7\ \text{ft}^2$$

Add a margin of 50% = 41.7 x 1.5 = 62.5 ft^2

Assume that the 4 sides are square, the sides of the reservoir will be:

One side of a square reservoir = $((62.5\ ft^2)/4)^{1/2}$ = 3.95 ft x 3.95 ft

Winter Time

The winter outside air temperature is 20° F and you wish to maintain the oil at a minimum of 140° F.

Winter ΔT = 140-20 = 120° F

Project a vertical line (line 2) from 120° F to the Black Body line. Read across to the "Watts per Square Foot" to find 80 watts per square foot.

Project a vertical line (line 1) from 80° F to the painted metal line. Read across to the "Watts per Square Foot" to find 40 watts per square foot. The difference between the two values is the amount of heat that must be added due to surface cooling of the reservoir to maintain the oil at 140° F.

Heat that must be added = 80-40 = 40 Watts per Square Foot

Multiply the heat that must be added, times the square footage of the tank to determine the heater size.

40 x 62.5 = 2500 watts

Add a margin of 50% = 2500 x 1.5
= 3750 watts total heater required

Example 2

Size a hydraulic reservoir with the same conditions as above except the size of the reservoir is limited to 3.0 feet wide x 2.5 feet tall and 3 feet long.

Summer Time

Total vertical surfaces are:

Ends = quantity 2 x 3 ft x 2.5 ft = 15 ft^2

Sides = quantity 2 x 2.5 ft x 3 ft = 15 ft^2

Total vertical surface = 15 + 15 = 30 ft^2

From example 1, the heat loss at the Summer ΔT is 40 Watts per Square foot. The heat loss in the summer, times the vertical surface area is the heat that can be removed by cooling of the surface.

Heat removal by cooling = 40 x 30 = 1200 watts (1.61 hp)

The heat required to be removed = 7.5 hp x .2 x 746 $\frac{\text{watts}}{\text{hp}}$ = 1119 watts

Add a margin of 50%

Total heat to be removed = 1119 x 1.5 = 1678 watts

The capacity of the reservoir in the summer time to remove heat is less than heat generated so a cooler must be added that can remove:

Heat that must be removed by an external cooler = 1678-1200 = 478 watts

Winter Time

From example 1, the heat loss at the winter ΔT is 80 watts per square foot. The heat loss in winter, times the vertical surface area is the heat that is removed by cooling of the surface.

Heat removal by cooling = 80 x 30 = 2400 watts (3.22 hp)

From above, the heat required to be removed is 1119 watts (to stay conservative do not add the margin in this case).

The capacity of the reservoir in the winter to remove heat is greater than the heat generated, so a heater must be added to keep the oil at 140° F.

The heat that must be added by a heater = 2400 - 1119 = 1281 watts

To determine the oil temperature in winter time if a heater is not used, divide the wattage by the vertical surface area to determine the watts per square foot.

Equilibrium heat loss = 1119/30 = 37.3 watts/ft^2

Then project a horizontal line on the graph from the calculated equilibrium heat loss to the black body line. The ΔT from the graph is approximately 78°F. Drop a vertical line to the X axis and read the resulting ΔT. Final temperature of the oil without a heater = 20 + 78 = 98°F.

Cooling In Hydraulic Systems

Heat Exchanger Size Selection

The exact size of a heat exchanger needed on a new system cannot be accurately calculated because of too many unknown factors. On existing systems, by making tank measurements and measuring air and oil temperatures, a rather accurate calculation can be made of the heat exchanger capacity needed to reduce the maximum oil temperature. Consult your local distributor for more information. For new systems, the following rules-of-thumb will give acceptable results for sizing:

The theoretical maximum cooling capacity of a heat exchanger for a hydraulic system will never have to be greater than the horsepower input to the system even if all the input power had to be removed. Usually its capacity can be considerably less based on the calculated input horsepower to drive the system. A rule-of-thumb is to provide a heat exchanger removal capacity of about 25% of the input horsepower. Please remember that this is a general rule and might not be accurate for unusual applications.

When ordering a heat exchanger, the information furnished must include the maximum rate of oil flow, in gpm, through the heat exchanger, and the horsepower or BTU/hr of heat to be removed. On water cooled models, state the maximum rate of water flow which will be available. Specify the temperature of the cooling water.

Another rule-of-thumb for size selection is that each square foot of cooling surface on shell and tube heat exchangers will remove approximately 2.16 hp, or 5500 BTU/hr under conditions of 50°F temperature difference between oil and water, using oil with viscosity of 150 SSU. Be sure that the model selected will handle the volume of oil flow with very low pressure loss.

Formulas For Electrical Standards

ELECTRICAL			
Single Phase		**Three-Phase**	
Current (Amperes) (hp known)	$I = \frac{hp \times 746}{E \times eff \times pf}$	Current (Amperes) (hp known)	$I = \frac{hp \times 746}{1.73 \times E \times eff \times pf}$
Current (Amperes) (KW known)	$I = \frac{KW \times 1000}{E \times pf}$	Current (Amperes) (KW known)	$I = \frac{KW \times 1000}{1.73 \times E \times pf}$
Current (Amperes) (Kva known)	$I = \frac{Kva \times 1000}{E}$	Current (Amperes) (Kva known)	$I = \frac{Kva \times 1000}{1.73 \times E}$
KW	$KW = \frac{I \times E \times pf}{1000}$	KW	$KW = \frac{1.73 \times I \times E \times pf}{1000}$
Kva	$Kva = \frac{I \times E}{1000}$	Kva	$Kva = \frac{1.73 \times I \times E}{1000}$
hp	$hp = \frac{I \times E \times eff \times pf}{746}$	hp	$hp = \frac{1.73 \times I \times E \times eff \times pf}{746}$

Electrical Rules of Thumb

Synch Speed rpm	Approx. Torque lb-ft per hp
3600	1.4
1800	3
1200	4.5
900	5.8

Rated Voltage	Approximate Amps/hp	
	Single-Phase	Three-Phase
115	10	
230	5	2.5
460		1.25
575		1.00

Electrical Standards

Resistor Color Codes

Color	1st Digit (A)	2nd Digit (B)	Multiplier (C)	Tolerance (D)
Black	0	0	1	
Brown	1	1	10	1%
Red	2	2	100	2%
Orange	3	3	1,000	3%
Yellow	4	4	10,000	4%
Green	5	5	100,000	
Blue	6	6	1,000,000	
Violet	7	7	10,000,000	
Gray	8	8	100,000,000	
White	9	9	10^9	
Gold			0.1 (EIA)	5%
Silver			0.01 (EIA)	10%
No Color				20%

Example: Red — Red — Orange = 22,000 ohms, 20%

Additional information concerning the Axial Lead resistor can be obtained if Band A is a wide band. Case 1: If only Band A is wide, it indicates that the resistor is wirewound. Case 2: If Band A is wide and there is also a blue fifth band to the right of Band D on the Axial Lead Resistor, it indicates the resistor is wirewound and flame proof.

Resistor

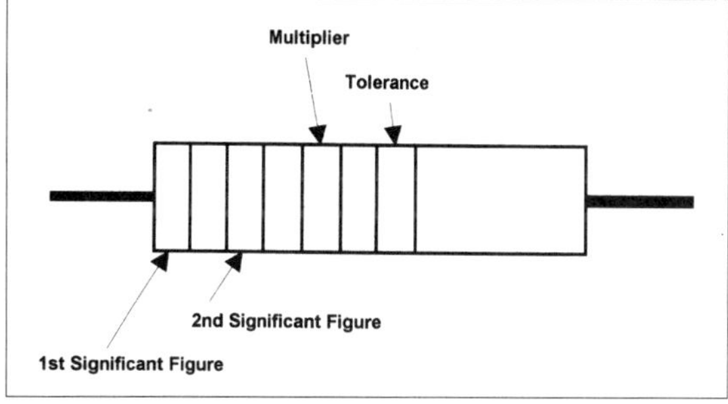

Capacitor Color Codes

Color	Capacitance: 1st & 2nd Significant Figures	Capacitance: Multiplier	Capacitance Tolerance
Black	0	1	±20% (E1A)
Brown	1	10	±1%
Red	2	100	±2%
Orange	3	1000	
Yellow	4	10000 (E1A)	
Green	5		±5%
Blue	6		
Purple	7		
Gray	8		
White	9		
Gold		0.1	±1/2% (E1A)
Silver		0.01 (E1A)	±10%

Resistor Standard Values

Standard Resistor Values for 5% class			
1	62	3.9k	240k
1.1	68	4.3k	270k
1.2	75	4.7k	300k
1.3	82	5.1k	330k
1.5	91	5.6k	360k
1.6	100	6.2k	390k
1.8	110	6.8k	430k
2.0	120	7.5k	470k
2.2	130	8.2k	510k
2.4	150	9.1k	560k
2.7	160	10k	620k
3.0	180	11k	680k
3.3	200	12k	750k
32.6	220	13k	820k
3.9	240	15k	910k
4.3	270	16k	1.0M
4.7	300	18k	1.1M
5.1	330	20k	1.2M
5.6	360	22k	1.3M
6.2	390	24k	1.5M
6.8	430	27k	1.6M
7.5	470	30k	1.8M
8.2	510	33k	2.0M
9.1	560	36k	2.2M
10	620	39k	2.4M
11	680	43k	2.7M
12	750	47k	3.0M
13	820	51k	3.3M
15	910	56k	3.6M
16	1.0k	62k	3.9M
18	1.1k	68k	4.3M
20	1.2k	75k	4.7M
22	1.3k	82k	5.1M
24	1.5k	91k	5.6M
27	1.6k	100k	6.2M
30	1.8k	110k	6.8M
33	2.0k	120k	7.5M
36	2.2k	130k	8.2M
39	2.4k	150k	9.1M
43	2.7k	160k	10.0M
47	3.0k	180k	
51	3.3k	200k	
56	3.6k	220k	

k = kohms = 1,000 ohms M = megohms = 1,000,000 ohms

Capacitor Standard Values

pF	mF	mF	mF	mF
10	0.001	0.1	10	1,000
12	0.0012			
13	0.0013			
15	0.0015	0.15	15	
18	0.0018			
20	0.002			
22	0.0022	0.22	22	2200
24				
27				
30				
33	0.0033	0.33	33	3,300
36				
43				
47	0.0047	0.47	47	4,700
51				
56				
62				
68	0.0068	0.68	68	6,800
75				
82				
100	0.01	1.0	100	10,000
110				
120				
130				
150	0.015	1.5		
180				
200				
220	0.022	2.2	220	22,000
240				
270				
300				
330	0.033	3.3	330	
360				
390				
430				
470	0.047	4.7	470	47,000
510				
560				
620				
680	0.068	6.8		
750				
820				82,000
910				

pF = picofarads = 1 x 10^{-12} farads
mF = microfarads = 1 x 10^{-6} farads

Electric Motor Selection and Application

The use of electric motors, like that of all other utilization of concentrated power, is potentially hazardous. The degree of hazard can be greatly reduced by proper design, selection, installation and use, but hazards cannot be completely eliminated.

Some of the factors that must be considered for the safe application of motors are as follows:

- Enclosure
- Service conditions
- Proper matching of the motor to the load

Enclosure

The selection of the proper enclosure is vital to the successful safe operation of A-C motors. Machine performance and life can be materially reduced by using an enclosure inappropriate for the application. The user must recognize the specific environmental conditions and specifiy the correct enclosure.

- Protected
- Totally Enclosed (Non-Ventilated, Fan-Cooled)
- Explosion-Proof
- Dust-Ignition-Proof

Power Supply

Phase — Motors are designed for either single-phase or three-phase operation. Single-phase motors may be operated on one-phase of a three-phase power supply with the rated voltage. Operation of a three-phase motor on a single-phase supply will result in motor damage.

Voltage

Single-Phase Motors — standard voltages are 115/230.

Three-Phase Motors — standard voltage are 190/380, 200, 200/400, 230/460, 208-230/460, 460 and 575. Motors rated at 230 or 230/460 volts will operate on a 208 volt supply except that starting and peak running torques at 208 volts will be 20-25% less than at the 230 volt rating. Motors nameplated 208-230/460 will operate satisfactorily at 208 volts.

Voltage Variation — In accordance with NEMA standards, motors will operate with ±10% variation in rated voltage with rated frequency. Performance within this voltage variation will not necessarily be in accordance at rated voltage.

Dual Voltage Motors — easily reconnectable by referring to the information on the motor nameplate.

Frequency — Motors are designed for operation on 50 or 60 Hz as specified. Many motors designed for operation on 60 Hz power may be operated on 50 Hz power of a lower voltage rating with resulting decrease in speed and horsepower. In accordance with NEMA standards, motors will operate with ±5% variation in rated frequency with rated voltage. Performance with this frequency variation will not necessarily be in accordance with performance at rated frequency.

Service Factor

The service factor of a motor is a multiplier which, when applied to the rated horsepower, indicates a permissible horsepower loading which may be carried under the conditions specified for the service factor.

Torque/NEMA Design Classes

The full load torque of a motor is expressed as follows:

$$\text{Full load torque (ft. lbs)} = \frac{\text{hp} \times 5250}{\text{Full load speed (rpm)}}$$

Since different loads present different torque requirements at starting (breakaway), minimum (pull-up), breakdown (pull-out) and full load, the National Electrical Manufacturers Association (NEMA) has defined four standard design classes A, B, C and D of squirrel cage polyphase induction motors. Refer to the following chart for summation of these design classes.

Characteristics of NEMA Design Classes For Squirrel-Cage Polyphase Induction Motors

NEMA Design	Starting Torque	Starting Current	Break-down Torque	Full Load Slip	Typical Applications
A	Normal	High	High	Low	Mach. Tools, Fans
B	Normal	Normal	Normal	Normal	General Industrial
C	High	Normal	Normal	Normal	Loaded compressor, Loaded conveyor
D	Very High	Low		High	Punch Press or Hoists

Speed/Slip

The synchronous speed of an A-C motor is as follows:

$$\text{Synchronous Speed (rpm)} = \frac{120 \times \text{Power Supply Frequency (HZ)}}{\text{Number of Poles}}$$

The number of poles is a function of motor design.

The actual operating speed of an A-C induction motor is determined by the synchronous speed and the slip. Slip is the difference between the speed of the rotating magnetic field (which is always synchronous) and the rotor speed. Slip generally increases with an increase in motor torque; therefore, actual operating speed generally decreases with an increase in motor torque.

Motor torque is defined at four points as shown.

1. Breakaway or starting
2. Minimum or "pull-up"
3. Breakdown or "pull-out"
4. Full load

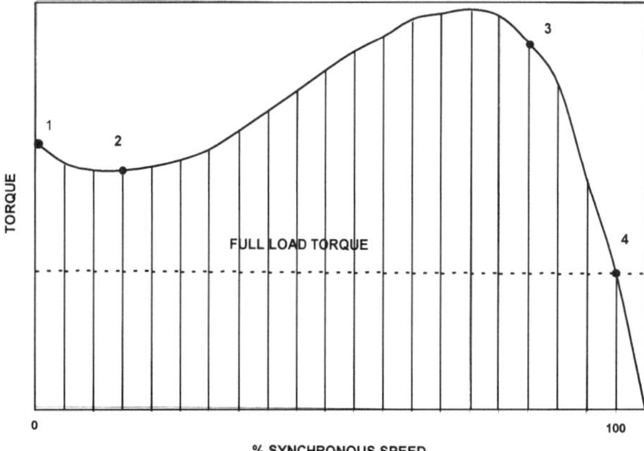

General Speed/Torque relationship for squirrel-cage polyphase induction motors

ELECTRICAL

Motor Starter & Wire Sizes

Motor Wiring — 3-Phase Squirrel-Cage Induction Motors

Not More Than 3 Conductors in Raceway, Cable, or Direct Burial. Based on Ambient Temperature of 30°C (86°F)

	230 Volts							460 Volts						
		Approx. Full-load Current Amps[1]		Minimum Wire Size THW or RHW AWG[2]		Branch Circuit Protection Amps			Approx. Full-load Current Amps[1] AWG[2]		Minimum Wire Size THW or RHW		Branch Circuit Protection Amps	
hp	NEMA Starter	x 1.0	x 1.25	Cu	Al	Fuse[3]	Break	Mag™ NEMA Starter	x 1.0	x 1.25	Cu	Al	Fuse[3]	Mag™ Break
1	00	3.6	4.5	14	12	10	7	00	1.8	2.3	14	12	3	3
1 1/2	00	5.2	6.5	14	12	15	7	00	2.6	3.3	14	12	6	3
2	0	6.8	8.5	14	12	20	15	00	3.4	4.3	14	12	10	7
3	0	9.6	12	14	12	25	15	0	4.8	6	14	12	10	7
5	1	15.2	19	12	10	40	30	0	7.6	9.5	14	12	20	15
7 1/2	1	22	27.5	10	8	60	30	1	11	13.8	14	12	30	15
10	2	28	35	8	8	80	50	1	14	17.5	12	10	40	30
15	2	42	52.5	6	4	125	50	2	21	26.3	10	8	60	30
20	3	54	67.5	4	3	150	100	2	27	33.8	8	8	80	50
25	3	68	85	4	2	200	100	2	34	42.5	8	6	100	50
30	3	80	100	3	1	225	100	3	40	50	6	6	100	50
40	4	104	130	1	2/0	300	150	3	52	65	6	4	150	100
50	4	130	162.5	2/0	4/0	350	150	3	65	81.3	4	2	175	100
60	5	154	192.5	3/0	250	450	225	4	77	96.3	3	1	225	100

Motor Wiring — Single-Phase Induction Motors

Not More Than 3 Conductors in Raceway, Cable, or Direct Burial. Based on Ambient Temperature of 30°C (86°F)

	230 Volts						460 Volts					
	Approx. Full-load Current Amps[1]		Minimum Wire Size THW or RHW AWG[2]		Branch Circuit Protection Amps		Approx. Full-load Current Amps[1]		Minimum Wire Size THW or RHW AWG[2]		Branch Circuit Protection Amps	
hp	x 1.0	x 1.25	Cu	Al	Fuse[3]	Mag™ Break	x 1.0	x 1.25	Cu	Al	Fuse[3]	Mag™ Break
1/2	9.8	12.3	14	12	30	15	4.9	6.1	14	12	10	7
3/4	13.8	17.3	12	10	40	30	6.9	8.6	14	12	20	15
1	16	20	12	10	45	30	8	10	14	12	20	15
1 1/2	20	25	10	10	60	30	10	12.5	14	12	30	15
2	24	30	10	8	70	30	12	15	14	12	35	15
3	34	42.5	8	6	100	50	17	21.3	10	10	50	30
5	56	70	4	3	150	100	28	35	8	8	80	50
7 1/2	80	100	3	1	225	100	40	50	6	6	110	50
10	100	125	1	2/0	300	150	50	62.5	6	4	150	100

[1] Full load currents shown from National Electric Code. NFPA 1978. Table 430-150. Column entitled "x 1.25" multiplies full-load current by 1.25 to aid in sizing motor branchcircuit conductors. See section 430-22 National Electric Code. Values of full-load current are for motors running at speeds usual for belted motors and motors with normal torque characteristics. Data does not apply to multispeed motors, or motors built for especially low speeds or high torques.
[2] Wire size data given for conductors having 75 C insulation. E.G. Types THW and RHW. Cu — Copper. Al — Aluminum.
[3] Maximum fuse ampere ratings for nontime delay fuses. Code F to Code V motors.

Motor Terminal Amperes at Full Load[1]

(Average Values for All Speeds and Frequencies)

	Single-Phase A-C		Polyphase A-C (Induction Type) Squirrel-Cage and Wound Motor								Direct Current		
			115 Volts		230 Volts		460 Volts		575 Volts				
hp	115 volts	230 volts[2]	3-ph.	2-Ph. 4-Wire[3]	3-ph.	2-Ph. 4-Wire[3]	3-Ph.	2-Ph. 4-Wire[3]	3-Ph.	2-Ph. 4-Wire[3]	180 volts	240 volts	500 volts
1/4	5.8	2.9	N/A	N/A	N/A	N/A	N/A	N/A	N/A	N/A	2.0	N/A	N/A
1/3	7.2	3.6	N/A	N/A	N/A	N/A	N/A	N/A	N/A	N/A	2.6	N/A	N/A
1/2	9.8	4.9	4.0	4.0	2.0	2.0	1.0	1.0	.8	.8	3.4	2.7	N/A
3/4	13.8	6.9	5.6	4.8	2.8	2.4	1.4	1.2	1.11	1.0	4.8	3.8	N/A
1	16.0	8.0	7.2	6.4	3.6	3.2	1.8	1.6	1.4	1.3	6.1	4.7	N/A
1 1/2	20.0	10.0	10.4	9.0	5.2	4.5	2.6	2.3	2.1	1.8	8.3	6.6	N/A
2	24.0	12.0	13.6	11.8	6.8	5.9	3.4	3.0	2.7	2.4	10.8	8.5	N/A
3	34.0	17.0	N/A	N/A	9.6	8.3	4.8	4.2	3.9	3.3	16.0	12.2	N/A
5	56.0	28.0	N/A	N/A	15.2	13.2	7.6	6.6	6.1	5.3	27.0	20.0	N/A
7 1/2	80.0	40.0	N/A	N/A	22.0	19.0	11.0	9.0	9.0	8.0	N/A	29.0	13.6
10	100.0	50.0	N/A	N/A	28.0	24.0	14.0	12.0	11.0	10.0	N/A	38.0	18.0
15	N/A	N/A	N/A	N/A	42.0	36.0	21.0	18.0	17.0	14.0	N/A	55.0	27.0
20	N/A	N/A	N/A	N/A	54.0	47.0	27.0	23.0	22.0	19.0	N/A	72.0	34.0
25	N/A	N/A	N/A	N/A	68.0	59.0	34.0	29.0	27.0	24.0	N/A	89.0	43.0
30	N/A	N/A	N/A	N/A	80.0	69.0	40.0	35.0	32.0	28.0	N/A	106.0	51.0
40	N/A	N/A	N/A	N/A	104.0	90.0	52.0	45.0	41.0	36.0	N/A	140.0	67.0
50	N/A	N/A	N/A	N/A	130.0	113.0	65.0	56.0	52.0	45.0	N/A	173.0	83.0
60	N/A	N/A	N/A	N/A	154.0	133.0	77.0	67.0	62.0	53.0	N/A	206.0	99.0
75	N/A	N/A	N/A	N/A	192.0	166.0	96.0	83.0	77.0	66.0	N/A	255.0	123.0

[1] These values of full-load currrent are for motors running, at speeds usual for belted motors with normal torque characteristics. Motors built for especially low speeds or high torques may require more running current, in which case the nameplate current rating should be used.
[2] For full-load currents of 208 and 200-volt motors increase the corresponding 230-volt motor full-load current by 10 and 15 percent respectively.
[3] Current in common conductor of 2-phase, 3 wire system will be 1.41 times value given.

Typical Dimensions
TEFC
48 — 140T

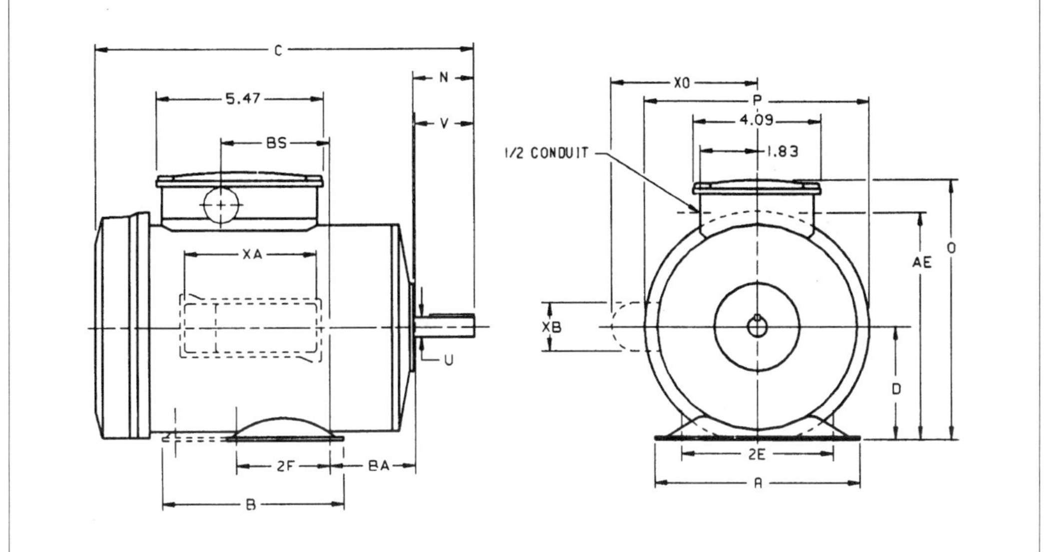

Frame Size	A (1)	2E	D	O	AE	P	H (4)	BA
E(*)48	5.62	4.24	3.00	7.33	6.22	6.20	0.34	2.50
E(*)56	6.38	4.88	3.50	7.83	6.72	7.29 (3)	0.34	2.75
F(*)56	6.62	4.88	3.50	8.19	7.13	7.28	0.34	2.75
F(*)140T	6.62	5.50	3.50	8.19	7.13	7.28	0.34	2.25

Frame Size	B (1)	2F	N	U	V	Key	
						Sq.	Lgth.
E(*)48	3.62	2.75	1.56	0.500	1.50	(2)	(2)
E(*)56	4.10	3.00	1.94	0.625	1.88	0.188	1.25
F(*)56	5.88	3.00	1.94	0.625	1.88	0.188	1.25
F(*)140T	5.88	4.00	2.31	0.875	2.25	0.188	1.25

Frame Size	C	BS	XO (1)	XA (1)	XB (1)
EC48	10.26	2.22	4.68	4.25	1.90
ED48	10.76	2.72	4.68	4.25	1.90
EE48	11.26	3.22	4.68	4.25	1.90
EF48	12.26	4.22	4.68	4.25	1.90
EH48	11.76	3.72	4.68	4.25	1.90
EC56	10.63	1.98	4.68	4.25	1.90
ED56	11.13	2.48	4.68	4.25	1.90
EE56	11.63	2.98	4.68	4.25	1.90
EF56	12.63	3.98	4.68	4.25	1.90
EH56	12.13	3.48	4.68	4.25	1.90
FB56	11.17	2.47	5.08	4.26	1.91
FC56	12.17	3.47	5.85	5.82	2.65
FD56	13.23	4.53	5.85	5.82	2.65
FE56	14.23	5.53	5.85	5.82	2.65
FF56	15.23	6.53	5.85	5.82	2.65
FJ56	12.61	3.91	5.85	5.82	2.65
FK56	11.67	2.97	5.08	4.26	1.91
FB140T	11.55	2.97	5.08	4.26	1.91
FC140T	12.55	3.97	5.80	5.82	2.65
FD140T	13.61	5.03	5.80	5.82	2.65
FE140T	14.61	6.03	5.80	5.82	2.65
FF140T	15.61	7.03	5.80	5.82	2.65
FJ140T	12.99	4.41	5.80	5.82	2.65
FK140T	12.05	3.47	5.08	4.26	1.91

NOTES: (*) Second letter defines "C" Dim.
(1) Maximum value
(2) No key — 1.25" long flat
(3) Single-phase P value is 6.20"
(4) Mounting hole diameter
Dimensions are for estimating purposes only

Typical Dimensions
TENV-XP & TEFC-XP
48 — 140T

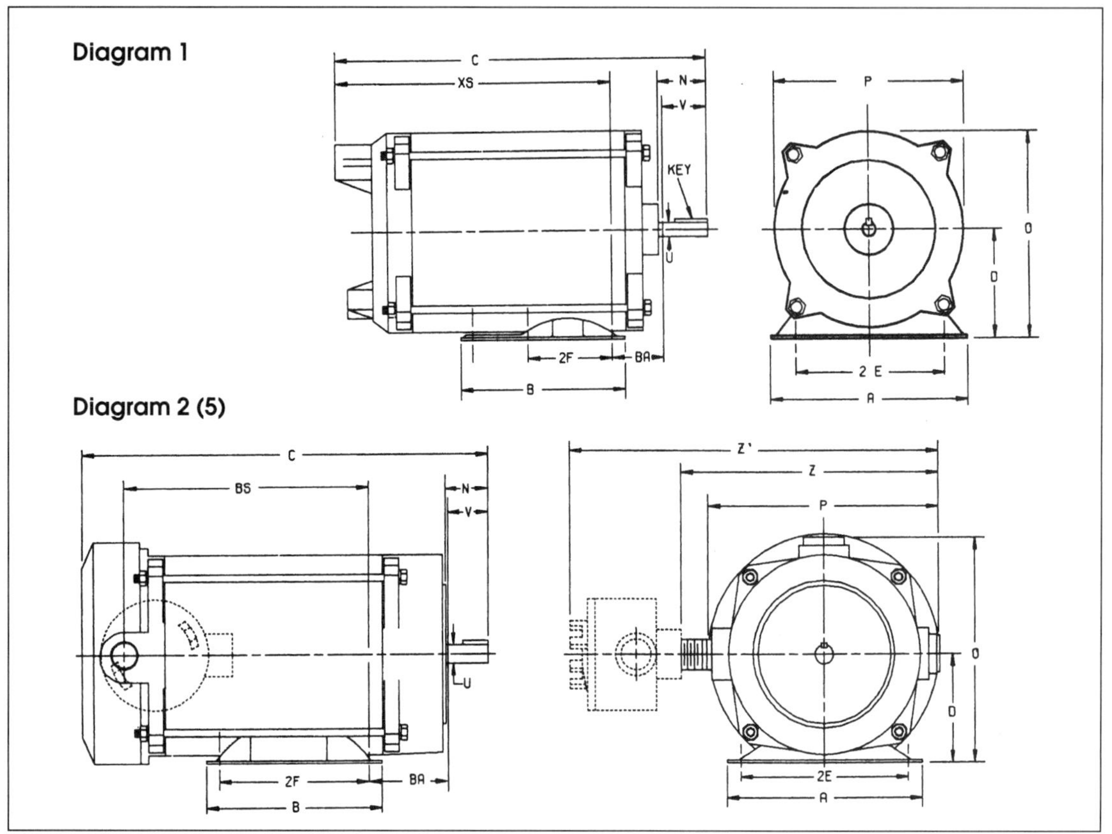

Frame Size	Refer to Diagram	A (1)	2E	D	O	BA	P	Z'
G(*)48 (2)	1	5.62	4.24	3.00	5.82	2.50	5.65	—
G(*)56 (2)	1	6.38	4.88	3.50	6.32	2.75	5.65	—
H(*)56 (2)	2	6.38	4.88	3.50	6.75	2.75	—	11.01
H(*)56	2	6.62	4.88	3.50	6.75	2.75	—	11.01
H(*)140T	2	6.62	5.50	3.50	6.75	2.75	—	11.01

Frame Size	B (1)	2F	H (4)	N	U	V	Key Sq.	Key Lgth.
G(*)48 (2)	3.62	2.75	0.34	1.56	0.500	1.50	(3)	(3)
G(*)56 (2)	5.72	3.00	0.34	1.94	0.625	1.88	0.188	1.25
H(*)56 (2)	5.72	3.00	0.34	2.07	0.625	1.88	0.19	1.25
H(*)56	5.88	3.00	0.34	2.07	0.625	1.88	0.19	1.25
H(*)140T	5.88	4.00	0.34	2.07	0.875	2.25	0.19	1.38

Frame Size	C	XS	Frame Size	C	BS	Frame Size	C	BS
GA48 (2)	10.38	6.38	GD56 (2)	12.25	7.62	HB56	12.69	4.06
GB48 (2)	11.12	7.12	GE56 (2)	12.25	7.62	HB56	12.69	4.06
GC48 (2)	11.38	7.38	GF56 (2)	13.75	9.12	HD56	14.75	6.12
GD48 (2)	11.88	7.88	GH56 (2)	13.25	8.62	HE56	15.75	7.12
GE48 (2)	12.38	8.38	HB56 (2)	11.39	4.06	HF56	16.75	8.12
GF48 (2)	13.38	9.38	HC56 (2)	12.39	5.06	HB140T	12.75	3.75
GH48 (2)	12.88	8.88	HD56 (2)	13.45	6.12	HC140T	13.75	4.75
GA56 (2)	10.75	6.12	HE56 (2)	14.45	7.12	HD140T	14.81	5.81
GB56 (2)	11.50	6.88	HF56 (2)	15.45	8.12	HE140T	15.81	6.81
GC56 (2)	11.75	7.12				HF140T	16.81	7.81

NOTES: (*) Second letter defines length "C" dim.
(1) Maximum value
(2) TENV-XP
(3) No key — 1.25" long flat
(4) Mounting hole diameter
(5) TENV-XP ratings do not include fan cover
Dimensions are for estimating purposes only.

Typical Dimensions TEFC & TEFC-XP 180T — 449T

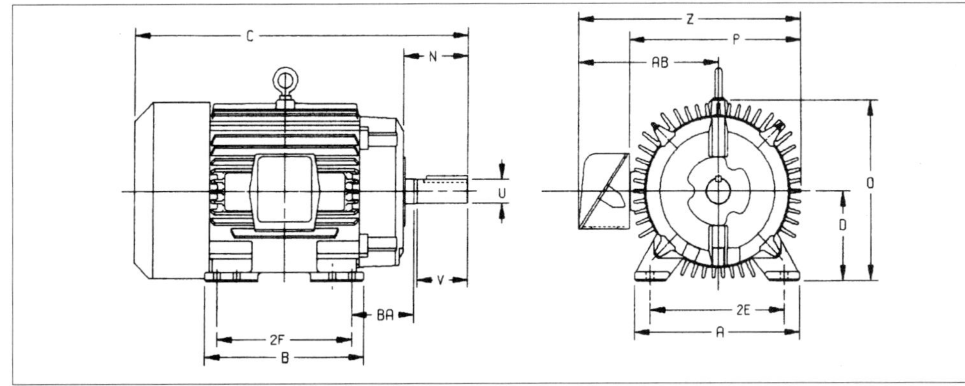

Frame Size	A	2E	D	O	AB TEFC	AB XO	BA	Z TEFC	Z XP	H (1)
182-184T	9.00	7.50	4.50	9.88	7.81	9.38	2.75	12.44	14.01	0.44
213-215T	10.50	8.50	5.25	11.25	8.69	10.25	3.50	13.94	15.50	0.44
254-256T	12.50	10.00	6.25	13.25	10.75	12.38	4.25	17.38	19.01	0.56
284-286T	13.75	11.00	7.00	14.75	12.75	13.25	4.75	20.19	20.69	0.56
324-326T	15.50	12.50	8.00	16.69	15.19	17.06	5.25	23.69	25.56	0.69
364-365T	17.00	14.00	9.00	18.50	18.06	18.81	5.88	27.81	28.56	0.69
404-405T	19.00	16.00	10.00	21.31	19.31	20.50	6.62	30.56	31.75	0.81
444-445T	21.00	18.00	11.00	23.38	23.38	26.25	7.50	36.01	38.88	0.81
447-449T	21.00	18.00	11.00	24.25	23.37	29.63	7.50	36.00	41.76	0.81

Frame Size	C	B	2F	N TEFC	N XP	U	V	Key Sq.	Key Lgth.
182T	15.62	7.00	4.50	3.00	2.81	1.125	2.50	0.250	1.75
L182T	17.12	8.50	4.50	3.00	2.81	1.125	2.50	0.250	1.75
184T	15.62	7.00	5.50	3.00	2.81	1.125	2.50	0.250	1.75
L184T	17.12	8.50	5.50	3.00	2.81	1.125	2.50	0.250	1.75
213T	19.25	8.50	5.50	3.62	3.44	1.375	3.12	0.312	2.38
L213T	20.12	9.12	5.50	3.62	3.44	1.375	3.12	0.312	2.38
215T	19.25	8.50	7.00	3.62	3.44	1.375	3.12	0.312	2.38
L215T	20.12	9.12	7.00	3.62	3.44	1.375	3.12	0.312	2.38
254T	24.56	12.00	8.25	4.12	4.06	1.625	3.75	0.375	2.88
256T	24.56	12.00	10.00	4.12	4.06	1.625	3.75	0.375	2.88
284T	27.44	13.00	9.50	5.00	4.69	1.875	4.38	0.500	3.25
284TS	26.06	13.00	9.50	3.62	3.31	1.625	3.00	0.375	1.88
286T	27.44	13.00	11.00	5.00	4.69	1.875	4.38	0.500	3.25
286TS	26.06	13.00	11.00	3.62	3.31	1.625	3.00	0.375	1.88
324T	30.44	14.75	10.50	5.62	5.62	2.125	5.00	0.500	3.88
324TS	28.94	14.75	10.50	4.12	4.12	1.875	3.50	0.500	2.00
326T	30.44	14.75	12.00	5.62	5.62	2.125	5.00	0.500	3.88
326TS	28.94	14.75	12.00	4.12	4.12	1.875	3.50	0.500	2.00
364T	33.44	15.00	11.25	6.25	6.00	2.375	5.62	0.625	4.25
364TS	31.31	15.00	11.25	4.12	3.88	1.875	3.50	0.500	2.00
365T	33.44	15.00	12.25	6.25	6.00	2.375	5.62	0.625	4.25
365TS	31.31	15.00	12.25	4.12	3.88	1.875	3.50	0.500	2.00
404T	38.31	16.00	12.25	7.50	7.50	2.875	7.00	0.750	5.62
404TS	35.31	16.00	12.25	4.50	4.50	2.125	4.00	0.500	2.75
405T	38.31	16.00	13.75	7.50	7.50	2.875	7.00	0.750	5.62
405TS	35.31	16.00	13.75	4.50	4.50	2.125	4.00	0.500	2.75
444T	44.62	19.00	14.50	8.94	8.94	3.375	8.25	0.875	6.88
444TS	40.88	19.00	14.50	5.19	5.19	2.375	4.50	0.625	3.00
445T	44.62	19.00	16.50	8.94	8.94	3.375	8.25	0.875	6.88
445TS	40.88	19.00	16.50	5.19	5.19	2.375	4.50	0.625	3.00
447T	48.13	22.50	20.00	8.50	8.50	3.375	8.25	0.875	6.91
447TS	44.37	22.50	20.00	4.75	4.75	2.375	4.50	0.625	3.03
449T	53.13	27.50	25.00	8.50	8.50	3.375	8.25	0.875	6.91
449TS	49.37	27.50	25.00	4.75	4.75	2.375	4.50	0.625	3.03

NOTES: (1) Mounting hole diameter
Dimensions are for estimating purposes only

Typical Dimensions Protected 180T — 449T

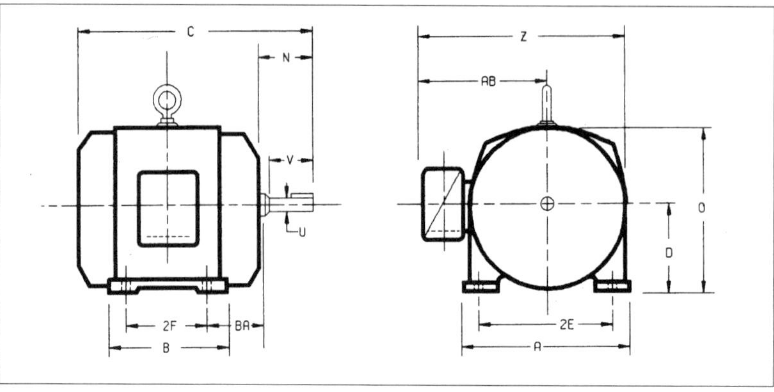

Frame Size	A	2E	D	O	AB	BA	Z	H (1)
182-184T(1)	9.00	7.50	4.50	9.31	7.38	2.75	11.88	0.44
21-215T	10.50	8.50	5.25	10.75	8.19	3.50	13.50	0.56
254-256T	12.50	10.00	6.25	12.56	10.19	4.25	16.45	0.56
284-286T	14.00	11.00	7.00	14.31	12.19	4.75	19.25	0.56
324-326T	15.75	12.50	8.00	16.10	14.25	5.25	22.46	0.69
364-365T	17.00	14.00	9.00	18.30	18.06	5.88	27.84	0.69
404-405T	19.00	16.00	10.00	20.56	19.06	6.62	29.97	0.81
444-445T	21.00	18.00	11.00	22.94	22.62	7.50	34.43	0.81
447-449T	21.00	18.00	11.00	24.25	22.75	7.50	34.88	0.81

Frame Size	C	B	2F	N	U	V	Key Sq.	Key Lgth.
182T	12.56	6.00	4.50	3.00	1.125	2.50	0.250	1.75
L182T	14.06	6.00	4.50	3.00	1.125	2.50	0.250	1.75
184T	13.56	7.00	5.50	3.00	1.125	2.50	0.250	1.75
L184T	15.06	7.00	5.50	3.00	1.125	2.50	0.250	1.75
213T	15.69	7.00	5.50	3.62	1.375	3.12	0.312	2.38
L213T	17.44	7.00	5.50	3.62	1.375	3.12	0.312	2.38
215T	17.19	8.50	7.00	3.62	1.375	3.12	0.312	2.38
L215T	18.06	8.50	7.00	3.62	1.375	3.12	0.312	2.38
254T	20.69	11.25	8.25	4.25	1.625	3.75	0.375	2.88
256T	22.44	11.50	10.00	4.25	1.625	3.75	0.375	2.88
284T	23.38	11.50	9.50	4.88	1.875	4.38	0.500	3.25
284TS	22.00	11.50	9.50	3.50	1.625	3.00	0.375	1.88
286T	24.88	13.00	11.00	4.88	1.875	4.38	0.500	3.25
286TS	23.50	13.00	11.00	3.50	1.625	3.00	0.375	1.88
324T	27.56	14.00	10.50	5.50	2.125	5.00	0.500	3.75
324TS	26.06	14.00	10.50	4.00	1.875	3.50	0.500	2.00
326T	27.56	14.00	12.00	5.50	2.125	5.00	0.500	3.75
326TS	26.06	14.00	12.00	4.00	1.875	3.50	0.500	2.00
364T	29.70	14.25	11.25	6.12	2.375	5.62	0.625	4.25
364TS	27.58	14.25	11.25	4.00	1.875	3.50	0.500	2.00
365T	29.70	14.25	12.25	6.12	2.375	5.62	0.625	4.25
365TS	27.58	14.25	12.25	4.00	1.875	3.50	0.500	2.00
404T	34.00	16.00	12.25	7.62	2.875	7.00	0.750	5.50
404TS	31.00	16.00	12.25	4.62	2.125	4.00	0.500	2.75
405T	34.00	16.00	13.75	7.62	2.875	7.00	0.750	5.50
405TS	31.00	16.00	13.75	4.62	2.125	4.00	0.500	2.75
444T	37.56	17.00	14.50	8.94	3.375	8.25	0.875	6.00
444TS	33.81	17.00	14.50	5.19	2.375	4.50	0.625	3.00
445T	39.56	19.00	16.50	8.94	3.375	8.25	0.875	6.00
445TS	35.81	19.00	16.50	5.19	2.375	4.50	0.625	3.00
447T	43.06	22.50	20.00	8.94	3.375	8.25	0.875	6.88
447TS	39.31	22.50	20.00	5.19	2.375	4.50	0.625	3.00
449T	48.06	27.50	25.00	8.94	3.375	8.25	0.875	6.88
449TS	44.31	27.50	25.00	5.19	2.375	4.50	0.625	3.00

NOTES: (1) Mounting hole diameter
Dimensions are for estimating purposes only

Typical Dimensions
C-Face
56C — 256TC

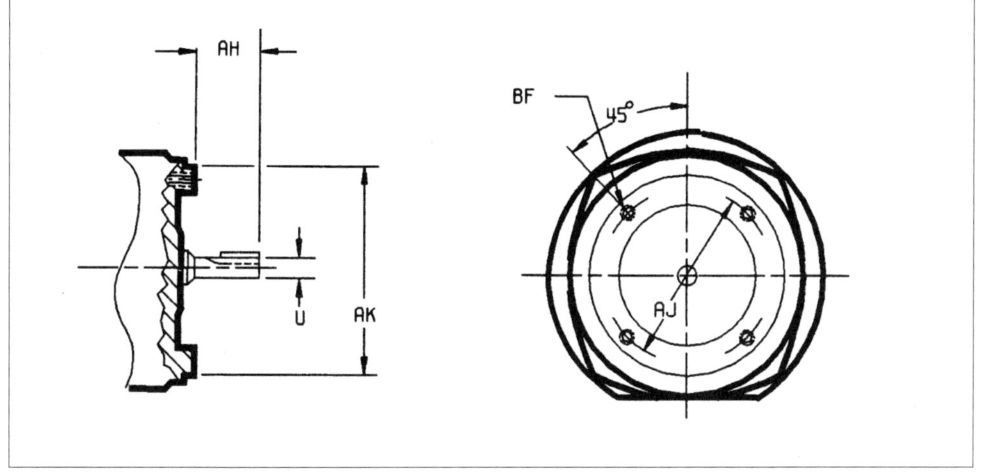

Frame Size	AH	AJ	AK	BF	U	Key Sq.	Key Length PROT	TEFC	XP
56C	2.06	5.88	4.500	3/8-16	0.625	0.188	1.25	1.25	1.25
140TC	2.12	5.88	4.500	3/8-16	0.875	0.188	1.38	1.25	1.38
182-184TC	2.62	7.25	8.500	1/2-13	1.125	0.250	1.78	1.75	1.75
213-215TC	3.12	7.25	8.500	1/2-13	1.375	0.312	2.41	2.38	2.38
254-256TC	3.75	7.25	8.500	1/2-13	1.625	0.380	2.91	2.91	2.91

Frame Size	BA (3)	C (3) PROT	TEFC	XP
56C (1)	2.56	same	same	same
56C (2)	2.75	+0.3	+0.3	+0.1
140C	2.75	same	same	-0.8
182TC	3.50	13.31	16.38	16.38
L182TC	3.50	14.81	17.88	17.88
184TC	3.50	14.31	16.38	16.38
L184TC	3.50	15.81	17.88	17.88
213TC	4.25	16.44	20.00	20.00
L213TC	4.25	18.19	20.88	20.88
215TC	4.25	17.94	20.00	20.00
L215TC	4.25	18.81	20.88	20.88
254TC	4.75	21.00	23.25	23.25
256TC	4.75	22.75	25.00	25.00

NOTES: (1) 5.5 frame, beginning with A, E or G
(2) 6.3 frame, beginning with B, F or H
(3) Refer to previous dimension pages
Dimensions are for estimating purposes only

NEMA Enclosure Descriptions[1]

Electrical Equipment (1000 volts maximum) • Non-classified locations

Type 1 Enclosures
Type 1 enclosures are intended for indoor use primarily to provide a degree of protection against contact with the enclosed equipment in locations where unusual service conditions do not exist. The enclosures shall meet the rod entry and rust-resistance design tests.

Type 2 Enclosures
Type 2 enclosures are intended for indoor use primarily to provide a degree of protection against limited amounts of falling water and dirt. These enclosures shall meet rod entry, drip, and rust-resistance design tests. They are not intended to provide protection against conditions such as dust or internal condensation.

Type 3 Enclosures
Type 3 enclosures are intended for outdoor use primarily to provide a degree of protection against wind-blown dust, rain, and sleet; and to be undamaged by the formation of ice on the enclosure. They shall meet rain, external icing, dust and rust-resistance design tests. They are not intended to provide protection against conditions such as internal condensation or internal icing.

Type 3R Enclosures
Type 3R enclosures are intended for outdoor use primarily to provide a degree of protection against falling rain; and to be undamaged by the formation of ice on the enclosure. They shall meet rod entry, rain, external icing, and rust-resistance design tests. They are not intended to provide protection against conditions such as dust, internal condensation, or internal icing.

Type 3S Enclosures
Type 3S enclosures are intended for outdoor use primarily to provide a degree of protection against wind-blown dust, rain, and sleet and to provide for operation of external mechanisms when ice laden. They shall meet rain, dust, external icing, and rust-resistance design tests. They are not intended to provide protection against conditions such as internal condensation or internal icing.

Type 4 Enclosures
Type 4 enclosures are intended for indoor use primarily to provide a degree of protection against wind-blown dust and rain, splashing water, and hose-directed water; and to be undamaged by the formation of ice on the enclsoure. They shall meet hose-down, external icing, and rust-resistance design tests. They are not intended to provide protection against conditions such as internal condensation or internal icing.

Type 4X Enclosures
Type 4X enclosures are intended for indoor or outdoor use primarily to provide a degree of protection against corrosion, windblown dust and rain, splashing water, and hose-directed water; and to be undamaged by the formation of ice on the enclosure. They shall meet the hosedown, external icing, and corrosion-resistance design tests. They are not intended to provide protection against conditions such as internal condensation or internal icing.

Type 5 Enclosures
Type 5 enclosures are intended for indoor use primarily to provide a degree of protection against dust and falling dirt. They shall meet the dust and rust-resistance design tests. They are not intended to provide protection against conditions such as internal condensation.

Tyep 6 Enclosures
Type 6 enclosures are intended for indoor or outdoor use primarily to provide a degree of protection against the entry of water during temporary submersion at a limited depth; and to be undamaged by the formation of ice on the enclosure. They shall meet submersion, external icing, and rust-resistance design tests. They are not intended to provide protection against conditions such as internal condensation, internal icing, or corrosive environments.

Type 6P Enclosures
Type 6P encosures are intended for indoor or outdoor use primarily to provide a degree of protection against the entry of water during prolonged submersion at a limited depth; and to be undamaged by the formation of ice on the enclosure. They shall meet air pressure, external icing, and corrosion-resistance design tests. They are not intended to provide protection against conditions such as internal condensation or internal icing.

Type 11 Enclosures
Type 11 enclosures are intended for indoor use primarily to provide a degree of protection against the corrosive effects of liquids and gases. In addition, they protect the enclosed equipment against the corrosive effects of fumes and gases by providing for immersion of the equipment in oil. They shall meet drip and corrosion-resistance design tests. They are not intended to provide protection against conditions such as internal condensation or internal icing.

Type 12 Enclosures
Type 12 enclosures are intended for indoor use primarily to provide a degree of protection against dust, falling dirt, and dripping noncorrosive liquids. They shall meet drip, dust, and rust-resistance design tests. They are not intended to provide protection against conditions such as internal condensation.

Type 12K Enclosures
Type 12K enclosures with knockouts are intended for indoor use primarily to provide a degree of protection against dust, falling dirt, and dripping noncorrosive liquids other than at knockouts. They shall meet drip, dust, and rust-resistance design tests. Knockouts are provided only in the top or bottom walls, or both. After installation of the enclosure, the knockout areas shall meet environmental characteristics listed above. They are not intended to provide protection against conditions such as internal condensation.

Type 13 Enclosures
Type 13 enclosures are intended for indoor use primarily to provide a degree of protection against dust, spraying of water, oil, and noncorrosive coolant. They shall meet oil exclusion and rust-resistance design tests. They are not intended to provide protection against conditions such as internal condensation.

Classified Location Enclosures

Type 7 Enclosures
Type 7 enclosures are for indoor use in locations classified as Class I, Groups A, B, C, or D, as defined in the *National Electrical Code.*

Type 8 Enclosures
Type 8 enclosures are for indoor or outdoor use in locations classified as Class I, Groups A, B, C, or D, as defined in the *National Electrical Code.*

Type 9 Enclosures
Type 9 enclosures are intended for indoor use in locations classified as Class II, Groups E or G, as defined in the *National Electrical Code* (Group F was reinstated in the 1987 NEC).

Type 10 Enclosures (MSHA)
Type 10 enclosures shall be capable of meeting requirements of the Mine Safety and Health Administration, *30 C.F.R.,* Part 18.

[1] Descriptions are excerpts from NEMA's "Standards Publications/No. 250—1985"

Hazardous Locations
Interpretation of NEC Definitions of Hazardous Locations: Class I, II and III, Groups A, B, C, D, E, F and G, Division 1 and 2

Classes — I, II and II: NEC 500-5, 500-6 and 500-7

Class I Locations (Gases)
An area where flammable gases or vapors are or may be present in the air in quantities sufficient to produce explosive or ignitible mixtures.

Class II Locations (Dust)
An area where presence of combustible dust presents a fire or explosion hazard.

Class III (Fibers)
An area made hazardous because of the presence of easily ignitible fibers or flyings, but on which such fibers or flyings are not likely to be in suspension in the air in quantities sufficient to produce ignitible mixtures.

Groups — A, B, C, D, (Class I); E, F and G (Class II): NEDC 500-3

Groups A, B, C and D (Class I)
Combustible and flammable gases and vapors are divided into four groups, the classification involving determinations of maximum explosion pressures, maximum safe clearance between parts of a clamped joint in an enclosure and the minimum ignition temperature of the atmospheric mixture.

There is no consistent relationship between the Group A, B, C or D classification and flash point/ignition temperature/explosive limits. Instead, the Groups are classified by chemical families. Certain chemicals create higher explosive pressures and heat when ignited. Generally speaking, Group A gas creates the most pressure during an explosion — and therefore is the most difficult to control. Group B is next highest in pressure; then Group C; and last Group D. This explains why a Group A or B listing is more difficult to obtain than a Group C or D listing for electrical equipment.

Groups E, F and G (Class II)
Combustible dusts are divided into these three Groups, the classification involving the tightness of the joints of assembly and shaft openings, the blanketing effect of layers of dust on the equipment that may cause overheating, the electrical conductivity of the dust and the ignition temperature of the dust.

Group E Atmospheres
Group E atmospheres contain metal dust, including aluminum, magnesium, and their commercial alloys, and other metals of similarly hazardous characteristics having resistivity of less than 10^5 ohm-cm.

Group F Atmospheres
Group F atmospheres contain combustible carbonaceous dusts, or other atmospheres containing these dusts sensitized by other hazardous materials, and having resistivity greater than 10^2 thru 10^8 ohm-cm.

Group G Atmospheres
Group G Atmospheres contain combustible dusts having resistivity of 10^5 ohm-cm, or greater.

Equipment used in these atmospheres must not only be approved for Class I, II or III, but also for the specific group (Class I and II).

Divisions — 1 and 2 (Class I, II and III)

Division I
NEC 500-5 (a), 500-6 (a) and 500-7 (a)

Class I, Division I
This is an area where the hazard exists under *normal* operating conditions. These situations include transferring flammable or combustible liquids from one container to another, open vats, paint spray booths or any location where ignitible mixtures are used. This also includes locations where the hazard is caused by *frequent* maintenance or repair work or *frequent* equipment failure.

Class II, Division 1
This is an area where combustible dust is normally in the air in sufficient quantities to produce ignitible mixtures or where mechanical failure or abnormal operation of equipment might produce ignitible mixtures. These locations also include (1) operations where this hazard exists because of frequent mechanical failure of machinery or equipment and (2) where electrically conductive combustible dusts (all Group E and some Group F) are present in hazardous quantities.

Class II, Division 1
This is an area where combustible dust is normally in the air in sufficient quantities to produce ignitible mixtures or where mechanical failure or abnormal operation of equipment might produce ignitible mixtures. These locations also include (1) operations where this hazard exists because of *frequent* mechanical failure of machinery or equipment and (2) where electrically conductive combustible dusts (all Group E and some Group F) are present in hazardous quantities.

Class III, Division I
This is an area where easily ignitible fibers or materials producing combustible flyings are handled, manufactured or used.

The Role of Partitions in Division I and Division II Locations
In most indoor areas with adequate partitions, Div. 1 and Div. 2 are self-contained areas. With partitions, a Div. 1 area may, for example, exist adjacent to a non-hazardous area.

However, in outdoor areas of large indoor areas where there are few or no partitions, Class I, Div. 1 and Class II, Div. 2 areas characteristically exist adjacent to each other — the Div. 1 location being near the point of vapor release and Div. 2 being at a given distance from the point of release from the flammable liquid. In these areas where the spread of flammable vapors and gases is not contained by adequate partitions, Class I, Div. 2 can be thought of as a "transition zone" between Class I, Div. 1 and the non-hazardous area. Class I, Div. 1 is a hazardous area where flammable gases or vapors are released from the liquid. Further away from the point of release, the gases or vapors are not normally of sufficient concentration to produce an ignitible mixture — and so such an area is designated Class I, Div. 2. This Class I, Div. 2 area is sometimes referred to as the "transition zone." Outside this Div. 2 "transition zone" is the non-hazardous area.

NOTE: Electrical Equipment Approved for Div. 1 is also suitable for Div. 2 per NEC 500-3 (A).

Noise Exposure & Decibels

Noise may be defined as any pressure variation (in air, water or other medium) that the human ear can detect. The number of pressure variations per second is called the frequency of the noise, and is measured in Hertz (Hz).

The second main quantity used to describe a noise is the size or amplitude of the pressure fluctuations. Noise amplitude can be described as the Sound Pressure Level (SPL) and is measured in decibels (dB).

Decibels are measured on three scales, A, B, and C, of intensity. The A scale is used in measuring noise intensity in the workplace and is expressed in decibels as dBA.

Environmental noise involves measurement of the total noise (irrespective of its source) at a particular location. The noise may thus be due to one or more sources and may also include reflections from walls, ceilings and other machines.

Noise at an employees work station is an example of environmental noise. The measurement is made where the person works.

Noise Exposure Chart

ISO Criterion level for maximum allowable noise dose per 40-hr week. (10–80 %)

OSHA Criterion level for maximum allowable exposure time per day. (16—0 to 9—11)

ISO Noise dose %	dBA	OSHA Max. time (hrs. - min.)
10	80	
15	82	
20	83	
25	84	
30	85	16—0
40	86	13—56
50	87	12—8
60	88	10—34
80	89	9—11
100	90	8—0
125	91	6—56
160	92	6—4
200	93	5—17
250	94	4—36
315	95	4—0
400	96	3—29
500	97	3—2
630	98	2—50
800	99	2—15
1,000	100	2—0
1,250	101	1—44
1,600	102	1—-31
2,000	103	1—19
2,500	104	1—9
3,150	105	1—0
4,000	106	0—52
5,000	107	0—46
6,300	108	0—40
8,000	109	0—34
10,000	110	0—30
12,500	111	0—26
16,000	112	0—23
20,000	113	0—20
25,000	114	0—17
31,500	115[1]	0—15 or less

[1] Maximum steady level permitted by OSHA

Dynamic Range of Human Ear

Sound pressure in dB	Common sounds
160	Medium jet engine
150	Large propeller aircraft
140	Air raid siren
130	Riveting and chipping
120	Loud music
110	Punch press
100	Canning plant
90	Heavy city traffic
80	Busy office
60	Normal speech
50	Private office
40	Quiet residential neighborhood
20	Whisper
0	Threshold of hearing

The dB chart above shows examples of various sources of common sounds and their dB equivalents over the dynamic range of the human ear.

Noise Reduction

Employers must now take the proper steps to avoid hearing loss, not just through limited exposure, but through machine and hydraulic equipment noise reduction. Listed below are some noise reduction strategies that have been effective for other companies and may be effective for you.

A major cause of noise exposure to machine operators is due to pumps. Noise is amplified by speed. An effective method of noise reduction, is to decrease pump speeds. Operating pumps at 1200 rpm will significantly reduce pump noise. Many manufacturers can now supply pumps with substantially reduced noise levels.

Isolate pump vibrations from other parts of the system to prevent sounding board effects. Attach the pump to an end-bell-mounting electric motor and support motor feet on vibration absorbing mounts. Install a short piece of flexible hose in the line to the pump (close to the pump) and also in the conduit to the electric motor. Short flexible hose "whip" connections isolate vibration as well as noise.

Fluid circulation through valves is another noise generator. Fluid circulation through these valves should be kept at manufacturer's recommended levels. Noise generated by fluid flow through valves or restrictions has been found proportional to the sixth or seventh power of the velocity. Oil velocity equals noise.

In order to keep noise from seeping through the floor, walls, etc., mount hydraulic machinery on vibration isolators. Piping should be stabilized to prevent rattling and noise transmission. Also, by isolating pump vibrations from other parts of the system, noise can be avoided.

Another source of noise from pumps is due to pump cavitation. Decreased suction lift and shorter suction line will avoid noise. Suction filter capacities are usually rated at normal operating temperatures but may be inadequate at lower temperatures due to increased viscosity.

Noise is somewhat related to the horsepower involved. These suggested practices ordinarily produce acceptable noise levels up to about 40 hp but they may not suffice for higher power systems. Shielding against airborne noise may then be necessary. This can be done with partitions or enclosures constructed of sound absorbers and/or sound barriers. Sound absorbers are lightweight and cellular or fibrous, such as glass wool or polyurethane foam. Sound barriers are flexible, high mass materials.

A sheet steel enclosure lined with a thin sheet of lead sandwiched between two layers of sound absorbent such as polyurethane can be used over a noisy pump. Ventilation openings may be necessary and they will require shielding or baffling.

A single wall or partition, with no openings, placed between operator and noise source, constructed of two sheet skins treated with auto undercoating, mastic or damping tape on the inside, and the space between them filled with sound absorbent, will reduce noise about 5 to 10 dBA at operator's position. Using 1/64" sheet lead laminated to inside of skins instead of mastic will give about 10 to 15 dBA reduction. However, sound can be reflected from walls and ceiling to bypass the partition. A three sided enclosure with top would give better shielding and still provide access and ventilation.

A well designed, complete enclosure with walls of plywood or sheet steel, including lead and sound absorbent, will give about 20 to 35 dBA improvement outside the enclosure. Ventilation and other openings should be through acoustically lined ducts. Access doors and windows must be tight fitting and of equivalent insulation. Any crack through which air can move will provide a substantial noise leak.

Background Noise (Subtracting Sound Levels)

One factor that may influence the accuracy of measurements is the level of the background noise compared to the level of the sound being measured. Obviously, the background noise must not "drown out" the sound of interest. In practice, this means that the level of the sound must be at least 3 dB higher than the background noise. However, a correction may still be necessary to get the correct result. The procedure for measuring the sound level from a machine under conditions of background noise is as follows:

1. Measure the total noise level (L_S) with the machine running.
2. Measure the background noise level (L_N) with the machine turned off.
3. Find the difference between the two readings (L_S - L_N). If this is less than 3 dB, the background noise level is too high for an accurate measurement. If it is between 3 and 10 dB, a correction will be necessary. No correction is necessary if the difference is greater than 10 dB.
4. To make corrections, the chart shown at the top of the next column may be used. Enter the bottom of the chart with the difference value (L_S - L_N) from step 3, go up until you intersect the curve and then go to the vertical axis on the left.
5. Subtract the value on the vertical axis (ΔL_N) from the total noise level in step 1. This gives sound level L_S of the machine.

Addition of Sound Levels

If the sound levels from two or more machines have been measured separately and you want to know the total SPL made by the machines when operating together, the sound levels must be added. However, dBs cannot just be added together directly (because of the logarithmic scale). Addition of dBs can be done simply using the chart at the top of the next column:

1. Measure the SPL of each machine separately (L_1, L_2).
2. Find the difference between these levels (L_2 - L_1).
3. Enter the bottom of the chart with this difference. Go up until you intersect the curve, then go to the vertical axis on the left.
4. Add the value indicated (ΔL) on the vertical axis to the level of the noisier machine (L_2). This gives the sum of the SPL's of the two machines.
5. If 3 machines are present, repeat steps 1 to 4 using the sum obtained for the first two machines and the SPL for machine 3.

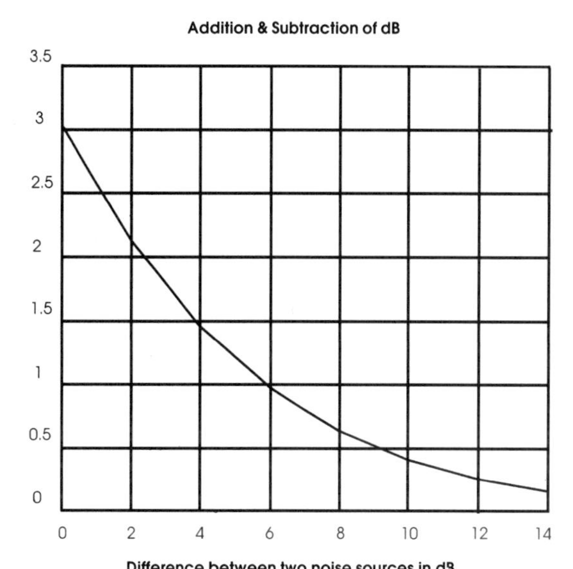

For addition, add this value to the higher of the two noise sources to obtain to the total dB.

For subtraction, subtract this value from the total noise to obtain the dB of the higher of the two noise sources.

Example:

1. Machine 1	L_1 = 82 dB
Machine 2	L_2 = 85 dB
2. Difference	L_2 - L_1 = 3 dB
3. Correction (from chart)	ΔL = 1.7 dB
4. Total Noise	= 85 + 1.7 = 86.7 dB

For addition of several equal noise sources use the chart below.

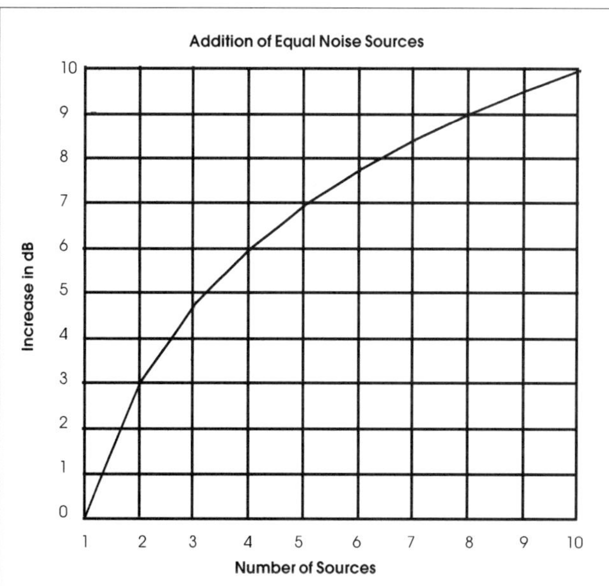

Example: 4 hydraulic motors each at 82 dB

From the chart, the increase in dB for 4 sources is 6 dB. Total dB for the 4 motors is 82 + 6 = 88 dB.

Noise Measurement Techniques

The following information on Noise Measurement Techniques, the Noise "Glossary of Terms" and the Explanation of Time-Weighted Sound Levels has been reproduced with permission of the "AMT — The Association For Manufacturing Technology".

10. MEASUREMENT

These techniques establish a minimum of control in acquiring accurate sound level data on a machine tool. They assume that measurements are made in a suitable test space under the control of the machine tool builder. Suitability of the test space is left to the builder since errors due to high ambient sound levels and acoustic reflections, while prejudicial to the sound level test results, may not warrant the effort needed to eliminate them.

10.1 GENERAL PROVISIONS

10.1 Ambient Noise Levels

To obtain an accurate measure of the sound level produced by a machine:

a. The ambient sound level at all microphone locations should preferably be at least 10 dB(A) lower than the level occurring when the machine is operating.

b. When the ambient sound level is less than 10 dB(A) lower than the level occurring when the machine is operating, correction factors may be applied to the sound level measurements in accordance with the chart below.

dB(A) Increase In Sound Level, Above Ambient, Produced By Machine Operation	dB(A) To Be Subtracted From Measured Sound Level
3 or less[1]	3
3 to 6	2
6 to 9	1
10 or more	0

[1] Caution must be exercised when working with this small differential. This can have a significant effect on the repeatability of the measurements.

c. If the application of correction factors is pertinent, the ambient sound level should remain steady within ±1 dB(A) for the duration of the test. Should the ambient level vary, the minimum applicable correction factor shall be applied.

10.1.2 Test Space

A plot plan may be included on the data sheet (Figure 2, page 128) following the example given in Figure 1. It should describe the test space, including all major reflecting surfaces such as walls, cabinets, control panels, etc., within 5 meters of the machine envelope. A brief description of the major reflecting surface materials may be provided.

10.1.3 Machine Operating Conditions

Measurements shall be made with the machine operating in one or more of the modes described below as agreed upon the builder and the purchaser:

a. At the required production rate and performing specified operations when the machine is purchased for a specified purpose.

b. In the unloaded mode of operation that generates maximum noise levels, when the machine is purchased without specified tooling.

c. At specified simulated load(s) typifying machine performance when the machine will be used for a broad range of purposes.

10.1.4 Measurement Locations

a. The measurement envelope shall be located 1 meter from the projected floor plan of the machine. Test points shall be located on the envelope at a height of 1.5 meters above floor level when the machine is installed. (See Figure 1).

b. Test points shall be located at the operator position(s) and on the measurement envelope.

c. Operator position(s) shall be defined as the operator(s) ear location(s) during normal machine operation.

10.1.5 Measurements

a. Sound level measurements shall be made at sufficient locations on the measurement envelope to determine the directional characteristics of the machine sound.

b. Data shall be taken at the measurement location at which the highest sound level exists. (Mandatory data)

c. Additional data may be taken of sound levels at operator position(s) and at other locations on the boundary of the measurement envelope to satisfy particular requirements.

d. The observer(s) shall remain a minimum of 0.5 meter away from the microphone.

10.1.6 Instrumentation Calibration

a. Field calibrations of the measuring system shall be performed at the beginning and at the end of each day. If the difference between the two calibrations shows more than a 1 dB deviation, the data acquired that day shall be invalid.

b. It is recommended that the accuracy of the calibrator be certified annually in accordance with the manufacturer's specifications.

10.1.7 Microphone

The microphone shall be oriented to have the incidence, either normal or grazing, that provides the flattest frequency response for the microphone being used. No microphone frequency response corrections shall be used.

10.1.8 Tape Recording

When sound levels are tape recorded for record or measurement purposes, a comparison between the original and recorded sound levels shall be made, preferably with head phones during recording, to determine that clipping does not occur and that recording noises do not distort the desired signal.

10.2 STEADY NOISE

Noise that is essentially constant in level, having a range of variation of 3 dB(A) or less with the meter set on slow response, shall be considered to be steady noise.

The average A-weighted sound level, dB(A), shall be measured with the meter on slow response at each microphone location.

10.3 NON-STEADY NOISE

10.3.1 Noise that varies in level over a range greater than 3 dB(A), with the meter set on slow response, shall be considered to be non-steady noise.The minimum and maximum A-weighted sound levels, dB(A), shall be measured with the meter on slow response at each microphone location.

Where noise levels repetitiously change in discrete increments, as during a machine duty cycle, the levels, durations, and associated events should be recorded for each distinct interval in order that a time-weighted average can be calculated. The time-weighted average may also be measured directly.

10.4 IMPULSIVE NOISE

Noise that consists of bursts of less than one-half second duration (with a rise time of not more than 35 milliseconds), either as single events or as repetitve events with greater than one second intervals between them, shall be considered to be impulsive noise.

10.4.1 The peak sound pressure level and frequency of occurrence of at least 10 noise bursts shall be measured and the maximum level reported.

10.4.2 Weighting networks or filters shall not be used when measuring impulsive peak levels.

11. INSTRUMENTATION

11.1 GENERAL

11.1.1 A condensor microphone or its equivalent in accuracy, stability, and frequency response is recommended. Where a cable length of 3 meters or more is required to couple a ceramic type microphone to the instrumentation, a preamplifier, located at the microphone, shall be used. A microphone providing the flattest response to a noise source located normal to it, in a free field, is preferable because it reduces the probability of errors due to extraneous noise sources.

11.1.2 When a tape recorder is used for impulsive noise measurement, a recorder of instrumentation quality is recommended. A playback and recording system frequency response that is uniform over the range of 45 to 11,200 Hz is preferred.

11.2 STEADY OR NON-STEADY NOISE

Sound levels shall be measured with a Type 1 or Type 2 Sound Level Meter meeting the requirements of ANSI S1.4-1971 except that it is not required to have a B-weighted filter. Suitable meters for measuring time-weighted averages shall have similar accuracy.

11.3 IMPULSIVE NOISE

11.3.1 Peak sound pressure level measurements may be taken with any peak reading sound level meter system having a rise time of 100 microseconds or less.

11.3.2 Peak measurements may also be taken with an oscilloscope. The oscilloscope should be of the memory type, or photographs should be taken of the oscilloscope trace.

12. REPORTING

12.1 DATA FORM

The sample Data Form illustrated (Figure 2, page 128) is intended as a guide for recording sound level measurement data.

12.2 REPORTED DATA

12.2.1 The data form is general in nature and may be used to report Mandatory or Supplemental Data as described in the Measurements section. The reporting of Mandatory Data fulfills the minimum requirements of this document.

12.2.2 Load conditions will be one of those described in the Machine Operating Conditions section.

12.2.3 Test points shall correspond with those labeled on the sketch.

12.2.4 The ambient sound level should be recorded at each test point since it indicates whether a correction factor may be applied to the tabulated data.

12.2.5 The minimum and maximum sound levels occurring during a machine cycle should be recorded. The character of the noise (steady or non-steady) will thus be shown.

12.2.6 When correction factors are applied in accordance with the second bulleted point in the Ambient Noise Level section, corrected minimum and maximum dB(A) shall be recorded.

12.2.7 Where impulsive noise exists, its amplitude and frequency of occurrence should be recorded.

12.2.8 When non-steady noise exists, a time-weighted average may be measured directly or may be calculated from data obtained during the machine cycle(s).

12.3 EXPLANATION OF SAMPLE DATA FORM

Page 128 summarizes Machine Specifications, Instrumentation and Certification

12.4 MEASUREMENT ENVELOPE

The machine outline and measurement envelope shall be indicated with a sketch on page 128 of the report form. All microphone locations shall be identified on the sketch. Major reflecting surfaces in the test area may be described at the builder's option.

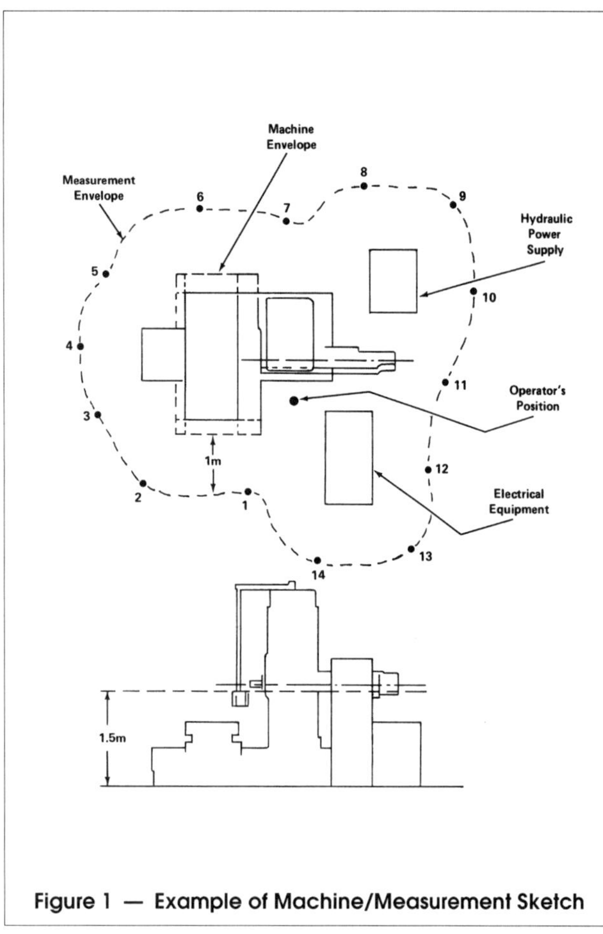

Figure 1 — Example of Machine/Measurement Sketch

Machine Identification No. ____________

Company Name ____________

NOISE MEASUREMENT DATA

A. MACHINE SPECIFICATIONS

Builder ______ Builder's No. ______ Buyer's P.O. No. ______

Equipment Specification: Type ________ Model ________

Serial No. ______ Size ________ Capacity ________

Speed ______ Horsepower ______ Auxiliaries ________

B. INSTRUMENTATION

Instrument	Model	Serial No.
Sound Level Meter		
Microphone		
Calibrator		
Tape Recorder		
Impact Meter		
Oscilloscope		
Graphic Level Recorder		

C. COMMENTS

____

____

____

By:

Name ____________

Position ____________

Company ____________

Date ____________

Figure 2 — Sample Form (Page 1 of 3)

Machine Identification No. ________

D. SKETCH OF MACHINE AND MEASUREMENT ENVELOPES

Test Location ____________

Test Space Description ____________

____

____

Figure 2 — Sample Form (Page 2 of 3)

Machine Identification No. ________

E. TEST DATA — TYPE: ________

Observer: ____________ Date: ________

Load Conditions (Describe) ____________

____

____

____

Test Point	Ambient dB(A) (slow)	Minimum dB(A) (slow)	Corrected Minimum dB(A)	Maximum dB(A) (slow)	Corrected Maximum dB(A)	Impulse	Frequency Impulses Per Hour	dB(A) Steady State Time-Weighted Average*

Remarks: ____________

____

____

____

____

*Basis For Time-Weighting: ________ dB(A) per doubling of time.

Figure 2 — Sample Form (Page 3 of 3)

Glossary of Terms

ACOUSTIC REFLECTIONS
Sound waves reflected by surfaces in the area of the machine being measured and having a direct effect on the measurement levels.

AMBIENT SOUND LEVEL (Background sound)
The sound level in the area surrounding the machine or component to be tested with machine being tested not operating.

CALIBRATE
To compare and adjust the accuracy of instrumentation by making comparative readings against a verified accurate source of known value.

CLIPPING (Recording)
Where the reproduced signal amplitude varies from the input signal effective amplitude by more than one dB, due to electronic compression or magnetic saturation.

DECIBEL (dB)
A non-dimensional number used to express sound pressure and sound power. It is a logarithmic expression of the ratio of a measured quantity to a reference quantity.

For example, two machines side by side, each separately producing a sound pressure level of 80 dB, do not produce a combined pressure levl of 160 dB when both are operating. They produce a combined sound pressure level of approximately 83 dB.

For sound pressure level the reference quantity is 20 microPascal[1] so that: Sound Pressure Level in dB = 20 $\log_{10}$ (measured pressure/20 x 10^{-6}).

For sound power level the reference quantity is 10^{-12} watt so that: Sound Power Level in dB = 10 $\log_{10}$ (measured power/ 10^{-12}).

Where a weighted network filter is employed in making sound pressure measurements this is indicated by a suffix added to the unit symbol, i.e., when an A-weighting network is used the symbol dB(A) is used.

[1] Confusion exists because the sound pressure reference is stated in several different forms in acoustic literature. The forms that follow are equivalent.

20 microPascal	.0002 dynes/cm squared
20 microNewtons/meter squared	.0002 microbar (microatmosphere)

dB(A)
A sound level reading in decibels made on the A-weighted network of a sound level meter.

DIRECTIONAL CHARACTERISTICS
The characteristics of sound or instrumentation that will determine the effective direction of maximum response from a fixed source.

FIELD CALIBRATION
The verification and adjustment of accuracy of the instrumentation used in the course of sound level measurement to be performed in a manner recommended by the manufacturer of the specific instrument.

FLAT RESPONSE
The characterization of a microphone, instrument or recorder having a sensitivity or response that is constant regardless of frequency.

FREE FIELD
An environment that, acoustically, is free of extraneous factors that will influence the result of the measurement of a specific sound generator.

GRAZING INCIDENCE
Microphone positioned so that its axis is perpendicular to a line from the microphone to the noise source.

HERTZ (Hz)
The unit of frequency which replaces the term "cycles per second." Most standardizing agencies have adopted "Hertz" as the preferred unit of frequency.

IMPACT NOISE
Same as impulsive noise.

IMPULSIVE NOISE
Noise that consists of bursts of less than one-half second duration (with a rise time of not more than 35 milliseconds), either as single events or as repetitive events with greater than one second intervals between them.

MACHINE AND MEASUREMENT ENVELOPE
The space occupied by a machine tool. The boundary of the measurement envelope is a reasonably uniform line around the periphery of the machine approximately 1 meter out from machine components. The measurement envelope extends up and over the top and down and under the bottom of machines. For machines mounted on long runways, the measurement envelope shall contain only "active" portions of the machine, excluding vacant portions of the runway.

MACHINE TOOL
A power-driven machine, not portable by hand, used to shape or form metal by cutting, impact, pressure, electrical techniques or a combination of these processes.

MACHINE TOOL BUILDER
Any individual, partnership, corporation or other form of enterprise which is engaged in the development, manufacture or assembly of a machine tool.

NORMAL INCIDENCE
Microphone positioned so it is pointing toward sound source.

OBSERVER
The person(s) assigned to measure and record data on the noise measurement data form.

OPERATOR
Any person(s) assigned to function within or at periphery of the measurement envelope shall be considered an operator.

OPERATOR'S NORMAL POSITION
The location of the operator's head with respect to the machine tool during the time the machine is operating. The position should be recorded for the particular machine installation. More than one position may be necessary.

REFLECTING SURFACE
A surface oriented to the sound so that sound waves are modified by or reflected from the surface.

RISE TIME
The time rate at which the sound pressure level (or other stated characteristic) increases.

SLOW RESPONSE
A selectable mode of operation of a sound level meter or analyzer in which the indicator has high damping and therefore responds slowly to a change in level. This mode tends to provide an average reading.

SOUND LEVEL (Noise level)
Weighted sound pressure level obtained by the use of a metering characteristic and the weightings as specified in the reference standards. The weighting employed must be indicated. Unit: decibel (dB).

STEADY NOISE
Noise that is essentially constant in level, having a range of variation of 3 dB(A) or less with the meter set on slow response.

TEST POINT
The actual physical location of measurement instrumentation during the measurement and recording of data on the noise measurement data form.

TEST SPACE
An area in the machine tool builder's facility occupied by the machine tool when tests are made.

TIME WEIGHTED AVERAGE
The numerical value determined by calculation or measurement that defines the effect of discrete increments or changing sound levels and duration in terms of equivalency to constant steady-state sound.

WEIGHTING FILTERS OR NETWORKS
Components of circuitry in sound level meters that provide prescribed frequency response.

Explanation of Time-Weighted Sound Levels

Manufacturers quite often use the A-weighted sound level of a machine to determine an employee's exposure to hazardous noise levels in relation to the requirements of the Occupational Safety and Health Administration. In these regulations the duration of exposure is considered as well as the level. How then should the varying noise levels of a cyclical machine be reported as a single number which can be related to noise exposure per OSHA? To solve some of the difficulties in reporting variable machine sound levels a time-weighted sound level can be a suitable descriptor.

A special time-weighted average, the equivalent sound level (L_{eq}), was developed as a solution to the problem of measurement of the variable sound levels associated with aircraft operations. The equivalent sound level is a steady-state sound level containing the same energy as a time-averaged variable sound level measured over some time period.

The equivalent sound level has received wide acceptance in Europe as a means of determining occupational noise exposure and is included in the International Organization for Standardization (ISO) Recommendation 1999 (Acoustics-Assessment of Occupational Noise Exposure for Hearing Conversation Purposes). Because energy is the basis for determining the equivalent sound level, a three decibel change in level represents a doubling or halving of energy. Hence in ISO R1999, if the allowable exposure to a A-weighted sound level of 90 dB is eight hours, an exposure to 93 dB would be allowed for four hours. This is known as a 3 dB "per doubling", or a 3 dB "trade-off". The 3 dB trade-off is also used by the U.S. Environmental Protection Agency (EPA).

The standard promulgated by OSHA does not use a 3 dB trade-off but uses, instead, a 5 dB trade-off for noise exposure. This means with an allowable exposure of eight hours to an A-weighted sound level of 90 dB, the allowable exposure at four hours would be 95 dB.

As a result, there are problems in properly describing fluctuating noise levels and in relating these data to noise exposure in terms of either the OSHA or EPA requirements.

To calculate the equivalent sound level of a noise that is varying over any time period, using the 3 dB trade-off, one of the following formulae may be used:

$$(1)\quad L_{eq} = 10 \log_{10}\left(\frac{1}{T}\int_0^T \left(\frac{p}{p_0}\right)^2 dt\right)$$

or

$$(1a)\quad L_{eq} = 10 \log_{10}\left(\frac{1}{T}\int_0^T 10^{\frac{L_{pi}}{10}} dt\right)$$

where

L_{eq} = equivalent sound level
T = time over which the sound is measured
p = sound pressure
p_0 = reference sound pressure
L_p = sound level (A-weighted with slow meter response for the purpose of discussion in the section).

Formula (1) is exact and can be used to calculate the eqivalent sound level no matter how the level varies with time, while formula (1a) is only a close approximation. For many practical situations, the sound level can be approximated as a series of discrete steps and the following formula can be used:

$$(2)\quad L_{eq} = 10 \log_{10}\left(\frac{1}{T}\sum_{i=1}^{n} 10^{\frac{L_{pi}}{10}} \times t_i\right)$$

where

L_{pi} = sound level (constant) over time interval t_i
t_i = length of time interval i
T = $t_i + t_2 + \ldots + t_n$

These formulae give the EPA equivalent sound level, i.e., 3 dB trade-off. For example, assume a sound level of 97 dB existing for one hour followed by a level of 50 dB for a second hour. Substituting 97 into equation (2) for L_{pi}, 50 for L_{p2}, setting t1 = 1, t2 = 1 and T = 2, an equivalent sound level of 94 dB is found. This illustrates the halving and doubling and 3 dB trade-off since the second level is insignificant in relation to the first.

However, if one wants to evaluate machine tool noise in terms of the noise exposure as related to the OSHA standards, the previous formulae cannot be used because OSHA uses the 5 dB trade-off. To calculate the time-weighted average OSHA sound level the following formulae must be used:

$$(3)\quad L_{os} = 16.7 \log_{10}\left(\frac{1}{T}\int_0^T \left(\frac{p}{p_0}\right)^{3.32} dt\right)$$

or

$$(3a)\quad L_{os} = 16.7 \log_{10}\left(\frac{1}{T}\int_0^T 10^{\frac{L_{pi}}{16.7}} dt\right)$$

where

L_{os} = "OSHA time-weighted average" sound level

All of the other variables are the same as those defined above. Again, this can be approximated by a finite summation formula as follows:

$$(4)\quad L_{os} = 16.7 \log_{10}\left(\frac{1}{T}\sum_{i=1}^{n} 10^{\frac{L_{pi}}{10}} \times t_i\right)$$

Great caution must be used in using the term "equivalent" sound level. If one simply says "the equivalent sound level of a machine is 86 dB", it is understood to be the time-weighted average sound level with the EPA weighting which would be applicable to machines sold where ISO R1999 is used. Obviously there could be confusion when an identical machine is sold in the United States with one sound level (OSHA) weighted and sold in Europe with another sound level (ISO/EPA) weighted.

To calculate or determine the equivalent sound level, several methods are available. The first method is somewhat indirect but relatively simple. By using a dosimeter intended for measuring occupational safety and health exposure, it is possible to determine the percentage of allowable exposure by using the dosimeter to measure the fluctuating sound levels. Such a procedure may require a modification of the dosimeter to allow it to sense lower sound levels (below 85 dB) and also to respond to short time periods. After the allowable exposure has been measured for a number of machine cycles, the percent allowable exposure must be extrapolated to obtain the level for an eight-hour period. There are generally instructions with a dosimeter describing the extrapolation. It is then possible to determine the equivalent sound level of the machine on the EPA or OSHA bases from the graphs on pages 131 and 132.

An alternate method would be to use a meter designed to measure the equivalent sound level directly without the necessity of additional calculations. The meter would have to be designed for either a 3 dB or 5 dB trade-off or both.

The last method to determine equivalent sound level would be to calculate the sound level from the varying sound levels measured. This is fairly easy to do by using a graph which plots measured sound levels versus time. It can be done by using equations (2) or (4). Finally, the sound measuring system can be connected to a computer which is programmed to manipulate the levels measured and calculate the equivalent sound level.

The method of equivalent sound level measurement for fluctuating machine tool noise has great merit and is very useful. The only complication is the 3 versus 5 dB trade-off.

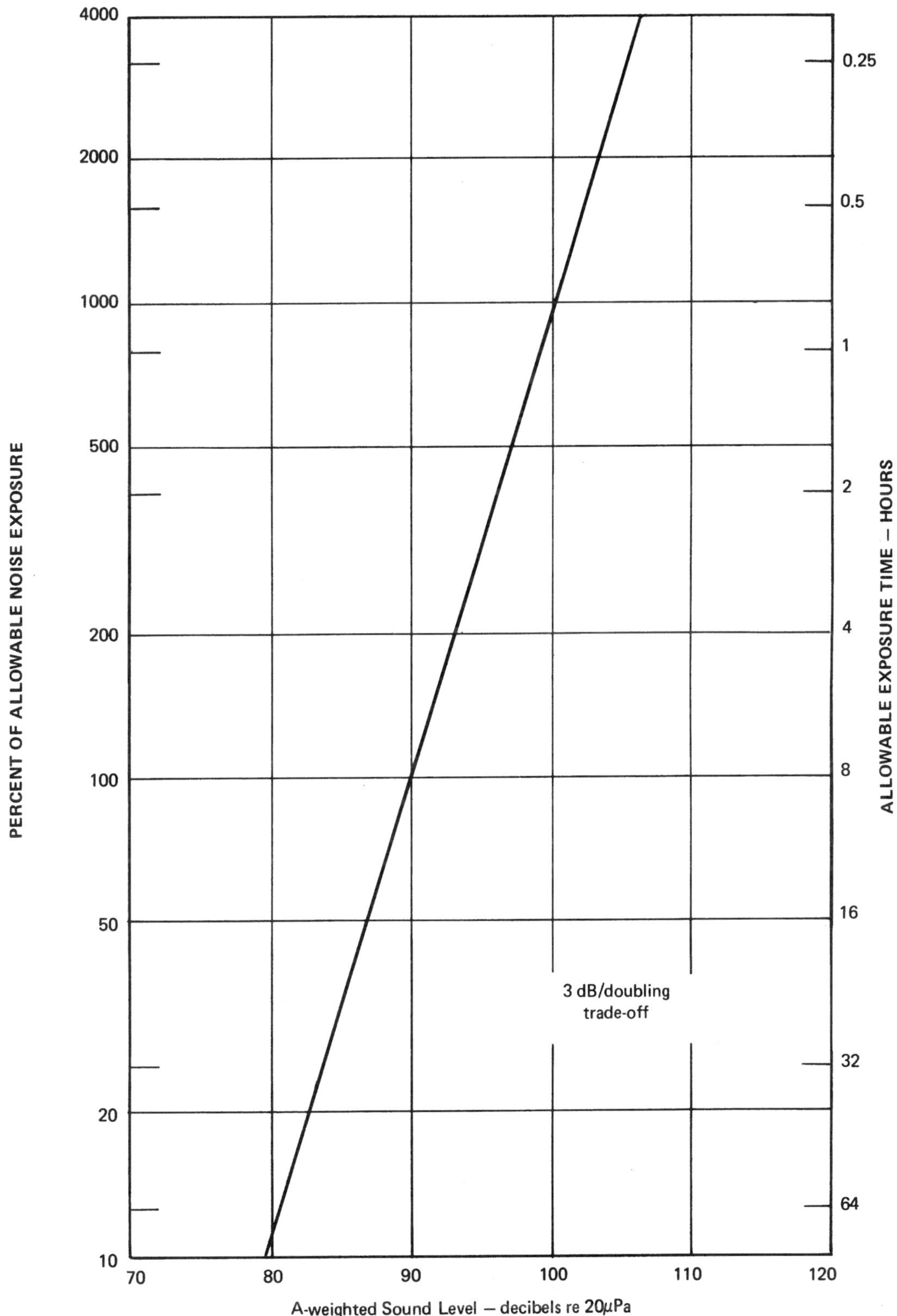

A-weighted Sound Level – decibels re 20μPa

Percent of allowable noise exposure and allowable exposure time based on 90 dB for eight hours

A-weighted Sound Level — decibels re 20μPa

Percent of allowable noise exposure and allowable exposure time based on 90 dB for eight hours

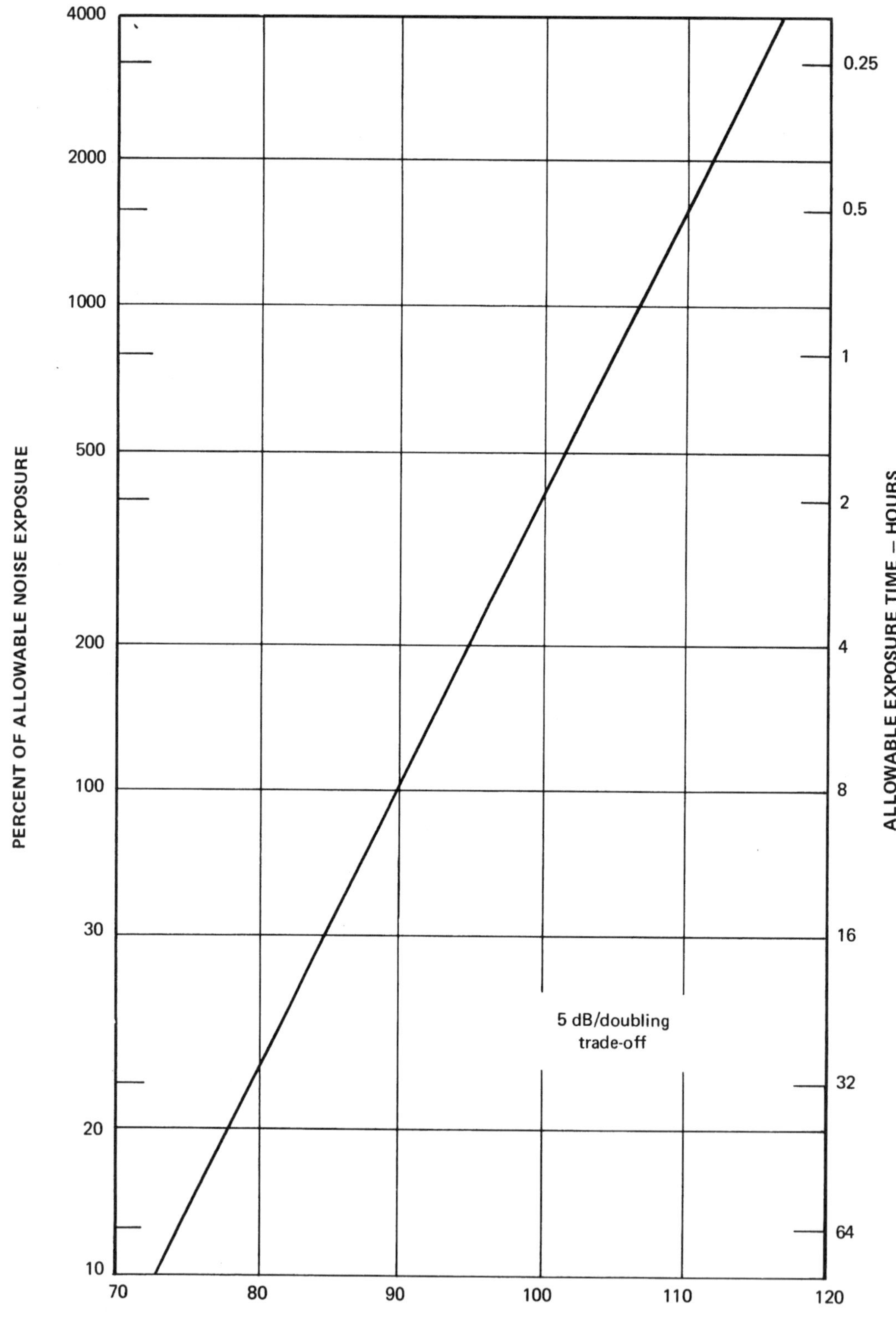

A-weighted Sound Level – decibels re 20μPa
Percent of allowable noise exposure and allowable exposure time based on 90 dB for eight hours

Safety Precautions Required For Hydraulic Machinery Operation and Maintenance

Although the scope of these instructions covers only the hydraulic operation of the equipment, these safety precautions also apply to pneumatically powered equipment and should be observed where appropriate.

Hydraulic equipment, like any other, may develop problems due to abuse, normal wear or unforeseeable circumstances. It, therefore, requires proper operation and maintenance. In the course of performing these functions, personnel may be required to work on or near the equipment. The following precautions are given to avoid injury to these personnel.

All safety requirements listed below are those generally applicable to hydraulically powered machinery, but do not pretend to be all inclusive. They are intended for qualified, experienced personnel who are capable of understanding the hazards of machinery operation and maintenance and, therefore, avoid injury by using the precautions. Particular types of machinery and hydraulic systems may require other precautions. Other precautions should be determined by someone in charge of the machinery who is capable of analyzing any hazards associated with operating and maintaining the equipment. These precautions should be included in the comprehensive safety program for the particular machinery, equipment, plant or process.

1. Return all movable machine members to their normal start-up condition if possible before starting the hydraulic power unit.

NOTE: In many types of equipment, parts of the machinery may start rotating, rising, falling, reciprocating, etc. out of their proper sequence as soon as the hydraulic (or pneumatic) circuit is filled and pressurized, which could result in injury to personnel or damage to machinery.

2. Be sure all personnel and product, workplace, etc. are clear of machinery before starting hydraulic power unit.

3. Be sure all hydraulic connections which may have been removed, replaced or disconnected during an equipment shutdown have been reconnected securely before starting hydraulic power unit.

4. Return all valves (manual and control system operated), which have been changed from their normal start-up conditions before starting hydraulic power unit.

5. Before shutting down hydraulic power unit, block up or lock in position any machine members which may move and cause damage to personnel, product or equipment upon loss of hydraulic flow and pressure.

6. Clear all personnel and product, work piece, etc. from machinery before shutting down hydraulic power unit.

7. If hydraulic system has oil accumulators in circuit, drain pressurized oil from all accumulators (if automatic drainage is not built into circuit) as soon as hydraulic unit is shut down. If accumulator has a shut-off valve, shut valve off also.

8. Shut down power unit and relieve pressure from all pressurized accumulators, actuators and line before removing, tearing down or performing maintenance on any remotely located actuators, hoses, filters, valves, piping, etc.

9. Maintain and keep in place any equipment guards, such as coupling guards, chain guards or protective cowlings. Do not wear loose clothing or jewelry that may get caught in moving parts.

10. If any personnel are required to work on equipment in the vicinity of the hydraulic system while the hydraulic power unit is running, they should always wear eye protection to prevent any eye injury in the event of a hydraulic line rupture and high velocity oil leak. (The above is in addition to any other personnel safety equipment needed for the work being performed).

11. Any personnel working near the equipment should wear ear protection if the noise level is high enough to require protection as set forth in OSHA regulations.

12. Any personnel observing or working on or adjacent to hydraulically powered equipment must never place themselves in a location or position that could produce an injury in the event of (a) a hydraulic line failure, (b) power blackout, (c) pump/motor failure or (d) movement of machine members during normal operating cycle as a result of component malfunction or failure.

13. Before removing or performing maintenance on any hydraulic system components that have an electrical interface (solenoid valves, switches, electric motors, etc.), shut off and padlock electrical power to power unit and/or control system. See paragraphs 5 and 12 before shut off of power. The above applies to pneumatically controlled equipment also.

14. Avoid locating equipment in any environment for which it was not designed and which may create a dangerous operating condition, such as explosive atmosphere (e.g., gas dust), high heat (e.g., molten metal, furnace), chemicals, extreme moisture, etc.

15. Certain hydraulic fluids may be irritating or injurious to the eyes and skin. Check with fluid suppliers for the effects of the fluid that you use and avoid bodily contact with those harmful fluids. Fire resistant or synthetic fluids should be especially guarded against.

16. Avoid the use of unauthorized or substitute parts and materials in servicing the equipment. Substitute parts or materials could produce a hazardous operating condition.

17. In piping the hydraulic system on the equipment, use only material of adequate size and strength to suit the flows and pressures, which will be present in the operating system. Use safety factors in selecting materials for strength to allow for shock and over-pressure conditions which may occur.

WARNING!!! ACCUMULATORS CONTAIN A GAS PRECHARGE. ALWAYS BLEED OFF ALL OIL PRESSURE BEFORE BREAKING ANY LINE OR PERFORMING MAINTENANCE.

THE HYDRAULIC SYSTEM IS STILL ACTIVE WHEN ALL ELECTRICAL POWER IS OFF WHEN AN ACCUMULATOR IS IN THE CIRCUIT.

Maintenance Instructions

Hydraulic systems are precision units and their continued smooth operations depends on proper care, therefore, do not neglect your hydraulic systems. Keep them clean, change the oil and oil filter at established intervals and follow prescribed maintenance.

Periodic procedures are as follows:

1. Check the reservoir oil level and add oil as required. The level must be maintained between the high and low mark on the sight gauge.

2. Check the operating temperature and oil pressure. For most industrial applications, an operating temperature of 150ºF is considered maximum.

3. If an external suction filter is used, check filter indicator for dirty element every two (2) hours for the first eight (8) hours of operation and clean when necessary. Check once every day for the next five (5) days of operation and clean when necessary. Check periodically thereafter at intervals that will prevent the filter from bypassing and cavitating the pump.

4. Check return filters as in step 3. These are usually finer filters, however, and will require more frequent element changes or cleansing than the suction filter. Always change filters when oil is changed.

5. At least once a year or every 4,000 operating hours, the reservoir, pump suction filter (if one is used) and air vent filter should be cleaned. At this time, check the entire system for possible future difficulties. Some applications or environmental conditions may dictate such maintenance be performed at more frequent intervals.

6. Make visual checks of all hose and tube connections. Regular checking and tightening of all hydraulic connections will help to assure trouble-free operation.

7. Periodically check pressure setting. The system was designed to operate at a specific pressure and increasing the pressure above that will result in motor overload. The system should be operated at the minimum pressure required to do the intended function, the lower the system pressure, the longer the pump life will be.

8. Check pump/motor coupling periodically for misalignment. A flexible coupling should always be used and shafts accurately aligned parallel and angularly. Check setscrews in couplings for loosening and tighten as required.

9. The reservoir cover should remain tightly sealed at all times, except in case of in-tank maintenance and periodic checks for in-tank leaks, in order to prevent atmospheric contaminations from entering the system.

Power Unit Startup Information

1. While in transit or during installation, a power unit may be subjected to many unusual conditions. On systems with separately mounted pumps and motors, the alignment between shafts should be checked with an indicator or straight edge and adjustments made if necessary. Pump misalignment drastically reduces pump bearing life. Check set screws in couplings for loosening. Tighten as required.

2. Fill the reservoir with fluid as recommended by pump manufacturer (see pump data).

3. Connect the motor to the proper electrical source, checking the motor nameplate for proper wiring of dual voltage motors. Jog the motor to check rotation. Polyphase motors are bi-directional and proper rotation can be established by reversing any two power leads.

4. Pump noise and "crackle" is most often caused by air entering the pump suction. The tightening of suction fittings will usually eliminate such problems. If pump fails to prime, vent discharge pipe to atmosphere to establish fluid flow.

5. **IMPORTANT:** After power unit has been started and all the lines filled, replenish the oil in the reservoir to the proper level. The fluid level should be maintained so it always shows in the sight gauge. This is of utmost importance when an immersion type heat exchanger is used to prevent condensation from collecting on uncovered cooling coils.

6. System pressures should be set as low as possible to prevent unnecessary fluid heating. On some applications, this setting may be from 50 to 200 psi above necessary static pressure to overcome dynamic pressure drop or to achieve proper acceleration.

7. For most industrial applications, an operating temperature of 150ºF is considered maximum. At higher temperatures, difficulty is often experienced in maintaining reliable and consistent hydraulic control, component service life is reduced, hydraulic fluid deteriorates and a potential danger to operating personnel is created.

8. After the first few hours of operation, clean or install new elements in all filters to remove contamination from initial flushing of system plumbing.

Preventing Trouble In Your Hydraulic System

1. Dirt or contamination is the number one villain in a system and should be kept out. It causes wear and malfunctions. Meticulous care should be used on installation to prevent dirt from getting into pipe, tubing, hose, fittings and ports of components. After completion, lines should be flushed and filter elements changed. Proper continuous filtration should be used.

2. Proper sized pipe, hoses and tubing should be used to assure desired flow rate with limited pressure drop. Suction side supply line to pump should be short, as straight as possible and sized to give as little pressure drop as possible within the limits set by pump manufacturer for the rpm chosen.

3. Pumps and fluid motors should not be run faster than recommended.

4. System pressure must be limited by a relief valve or other means to stay within pressure limits of components. System should be run at lowest pressure possible to accomplish job as excessive pressure accelerates wear on components.

5. Reservoir must be sized sufficiently and designed properly to allow deceleration of fluid and cooling.

6. Proper fluid must be used for compatibility to seals and of viscosity range recommended by component manufacturers.

7. System should be designed properly to take care of shock generated by "water hammer" effect when stopping or excessive deceleration forces.

8. Coils on a double solenoid valve should not be energized at same time.

9. Power unit should be located so that it is protected from weather, accumulation of debris and heating effect of ambient conditions or direct sun.

10. When needed, heating or cooling of the reservoir by heat exchanger (or heater of proper design) should be considered so as not to oxidize oil.

11. On start-up, be sure pump rotates in correct direction.

12. After filling lines on start-up, be sure to add oil to reservoir to bring oil level to proper height.

13. Proper design of circuit must be accomplished to prevent dead-heading of pump when not performing work as a pump. Pumping over a relief valve generates considerable heat and may ruin components.

14. The pre-charge on accumulators must be maintained or the accumulator becomes ineffective.

15. Crankcases of pumps and fluid motors must be filled when required before start-up.

NOTE: The above are not intended to be all inclusive, but cover some of the most common reasons for trouble in a hydraulic system.

Troubleshooting Your Hydraulic System

Problem	Possible Causes	Solution
No pressure in the system	Pump does not deliver fluid	Follow the solutions given for "Failure of Pump to Deliver Fluid"
	Relief valve malfunction: 1. Incorrect valve setting 2. Valve leaking or bypassing 3. Valve spring broken	1. Reset to specifications 2. Check main valve seat and pilot valve for scoring and dirt. Replace and clean 3. Replace spring and re-adjust
	Free re-circulation of oil to tank being allowed through system	Check or relief valve may be stuck in open position or return line open intentionally Vent valve[1] is dumping flow through relief valve at low pressure Bypass valve to tank is open Open center 4-way valve[1] is not shifted, dumping oil to tank
	Pressure-compensated — pump not compensating properly	[1]See "Solenoid Valve" if valve is solenoid operated Check compensator for setting and re-adjust if required Check compensator for broken spring or contamination and repair as required
Failure of pump to deliver fluid	Low fluid level in reservoir	Add recommended oil and check level to be certain pump suction line inlet is submerged deeply enough to prevent air entering directly or by the formation of a vortex
	Oil intake pipe or strainer plugged	Clean or replace
	Air leak in suction line, preventing priming or causing noise and irregular action of control circuit	Repair leaks
	Oil viscosity too heavy to pick up prime (especially in cold weather)	Use lighter viscosity oil or install a low density immersion heater (with steel element only)
	Wrong direction of rotation	Must be reversed to prevent damage to pump
	Broken pump shaft or parts broken inside pump	Replace
Pump making noise	Intake line, suction filter of pipe restricted	Clean intake, filter or eliminate restriction. Be sure suction line is completely open
	Air leaks: 1. At pump intake piping joints 2. At pump shaft packing (if present) or seals 3. Air drawn in through inlet pipe opening	1. Tighten as required 2. Repair or replace 3. Be sure suction and return lines are below oil level in reservoir

Troubleshooting Your Hydraulic System

Problem	Possible Causes	Solution
Pump making noise, cont'd	Reservoir air vent plugged	Air must be allowed to breathe in the reservoir. Clean or replace reservoir breather
	Too high viscosity	Use lower viscosity oil. Check recommendations start-up inform-ation
	Coupling misalignment	Re-align
	Worn or broken parts	Replace
Excessive wear of pump parts	Sustained high pressure above maximum pump rating	Check relief valve maximum setting
	Drive misalignment	Check and correct
	Air recirculation causing chatter in system	Check air problems in "Pump Making Noise" section
	Abrasive material in the oil	Clean or replace filter and change oil
	Viscosity of oil too low at working conditions	Check recommendations
Breakage of inside pump housing	Excessive pressure above pump rating	Check relief valve maximum setting
	Seizure due to lack of oil	Check reservoir level, oil filter and possibility of restriction in suction line
External oil leakage around pump shaft or housing	Shaft packing or seals worn	Replace
	Damaged head packing seals	Replace
	Excessive case pressure due to restricted case drain flow (back to tank)	Check case drain line for a restrict-ion and remove. Drain line may be too small; if small, line is creating back pressure, replace with larger line
	Excessive case pressure due to excessive drain flow	Repair pump for excessive leakage from pumping element to case
	Cracked housing	Replace all damaged parts and seals
Solenoid valve	Solenoid burned out	Replace solenoid coil. Check control voltage (high or low voltage will burn out coil)
	No pilot pressure for shifting main spool	For double solenoid valves, check to see if both solenoids are being energized at the same time. Correct control circuit if this occurs
	No pilot pressure for shifting main spool	If valve is externally piloted, check pilot pressure source for adequate pressure
		Check valve for proper internal plugs for external pilot
		If valve is internally piloted, check for proper internal plugs for internal pilot

Troubleshooting Your Hydraulic System

Problem	Possible Causes	Solution
Solenoid valve, cont'd	No pilot pressure for shifting main spool, cont'd	Clear all internal pilot passages and orifices of foreign particles and clogging
	No pilot drain	Check to see if valve has proper internal plugs and orifices for internal or external drain
		If externally drained, check drain line and clear of any clogging
	Spool jammed by foreign particles	Disassemble pilot valve and main valve and remove all foreign particles and silt. If a contamination or silting problem continues to re-occur, install filtration in the system to remove particles
	Internal breakage or damage	Disassemble, examine and replace any damaged parts in pilot valve or main valve
Cylinder will not develop full force or hold position when valve closes lines	Leaking piston packing	Apply pump flow to one end and stroke cylinder until it bottoms out at end of stroke. Loosen return line at opposite end (from pressurized end) and measure leakage flow coming out. Any flow more than slow dripping requires new piston packings. Repeat above for opposite ends and direction of stroke
	Worn 4-way control valve	Disconnect cylinder lines at 4-way valve and plug. Shift valve to either extreme position and check to see if full system pressure builds up. If possible, check to see how much flow is leaking across to the tank port (Note: this check cannot be made if system relief is built into 4-way valve)
Cylinder loses force at some intermediate point in stroke	Scored tube	Repeat above except mechanically stop piston rod so piston is at point where force drops off. If leakage is determined, tube must be repaired or replaced
	Dented tube	Examine tube for dents at point where it loses force. Replace if tube is dented.

Centrifugal Force

Centrifugal force in pounds is .0003409 times the weight of a revolving body in lbs. times the radius of gyration times the square of the number of revolutions per minute.

$$F = .0003409 \times W \times K \times N^2$$

Where F is centrifugal force in pounds. W is weight of revolving body in pounds, K is radius of gyration (in feet) and N is number of revolutions per minute.

Example — Find the centrifugal force of a flywheel where the outside diameter is 17" and the base or inside diameter is 8.6" and the thickness is 1.8" and W = 79, K = .5613 and N = 4000.

$$F = .0003409 \times 79 \times .5613 \times 4000^2 = 241{,}863 \text{ lbs.}$$

Kinetic Energy of Revolving Flywheel

The kinetic energy of a revolving flywheel is the weight of wheel in lbs., times the square of the velocity at radius of gyration in feet per second divided by the sum of 2 times the acceleration due to gravity.

$$E = \frac{W \times V^2}{2 \times g}$$

Where E is kinetic energy in lbs. feet, W is weight of wheel in lbs., V is velocity at radius of gyration in feet per second and g is acceleration due to gravity (32.16 ft. per second).

Example — Find the kinetic energy of a flywheel where W = 79 lb., V is 126.4 ft./sec. and g is 32.16 ft./sec^2.

$$E = \frac{79 \times 126.4^2}{2 \times 32.16} = 19{,}623 \text{ lbs. ft.}$$

Torsional Shaft Stress

Torsional shaft stress is the torque in pound-inches divided by the sum of .1963 times shaft diameter cubed.

$$S = \frac{T}{.1963 \times D^3}$$

Where S is torsional shaft stress in psi, T is torque in lbs.-inches, D is diameter of shaft in inches and .1963 = factor.

Example — What is the torsional stress of a shaft whose diameter is 1.50 and the torque is 2000 lb. in.

$$S = \frac{2000}{.1963 \times 1.50^3} = 30189 \text{ psi}$$

Shaft Torque

Torque is the product of the stresses times .1963 times the cube of shaft diameter.

$$T = S \times .1963 \times d^3$$

Example

What is the torque of a shaft whose diameter is 1.5 and the stress is 30189 psi.

$$T = 30189 \times .1963 \times 1.50^3 = 2000 \text{ lb.-in.}$$

Shaft Diameter Selection

Shaft diameter is the cube root of the torque divided by the sum of .1963 times the stress.

$$D = \sqrt[3]{\frac{T}{.1963 \times S}} \times FS$$

Example — What is the minimum shaft diameter where the torque is 20,000 in.-lbs. and the torsional stress is 30189 psi.

$$D = \sqrt[3]{\frac{20{,}000}{.1963 \times 30189}} \times 2 = 1.89 \text{ inches}$$

To Find Angles and Sides of Oblique Angle Triangles

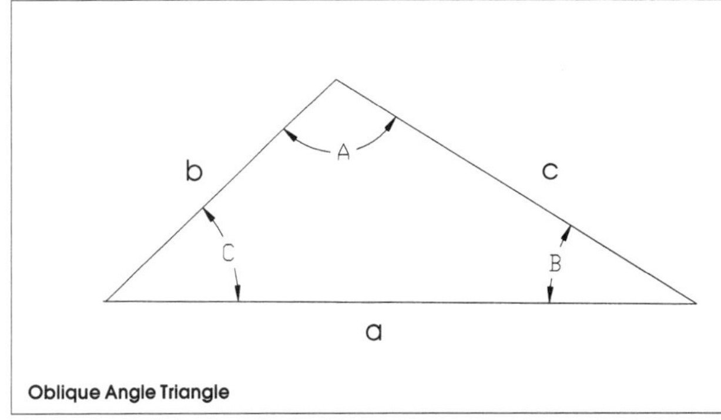

Oblique Angle Triangle

To Find	Known	Formulas
C	A,B	$180^0 - (A + B)$
b	a, B, A	$\frac{a \times \sin B}{\sin A}$
c	a, A, C	$\frac{a \times \sin C}{\sin A}$
Tan A	a, C, b	$\frac{a \times \sin C}{b - (a \times \cos C)}$
B	A, C	$180^0 - (A + C)$
Sin B	b, A, a	$\frac{b \times \sin A}{a}$
A	B,C	$180^0 - (B + C)$
Cos A	a, b, c	$\frac{b^2 + c^2 - a^2}{2bc}$
Sin C	c, A, a	$\frac{c \times \sin A}{a}$
Cot B	a, C, b	$\frac{a \times \text{Cosec } C}{b} - \text{Cot } C$
c	b, C, B	$b \times \sin C \times \text{Cosec } B$

GENERAL ENGINEERING

Uniform Linear Acceleration Formulas

To Find	Given These	Use This Equation
t	$\alpha\upsilon_0\upsilon$	$t = \dfrac{\upsilon - \upsilon_0}{\alpha}$
t	$\alpha\upsilon_0 s$	$t = \dfrac{\sqrt{2\alpha s + \upsilon_0^2} - \upsilon_0}{\alpha}$
t	$\upsilon_0\upsilon s$	$t = \dfrac{2s}{\upsilon_0 + \upsilon}$
α	$t\upsilon_0\upsilon$	$\alpha = \dfrac{\upsilon - \upsilon_0}{t}$
α	$t\upsilon_0 s$	$\alpha = \dfrac{2s - 2\upsilon_0 t}{t^2}$
α	$\upsilon_0\upsilon s$	$\alpha = \dfrac{\upsilon^2 - \upsilon_0^2}{2s}$
υ_0	$t\alpha\upsilon$	$\upsilon_0 = \upsilon - \alpha t$
υ_0	$t\alpha s$	$\upsilon_0 = \dfrac{s}{t} - \dfrac{1}{2}\alpha t$
υ_0	$\alpha\upsilon s$	$\upsilon_0 = \sqrt{\upsilon^2 - 2\alpha s}$
υ	$t\alpha\upsilon_0$	$\upsilon = \upsilon_0 + \alpha t$
υ	$\alpha\upsilon_0 s$	$\upsilon = \sqrt{\upsilon_0^2 + 2\alpha s}$
s	$t\alpha\upsilon_0$	$s = \upsilon_0 t + \dfrac{1}{2}\alpha t^2$
s	$\alpha\upsilon_0\upsilon$	$s = \dfrac{\upsilon^2 - \upsilon_0^2}{2\alpha}$
s	$t\upsilon_0\upsilon$	$s = \dfrac{1}{2} t (\upsilon_0 + \upsilon)$

t = Time
α = Acceleration, ft/sec^2
υ_0 = Initial Velocity, ft/ sec
s = Distance, ft
υ = Final Velocity, ft/sec

Uniform Angular Acceleration Formulas

To Find	Given These	Use This Equation
t	$\alpha\omega_0\omega$	$t = \dfrac{\omega - \omega_0}{\alpha}$
t	$\alpha\omega_0\theta$	$t = \dfrac{\sqrt{2\alpha\theta + \omega_0^2} - \omega_0}{\alpha}$
t	$\omega_0\omega\theta$	$t = \dfrac{2\theta}{\omega_0 + \omega}$
α	$t\omega_0\omega$	$\alpha = \dfrac{\omega - \omega_0}{t}$
α	$t\omega_0\theta$	$\alpha = \dfrac{2\theta - 2\omega_0 t}{t^2}$
α	$\omega_0\omega\theta$	$\alpha = \dfrac{\omega^2 - \omega_0^2}{2\theta}$
ω_0	$t\alpha\omega$	$\omega_0 = \omega - \alpha t$
ω_0	$t\alpha\theta$	$\omega_0 = \dfrac{\theta}{t} - \dfrac{1}{2}\alpha t$
ω_0	$\alpha\omega\theta$	$\omega_0 = \sqrt{\omega^2 - 2\alpha\theta}$
ω	$t\alpha\omega_0$	$\omega = \omega_0 + \alpha t$
ω	$\alpha\omega_0\theta$	$\omega = \sqrt{\omega_0^2 + 2\alpha\theta}$
θ	$t\alpha\omega_0$	$\theta = \omega_0 t + \dfrac{1}{2}\alpha t^2$
θ	$\alpha\omega_0\omega$	$\theta = \dfrac{\omega^2 - \omega_0^2}{2\alpha}$
θ	$t\omega_0\omega$	$\theta = \dfrac{1}{2} t (\omega_0 + \omega)$

t = Time
α = Acceleration, rad/sec^2
ω_0 = Initial Velocity, rad/sec
θ = Distance, rad
ω = Final Velocity, rad/sec

One revolution = 2 π rad = 360^0

Reference Design Formulas

PROPERTIES OF SECTIONS	MASS & MASS MOMENTS OF INERTIA OF GEOMETRIC SHAPES
A = area, in^2 I = moment of inertia, in^4 J = polar moment of inertia, in^4 Z = section modulus, in^3 k = radius of gyration, in $\bar{y}$ = centroidal distance, in ρ = density lb/in^3	**ROD** $m = \frac{\pi d^2 l \rho}{4g}$ $I_y = I_z = \frac{ml^2}{12}$
RECTANGLE $A = bh$ $k = 0.289h$ $I = \frac{bh^3}{12}$ $\bar{y} = \frac{h}{2}$ $Z = \frac{bh^2}{6}$	**ROUND DISK** $m = \frac{\pi d^2 t \rho}{4g}$ $I_x = \frac{md^2}{8}$ $I_y = I_z = \frac{md^2}{16}$
TRIANGLE $A = \frac{bh}{2}$ $k = 0.236h$ $I = \frac{bh^3}{36}$ $\bar{y} = \frac{h}{3}$ $Z = \frac{bh^2}{24}$	**RECTANGULAR PRISM** $m = \frac{abc\rho}{g}$ $I_x = \frac{m}{12}(a^2 + b^2)$ $I_y = \frac{m}{12}(a^2 + c^2)$ $I_z = \frac{m}{12}(b^2 + c^2)$
CIRCLE $A = \frac{\pi d^2}{4}$ $J = \frac{\pi d^4}{32}$ $I = \frac{\pi d^4}{64}$ $k = \frac{d}{4}$ $Z = \frac{\pi d^3}{32}$ $\bar{y} = \frac{d}{2}$	**CYLINDER** $m = \frac{\pi d^2 l \rho}{4g}$ $I_x = \frac{md^2}{8}$ $I_y = I_z = \frac{m}{48}(3d^2 + 4l^2)$
HOLLOW CIRCLE $A = \frac{\pi}{4}(d^2 - d_i^2)$ $J = \frac{\pi}{32}(d^4 - d_i^4)$ $I = \frac{\pi}{64}(d^4 - d_i^4)$ $k = \sqrt{\frac{d^2 + d_i^2}{16}}$ $Z = \frac{\pi}{32\,d}(d^4 - d_i^4)$ $\bar{y} = \frac{d}{2}$	**HOLLOW CYLINDER** $m = \frac{\pi l \rho}{4g}(d_o^2 - d_i^2)$ $I_x = \frac{m}{8}(d_o^2 + d_i^2)$ $I_y = I_x = \frac{m}{48}(3d_o^2 + 3d_i^2 + 4l^2)$

Reference Design Formulas

BEAM FORMULAS

Beam supported at both ends
Concentrated load at mid-span

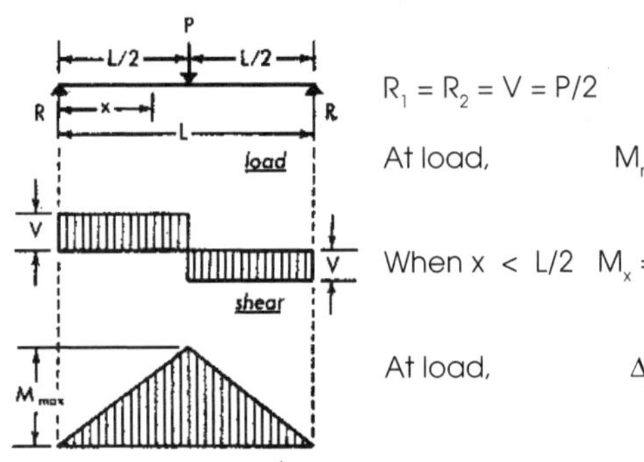

$R_1 = R_2 = V = P/2$

At load, $M_{max} = \frac{PL}{4}$

When $x < L/2$ $M_x = \frac{Px}{2}$

At load, $\Delta_{max} = \frac{PL^3}{48EI}$

When $x < L/2$ $\Delta_x = \frac{Px}{48EI}(3L^2 - 4x^2)$

At end, $\theta_1 = -\frac{PL^2}{16EI} = -\theta_2$

Beam supported at both ends
Two equal concentrated loads, equally spaced from ends

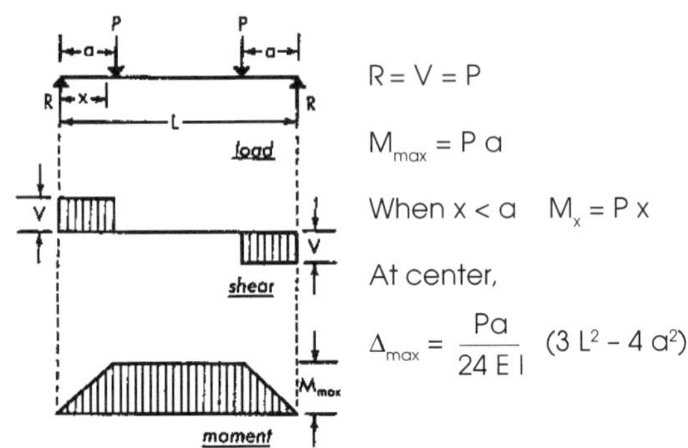

$R = V = P$

$M_{max} = Pa$

When $x < a$ $M_x = Px$

At center,

$\Delta_{max} = \frac{Pa}{24EI}(3L^2 - 4a^2)$

When $x < a$ $\Delta_x = \frac{Px}{6EI}(3La - 3a^2 - x^2)$

When $x > a$
but $x < (L - a)$ $\Delta_x = \frac{Pa}{6EI}(3Lx - 3x^2 - a^2)$

At ends, $\theta = \frac{Pa}{2EI}(L - a)$

Beam supported at both ends
Concentrated load at any point

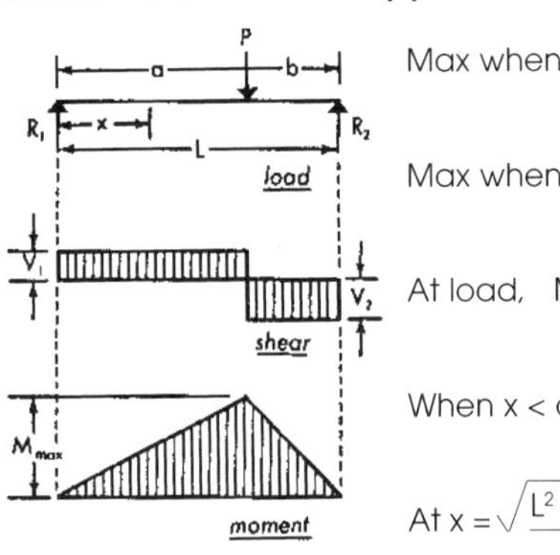

Max when $a < b$ $R_1 = V_1 = \frac{Pb}{L}$

Max when $a > b$ $R_2 = V_2 = \frac{Pa}{L}$

At load, $M_{max} = \frac{Pab}{L}$

When $x < a$ $M_x = \frac{Pbx}{L}$

At $x = \sqrt{\frac{L^2 - b^2}{3}}$

when $a > b$ $\Delta_{max} = \frac{Pb}{3EIL}\sqrt{\left(\frac{L^2 - b^2}{3}\right)^3}$

At load, $\Delta = \frac{Pa^2b^2}{3EIL}$

When $x < a$ $\Delta_x = \frac{Pbx}{6EIL}(L^2 - b^2 - x^2)$

When $a < b$ $\Delta_{L/2} = \frac{Pa}{48EI}(3L^2 - 4a^2)$

At ends,
$$\left\{\begin{array}{l} \theta_1 = -\frac{P}{6EI}\left(2aL + \frac{a^3}{L} - 3a^2\right) \\ \theta_2 = +\frac{P}{6EI}\left(aL - \frac{a^3}{L}\right) \end{array}\right.$$

Beam supported at both ends
Two equal concentrated loads, unequally spaced from ends

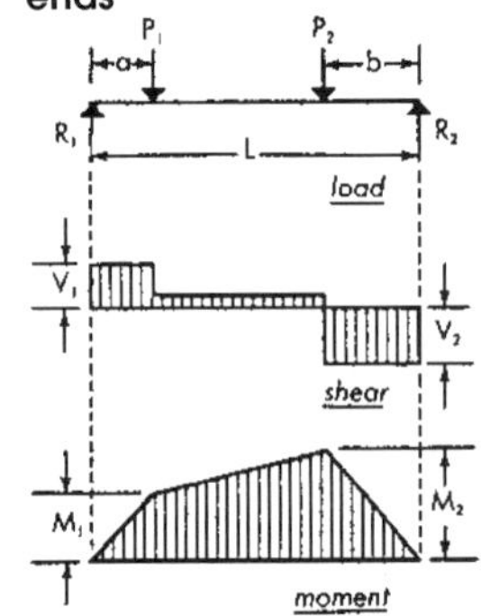

$R_1 = V_1 = \frac{P_1(L - a) + P_2 b}{L}$

$R_2 = V_2 = \frac{P_1 a + P_2(L - b)}{L}$

When $x > a$
but $x < (L - b)$ $V_x = R_1 - P_1$

Max when $R_1 < P_1$ $M_1 = R_1 a$

Max when $R_2 < P_2$ $M_2 = R_2 b$

When $x < a$ $M_x = R_1 x$

When $x > a$ but $x < (L - b)$ $M_x = R_1 x - P_1(x - a)$

BEAM FORMULAS

Beam supported at both ends
Uniform load partially distributed over span

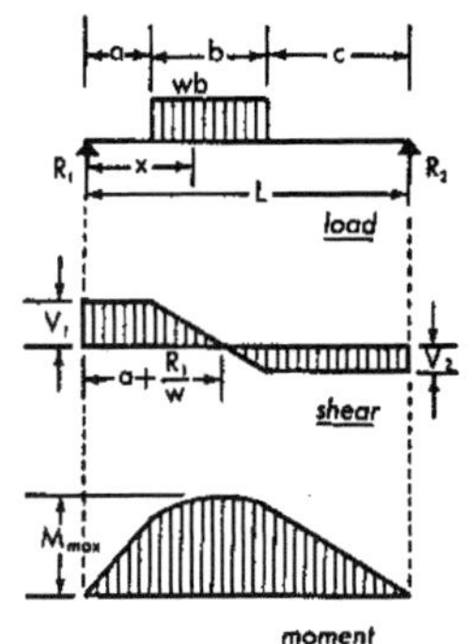

Max when a<c

$R_1 = V_1 = \frac{w\,b}{2\,L}(2c + b)$

Max when a > c

$R_2 = V_2 = \frac{w\,b}{2\,L}(2a + b)$

When x > a but x < (a + b)
$V_x = R_1 - w(x - a)$

At $x = a + \frac{R_1}{w}$ $\quad M_{max} = R_1\left(a + \frac{R_1}{2\,w}\right)$

When x < a $\quad M_x = R_1 x$

When x > a but x < (a + b) $\quad M_x = R_1 x - \frac{w}{2}(x - a)^2$

When x > (a + b) $\quad M_x = R_2(L - x)$

<u>When a = c</u>

$R = V = \frac{w\,b}{2}$ $\quad V_x = w\left(a + \frac{b}{2} - x\right)$

At center, $M_{max} = \frac{wb}{2}\left(a + \frac{b}{4}\right)$

When x < a $\quad M_x = \frac{w\,b\,x}{2}$

When x > a but x < (a + b) $\quad M_x = \frac{w\,b\,x}{2} - \frac{w}{2}(x - a)^2$

At center, $\quad \Delta_{L/2} = \frac{w\,b}{384EI}(-8L^3 + 4b^2L - b^3)$

Beam supported at both ends
Uniform load over entire span

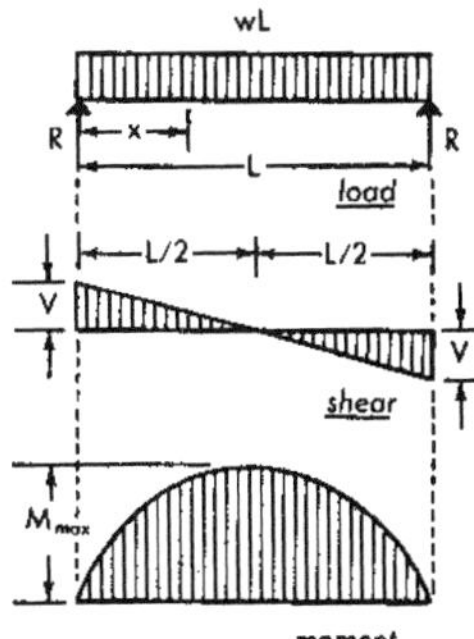

$R = V = \frac{w\,L}{2}$

$V_x = w\left(\frac{L}{2} - x\right)$

At center, $M_{max} = \frac{w\,L^2}{8}$

$M_x = \frac{w\,x}{2}(L - x)$

At center, $\Delta_{max} = \frac{5\,w\,L^4}{384\,E\,I}$ $\quad \Delta_x = \frac{w\,x}{24\,E\,I}(L^3 - 2Lx^2 + x^3)$

At ends, $\theta = \frac{w\,L^3}{24\,E\,I}$

Beam supported at both ends
Varying load, increasing uniformly to center

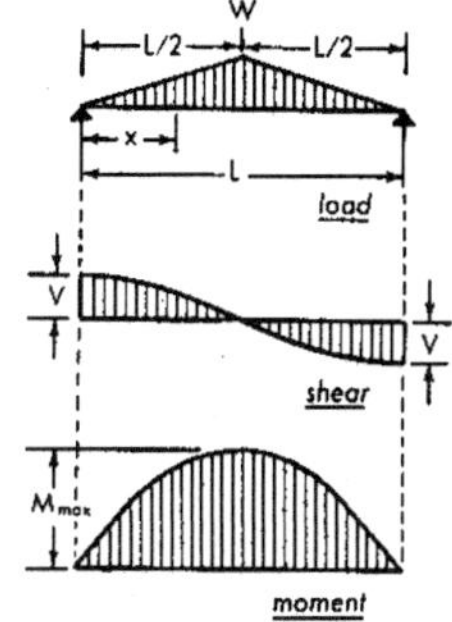

$R_1 = R_2 = V = \frac{W}{2}$

When x < L/2 $\quad V_x = \frac{W}{2L^2}(L^2 - 4x^2)$

At center, $M_{max} = \frac{W\,L}{6}$

When x < L/2 $\quad M_x = Wx\left(\frac{1}{2} - \frac{2x^2}{3L^2}\right)$

At center, $\Delta_{max} = \frac{W\,L^3}{60\,E\,I}$

$\Delta_x = \frac{W\,x}{480\,E\,I\,L^2}(5L^2 - 4x^2)^2$

At ends, $\theta = \frac{5\,W\,L^2}{96\,E\,I}$

Beam supported at both ends
Varying load, increasing uniformly to one end

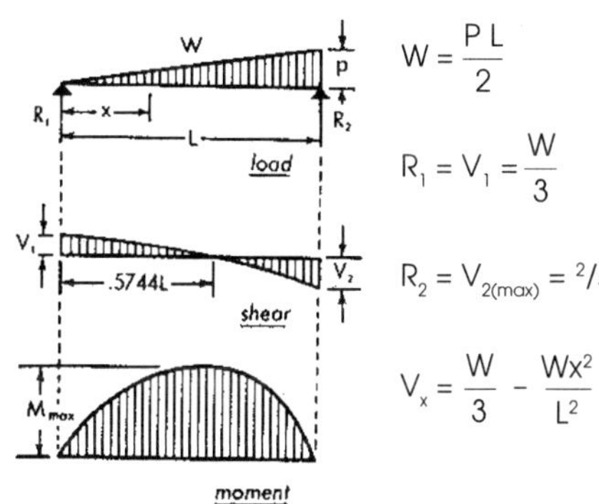

$W = \frac{P\,L}{2}$

$R_1 = V_1 = \frac{W}{3}$

$R_2 = V_{2(max)} = {}^2/_3\,W$

$V_x = \frac{W}{3} - \frac{Wx^2}{L^2}$

At $x = L/\sqrt{3} = .5744L$ $\quad M_{max} = \frac{2\,W\,L}{9\sqrt{3}} = .1283\,W\,L$

$M_x = \frac{W\,x}{3\,L^2}(L^2 - x^2)$

At $x = L\sqrt{1 - \sqrt{8/15}} = .5193L$ $\quad \Delta_{max} = .01304\,\frac{W\,L^3}{E\,I}$

$\Delta_x = \frac{W\,x}{180\,E\,I\,L^2}(3x^4 - 10L^2x^2 + 7L^4)$

At center, $\Delta_{L/2} = \frac{5\,W\,L^3}{384\,E\,I}$

At ends, $\quad \theta_1 = -\frac{7\,W\,L^2}{180\,E\,I}$, $\quad \theta_2 = +\frac{8\,W\,L^2}{180\,E\,I}$

Reference Design Formulas

BEAM FORMULAS

Beam supported at both ends
Moment applied at one end

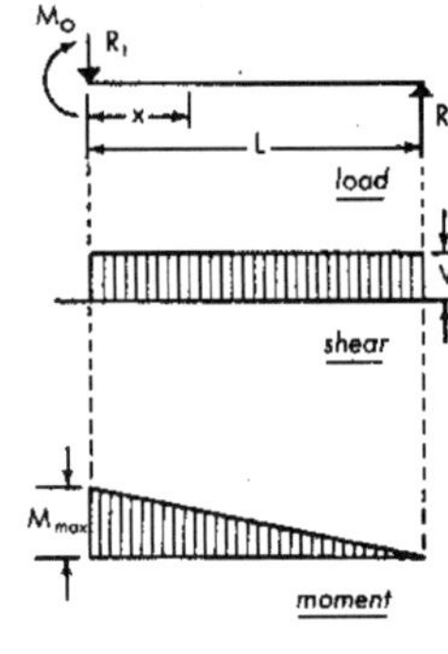

$$R_1 = R_2 = V = \frac{M_o}{L}$$

At R_1, $M_{max} = M_o$

$$M_x = M_o + R_1 X = M_o \left(1 - \frac{x}{L}\right)$$

When $x = .422L$

$$\Delta_{max} = .0642 \frac{M_o L^2}{E\,I}$$

$$\Delta_x = \frac{M_o}{6\,E\,I}\left(3x^2 - \frac{x^3}{L} - 2Lx\right)$$

At R_1, $\theta_1 = -\frac{M_o L}{3\,E\,I}$

At R_2, $\theta_2 = \frac{M_o L}{6\,E\,I}$

Beam supported at both ends
Moment applied at any point

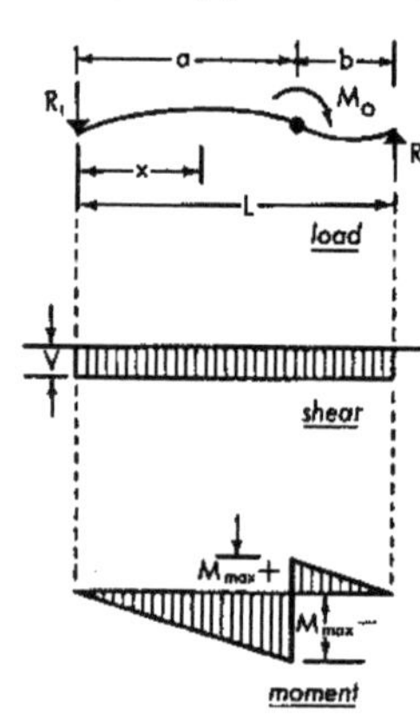

When a>b $R_1 = -\frac{M_o}{L} = V \quad R_2 = \frac{M_o}{L}$

At $x = a$, $M_{max}(-) = -\frac{M_o a}{L}$

At $x = a$, $M_{max}(+) = M_o\left(1 - \frac{a}{L}\right)$

When $x < a$ $\quad M_x = -\frac{M_o x}{L}$

When $x > a$ $\quad M_x = M_o\left(1 - \frac{x}{L}\right)$

When $x < a$ $\quad \Delta_x = +\frac{M_o x}{6\,E\,I\,L}(L^2 - 3b^2 - x^2)$

When $x > a$ $\quad \Delta_x = \frac{M_o(L - x)}{6\,E\,I\,L}(3a^2 - 2Lx + x^2)$

At $x = \sqrt{\frac{L^2 - 3b^2}{3}}$ if $a > .4226L$, $\quad \Delta_{max} = \frac{M_o}{3\,E\,I\,L}\left(\frac{L^2 - 3b^2}{3}\right)^{3/2}$

At $x = L - \sqrt{\frac{L^2 - 3a^2}{3}}$ if $a < .5774L$, $\quad \Delta_{max} = -\frac{M_o}{3\,E\,I\,L}\left(\frac{L^2 - 3a^2}{3}\right)^{3/2}$

At center, $\quad M_{L/2} = -\frac{M_o}{2}$

At center, $\quad \Delta_{L/2} = +\frac{M_o}{16\,EI}(L^2 - 4b^2)$

When a = b = L/2

At $x = \frac{\sqrt{3}}{6} L = .28867L$, $\quad \Delta_{max} = \frac{M_o L^2}{124.71\,EI}$

At center, $\quad \theta_{L/2} = \frac{M_o L}{12\,EI}$

Beam fixed at one end only (cantilever)
Concentrated load at free end

$$R = V = P$$

At support, $M_{max} = P\,L$

$M_x = Px$

At free end, $\Delta_{max} = \frac{PL^3}{3\,E\,I}$

$$\Delta_x = \frac{P}{6\,EI}(2L^3 - 3L^2 + x^3)$$

$$= \frac{P\,z^2}{6\,E\,I}(3L - z)$$

Beam fixed at one end only (cantilever)
Concentrated load at any point

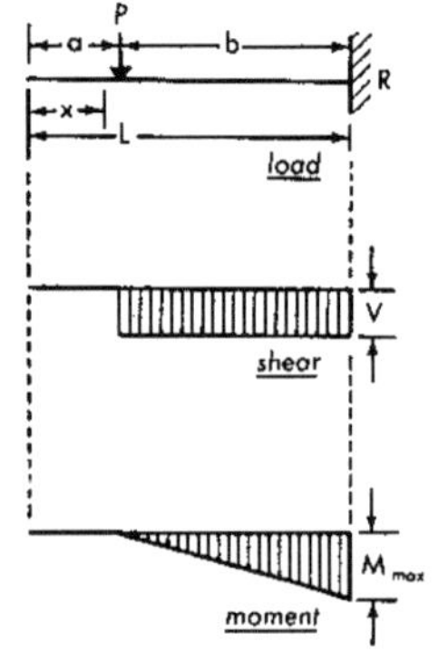

$$R = V = P$$

At support, $M_{max} = Pb$

When $x > a$ $\quad M_x = P(x - a)$

At free end, $\Delta_{max} = \frac{Pb^2}{6\,E\,I}(3L - b)$

At load, $\Delta = \frac{Pb^3}{3\,E\,I}$

When $x < a$ $\quad \Delta_x = \frac{Pb^2}{6\,E\,I}(3L - 3x - b)$

When $x > a$ $\quad \Delta_x = \frac{P(L - x)^2}{6\,EI}(3b - L + x)$

Beam fixed at one end only (cantilever)
Uniform load over entire span

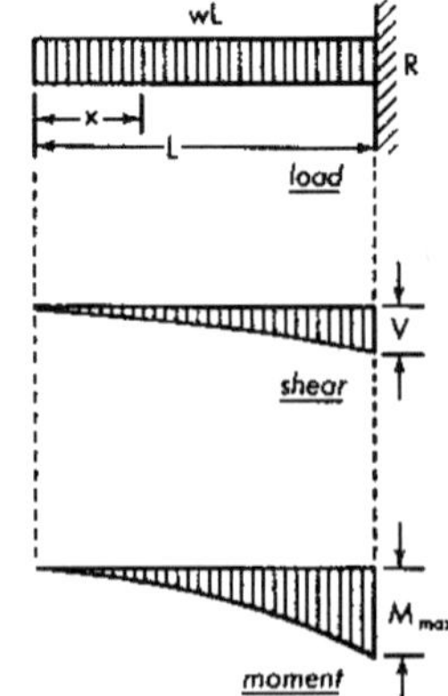

$$R = V = w\,L$$

$$V_x = w\,x$$

At support, $M_{max} = \frac{w\,L^2}{2}$

$$M_x = \frac{w\,x^2}{2}$$

At free end, $\Delta_{max} = \frac{wL^4}{8\,E\,I}$

$$\Delta_x = \frac{w}{24\,E\,I}(x^4 - 4L^3x + 3L^4)$$

BEAM FORMULAS

Beam fixed at one end only (cantilever)
Uniform load partially distributed at free end

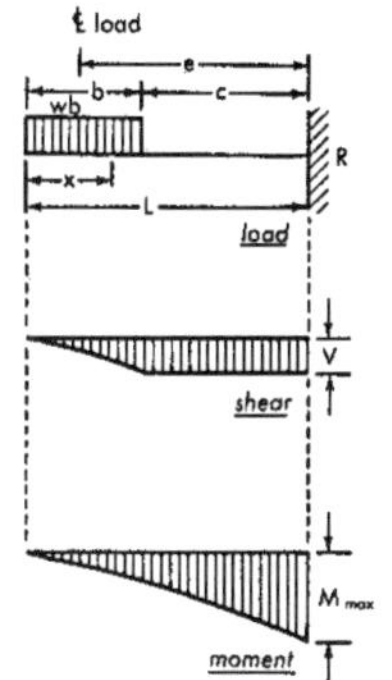

$R = V = w\,b$

At support, $M_{max} = wbe$

When $x < b$ $\quad M_x = \dfrac{w\,x^2}{2}$

When $x > b$ $\quad M_x = \dfrac{wb}{2}(b - 2x)$

At free end, $\Delta_{max} = \dfrac{wb}{48\,E\,I}(8e^3 - 24e^2L - b^3)$

When $x < b$ $\quad \Delta_x = \dfrac{w}{48\,E\,I}(8be^3 - 24be^2(L - x) + 2b^3x - b^4 - 2x^4)$

When $x > b$ $\quad \Delta_x = \dfrac{wb}{48\,E\,I}(8e^3 - 24e^2(L - x) - (2x - b)^3)$

At free end, $\quad \theta = +\dfrac{wb}{24\,E\,I}(b^2 + 12e^2)$

Beam fixed at one end only (cantilever)
Varying load increasing uniformly from support to free end

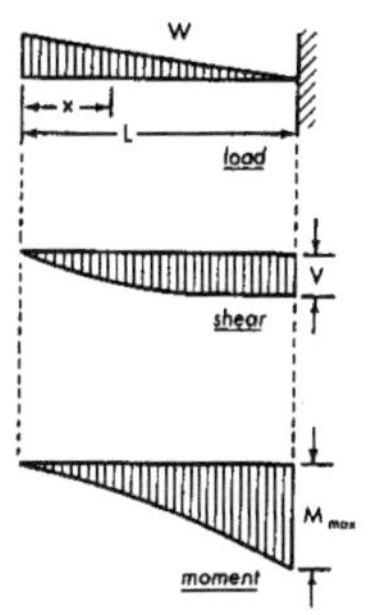

$R = V = W$

$V_x = \dfrac{2\,W\,x}{L^2}\left(L - \dfrac{x}{2}\right)$

At support, $\quad M_{max} = \dfrac{2\,W\,L}{3}$

$M_x = \dfrac{W\,x^2}{3\,L^2}(x - 3\,L)$

At free end, $\Delta_{max} = \dfrac{11\,W\,L^3}{60\,E\,I}$

$\Delta_x = \dfrac{W}{60\,E\,I\,L^2}(L^4(15x - 11L) - x^4(5L - x))$

At free end, $\quad \theta = +\dfrac{W\,L^2}{4\,E\,I}$

Beam fixed at one end only (cantilever)
Varying load increasing uniformly from free end to support

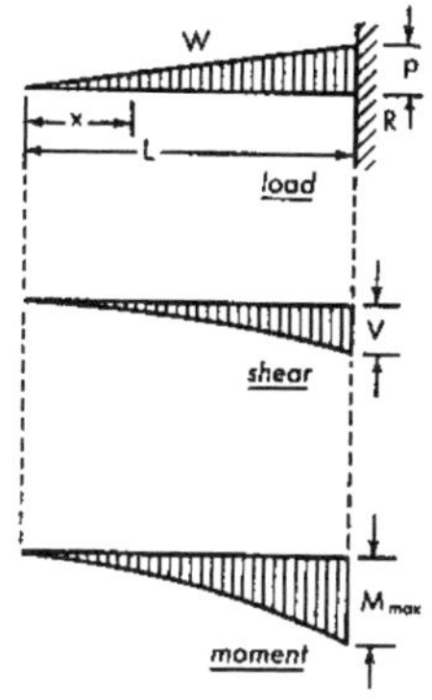

$W = \dfrac{p\,L}{2}$

$R = V = W$

$V_x = W\dfrac{x^2}{L^2}$

At support, $\quad M_{max} = \dfrac{w\,L}{3}$

$M_x = \dfrac{W\,x^3}{3\,L^2}$

At free end, $\quad \Delta_{max} = \dfrac{W\,L^3}{15\,E\,I}$

$\Delta_x = \dfrac{W}{60\,E\,I\,L^2}(x^5 - 5L^4x + 4L^5)$

At free end, $\quad \theta = +\dfrac{W\,L^2}{12\,E\,I}$

Beam fixed at one end only (cantilever)
Moment applied at free end

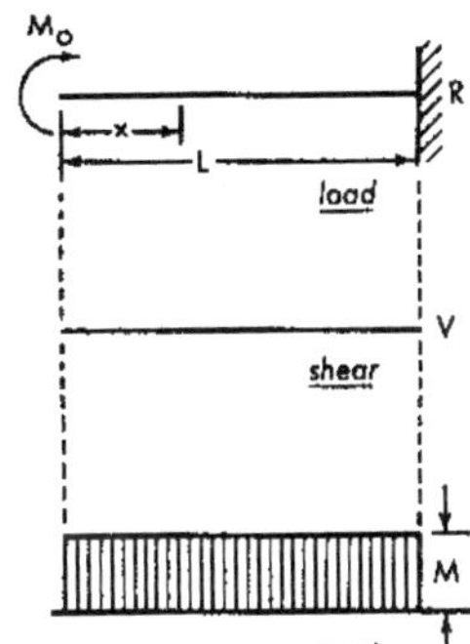

$R = V = O$

$M_x = M_o$

At free end, $\quad \Delta_{max} = \dfrac{M_o L^2}{2\,E\,I}$

$\Delta_x = \dfrac{M_o}{2\,E\,I}(L - x)^2$

At free end, $\quad \theta = -\dfrac{M_o L}{E\,I}$

Reference Design Formulas

BEAM FORMULAS APPLIED TO SIDE OF TANK

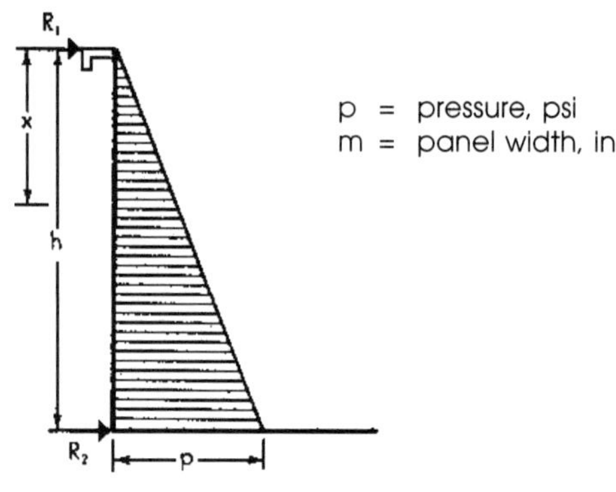

p = pressure, psi
m = panel width, in

$R_1 = \frac{p\,h\,m}{6}$ $\qquad R_2 = \frac{p\,h\,m}{3} = V_{max}$

$M_{max} = \frac{p\,h^2\,m}{9\sqrt{3}} = .0642\,p\,h^2 m$

$M_x = \frac{p\,x\,m}{6\,h}\,(h^2 - x^2)$

$\Delta_{h/2} = \frac{5\,p\,h^4 m}{768\,E\,I}$

$\Delta_x = \frac{p\,x\,m}{360\,E\,I\,h}\,(3x^4 - 10\,h^2 x^2 + 7h^4)$

$\Delta_{max} = .00652\,\frac{p\,h^4\,m}{E\,I}$ (at x = .5193h)

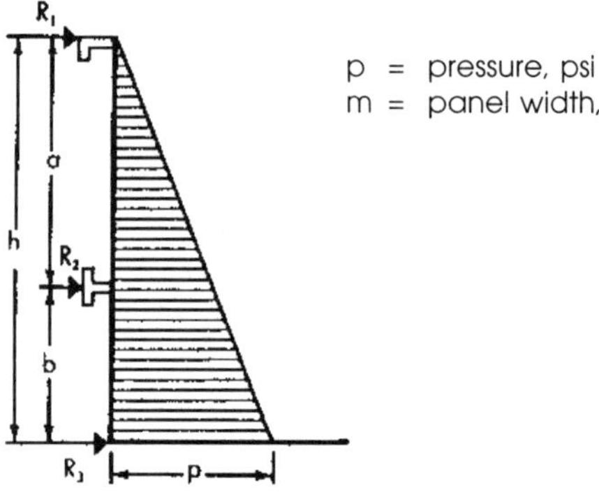

p = pressure, psi
m = panel width, in

Maximum bending moment is least when
a = .57 h
b = .43h

$M_{max} = .0147\,p\,h^2\,m$
(negative moment at middle support, 2)

$R_1 = +.030\,p\,h\,m$

$R_2 = +.320\,p\,h\,m$

$R_3 = +.150\,p\,h\,m$

$V_{max} = +.188\,p\,h\,m$
(at middle support, 2)

TORSIONAL MEMBERS

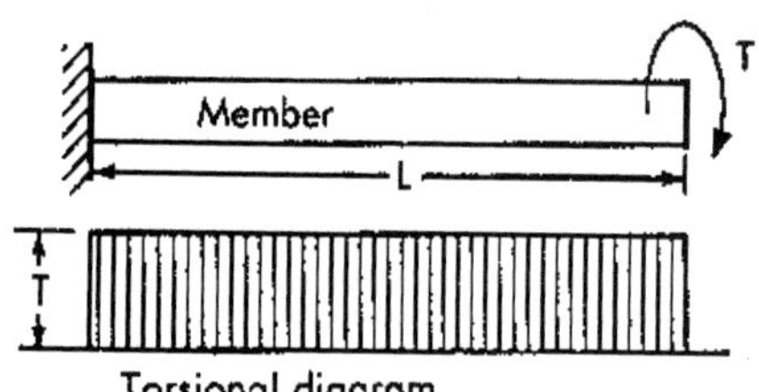

J = Polar moment of inertia for shaft

G = Shearing modulus of elasticity (See material section)

T = T

At end, $\theta = \frac{T\,L}{J\,G}$

J for shaft $= \frac{\pi d^4}{32}$

J for hollow shaft $= \frac{\pi(d^4 - d_i^4)}{32}$

CIRCULAR FLAT PLATE FORMULAS

The following table of formulas is for stress and deflection of circular flat steel plates. (Poisson's ratio = .3)

σ_r = tensile stress in radial direction (psi)
σ_t = tensile stress in tangential direction (psi)
t = thickness of plate (inches)
r = outer radius of plate (inches)
E = modulus of elasticity (for steel = 30,000,000 psi)
W = total load on plate (pounds)
p = uniform load on plate (psi)
M = couple or moment applied to central portion (inch-lbs)
log to the base (e) (Natural or Naperian logarithms):
$\log_e x = 2.3026 \log_{10} x$

A positive sign for the stress indicates tension on the top surface and compression on the lower surface.

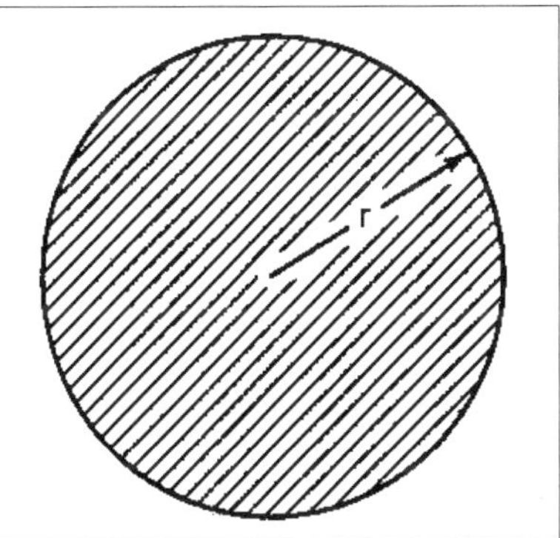

Adapted from Roark "Formulas for Stress and Strain", and Timoshenko "Theory of Plates and Shells".

Outer edge fixed and supported
Uniform load over entire area

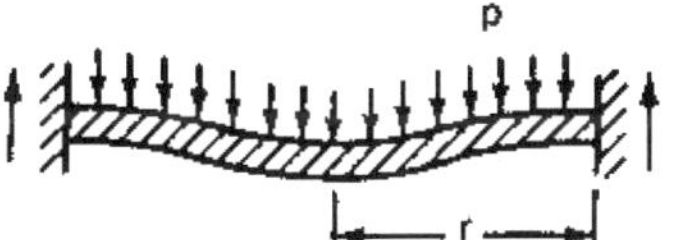

At center, $\sigma_r = \sigma_t = -\dfrac{39\,p}{80}\left(\dfrac{r}{t}\right)^2$

At edge, Max $\sigma_r = \dfrac{3\,p}{4}\left(\dfrac{r}{t}\right)^2$

At center, Max$\Delta = -\dfrac{273\,p\,r^4}{1600\,E\,t^3}$

Outer edge supported
Uniform load over entire area

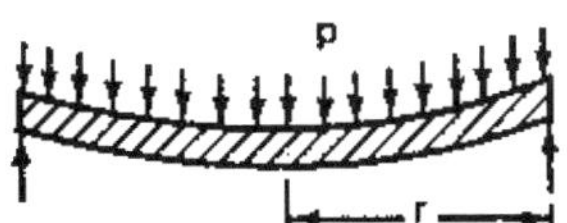

At center, Max $\sigma_r = \sigma_t = -\dfrac{99\,p}{80}\left(\dfrac{r}{t}\right)^2$

At center, Max$\Delta = -\dfrac{1113\,p\,r^4}{1600\,E\,t^3}$

At edge, $\theta = \dfrac{21\,p}{20\,E}\left(\dfrac{r}{t}\right)^3$

Outer edge fixed and supported
Uniform load over concentric circular area of radius r_1

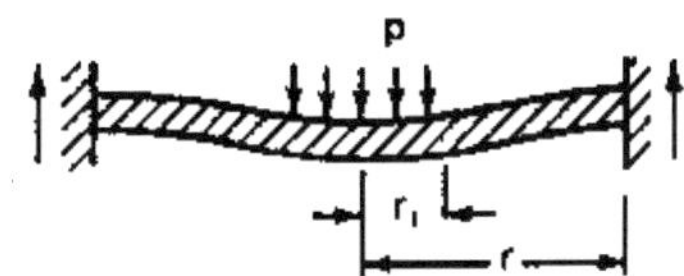

At center (max when $r_1 < .588r$),

$$\sigma_r = \sigma_t = -\frac{3\,p}{80}\left(\frac{r_1}{t}\right)^2\left[\ 52\log\frac{r}{r_1} + 13\left(\frac{r_1}{r}\right)^2\right]$$

At edge (max when $r_1 > .588r$),

$$\sigma_r = \frac{3\,p}{4}\left(\frac{r_1}{t\,r}\right)^2(2\,r^2 - r_1^2)$$

At edge, $\sigma_t = \dfrac{9\,p}{40}\left(\dfrac{r_1}{t\,r}\right)^2\ (2\,r^2 - r_1^2)$

At center, Max$\Delta = -\dfrac{273\,p\,r_1^2}{1600\,E\,t^3}\ (4r^2 - 3r_1^2 - 4r_1^2\log\dfrac{r}{r_1})$

<u>When r_1 is very small (concentrated load)</u>

At center, Max$\Delta = -\dfrac{273\,W\,r^2}{400\,\pi\,E\,t^3}$

For stress, Roark suggests using the following in which r_e is equal to either r_1 or the effective radius ($\sqrt{1.6\,r_1^2 + t^2} - .675t$) whichever is largest:

At center, Max $\sigma_r = \sigma_t = -\dfrac{3\,W}{80\,\pi\,t^2}\left[\,52\log\dfrac{r}{r^e} + 13\left(\dfrac{r_e}{r}\right)^2\right]$

At edge, $\sigma_t = \dfrac{9\,W}{40\,\pi\,t^2\,r^2}\ (2r^2 - rb^2)$

Reference Design Formulas

CIRCULAR FLAT PLATE FORMULAS

Outer edge supported
Uniform load over concentric circular area of radius r_1

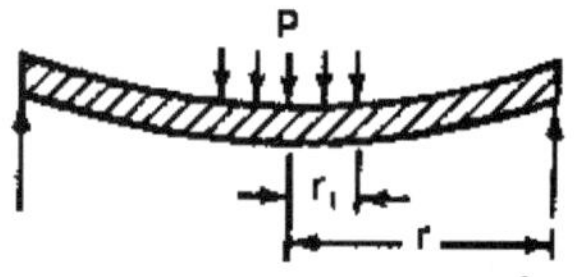

At center, $\text{Max}\sigma_r = \sigma_t = \frac{3p}{80}\left(\frac{r_1}{t}\right)\left[40 + 52\log\frac{r}{r_1} - 7\left(\frac{r_1}{r}\right)^2\right]$

At center, $\text{Max}\,\Delta = -\frac{21\,p\,r_1^2}{1600E\,t^3}\left(132r^2 - 52r_1^2\log\frac{r}{r_1} - 73r_1^2\right)$

When r_1 is very small (concentrated load)

At center, $\text{Max}\Delta = -\frac{693\,Wr^2}{400\,\pi\,E\,t^3}$

At edge, $\theta = \frac{21\,W\,r}{10\,\pi\,E\,t^3}$

For stress, Roark suggests using the following in which r_e is equal to either r_1 or the effective radius $(\sqrt{1.6\,r_1^2 + t^2} - .675t)$ whichever is largest:

At center, $\text{Max}\,\sigma_r = \sigma_t = \frac{3\,W}{80\,\pi\,t^2}\left[40 + 52\log\frac{r}{r_e} - 7\left(\frac{r_e}{r}\right)^2\right]$

Outer edge supported
Uniform load on concentric circular ring of radius r_1

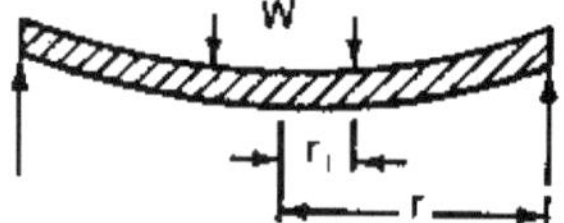

At $< r_1$, $\text{Max}\,\sigma_r = \sigma_t = -\frac{3\,W}{40\,\pi\,t^2}\left[7 + 26\log\frac{r}{r_1} - 7\left(\frac{r_1}{r}\right)^2\right]$

At center, $\text{Max}\Delta = -\frac{273\,W}{5200\,\pi\,E\,t^3}\left[33(r^2 - r_1^2) - 26r_1^2\log\frac{r}{r_1}\right]$

Outer edge fixed and supported
Uniform load on concentric circular ring of radius r_1

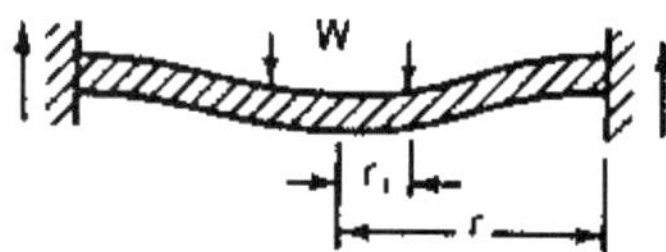

At $< r_1$ (max when $r_1 < .31r$),

$\sigma_r = \sigma_t = -\frac{39\,W}{40\,\pi\,t^2}\left[2\log\frac{r}{r_1} + \left(\frac{r_1}{r}\right)^2 - 1\right]$

At edge (max when $r_1 > .31r$),

$\sigma_r = \frac{3\,W}{2\,\pi\,t^3}\left[1 - \left(\frac{r_1}{r}\right)^2\right]$

At edge $\sigma_t = \frac{9\,W}{20\,\pi\,t^2}\left[1 - \left(\frac{r_1}{r}\right)^2\right]$

At center, $\text{Max}\Delta = -\frac{273\,W}{400\,\pi\,E\,t^3}\left(r^2 - r_1^2 - 2r_1^2\log\frac{r}{r_1}\right)$

Plating Information

Material	Approximate BHN	Wear Resistance	Process Cost
Cadmium	35-50	Poor	Medium
Chromium	500-1000	Excellent	High
Copper	50-150	Fair	Medium
Nickel	20-500	Good	Medium
Tin	5-15	Poor	Low-Medium
Zinc	35-55	Poor	Low

Cadmium
An electronically deposited plating which provides an exceptionally good protection against corrosion, particularly in a salt atmosphere. The plating is silver-gray in color. The cadmium surfaces can be passivated and the cyanide from plating neutralized with a clear Chromate overcoat. The overcoat provides a bright shiny appearance which resists staining. Another post plating process for cadmium plated parts is a Phosphate coating. The phosphate coating adheres to the plated surface very well and is porous, making it an excellent base for parts that will have a final coat of paint applied. Cadmium is normally applied in a thickness of .0002 to .0004 inches.

Chromium
An electronically deposited plating which provides a very wear resistant coating. For corrosion protection, an undercoat of nickel is normally deposited prior to the plating by chromium. No overcoats are normally used. Chromium is normally applied in a thickness of .0002 to .0004 inches.

Copper
An electronically deposited plating which provides a corrosion resistant surface but is subject to rapid tarnishing and is therefore rarely used by itself. Copper is normally used as an undercoat for nickel and applied in a thickness of .0002 to .0004 inches.

Nickel
An electronically deposited plating which provides corrosion resistance in mildly corrosive atmospheres. It is normally used for decorative finishes only or when parts are subject to mild acids. Nickel is normally applied in a thickness of .0002 to .0004 inches.

Tin
Tin plating is resistant to organic acids leading to extensive use in the food industry. It resists normal atmospheric corrosion and acts as a lubricant on steel, preventing galling. Tin is normally applied in a thickness of .0002 to .0004 inches.

Zinc
An electronically deposited plating which provides an excellent resistance to corrosion. Unlike cadmium it can be used in the presence of food. Zinc is normally applied in a thickness of .0002 to .0004 inches.

Typical Properties of Selected Materials

Material	Modulus of Elasticity, E		Modulus of Rigidity, G		Poisson's ratio	Unit Weight		
	Mpsi	GPa	Mpsi	GPa		lb/in^3	lb/ft^3	kN/m^3
Aluminum-all alloys	10.3	71.0	3.80	26.2	0.334	0.098	169	26.6
Beryllium copper	18.0	124.0	7.0	48.3	0.285	0.297	513	80.6
Brass	15.4	106.0	5.82	40.1	0.324	0.309	534	83.8
Carbon steel	30.0	207.0	11.5	79.3	0.292	0.282	487	76.5
Cast iron, gray	14.5	100.0	6.0	41.4	0.211	0.260	450	70.6
Copper	17.2	119.0	6.49	44.7	0.326	0.322	556	87.3
Douglas fir	1.6	11.0	0.6	4.1	0.33	0.016	28	4.3
Glass	6.7	46.2	2.7	18.6	0.245	0.094	162	25.4
Inconel	31.0	214.0	11.0	75.8	0.290	0.307	530	83.3
Lead	5.3	36.5	1.9	13.1	0.425	0.411	710	111.5
Magnesium	6.5	44.8	2.4	16.5	0.350	0.065	112	17.6
Molybdenum	48.0	331.0	17.0	117.0	0.307	0.368	636	100.0
Monel metal	26.0	179.0	9.5	65.5	0.320	0.319	551	86.6
Nickel silver	18.5	127.0	7.0	48.3	0.322	0.316	546	85.8
Nickel steel	30.0	207.0	11.5	79.3	0.291	0.280	484	76.0
Phosphor bronze	16.1	111.0	6.0	41.4	0.349	0.295	510	80.1
Stainless steel (18-8)	27.6	190.0	10.6	73.1	0.305	0.280	484	76.0

Typical Properties of Wrought Aluminum Alloys

	AA Designation								
	1100		2024		6013	6061		7075	
	Temper								
	-0	-H18	-0	-T4	-T8	-0	-T6	-0	-T6
Yield Strength (10^3 psi)	5	22	11	47	62	8	40	15	73
Tensile Strength (10^3 psi)	13	24	27	68	64	18	45	33	83
Fatigue Endurance Limit (10^3 psi)	5	9	13	20	16	9	14	—	23
Elongation in 2 in. (%)	45	15	22	19	11	30	17	16	11
Modulus of Elasticity (10^6) psi)	10.0	10.0	10.6	10.6	10.0	10.0	10.0	10.4	10.4
Coeff. of Therm. Expansion (in./in.-°F x 10^{-6})	13.1	13.1	12.9	12.9	13.0	13.1	13.1	13.1	13.1
Density (lb/in^3)	0.098	0.098	0.100	0.100	.098	0.098	0.098	0.101	0.101

Typical Properties of Wrought Stainless Steels

	Austenitic								Ferritic			Martensitic (annealed)			
	AISI Grades								AISI Grades			AISI Grades			
	202	301	302	304/ 304L	309/ 309S	310/ 310S	316/ 316L	347/ 348	405	430	446	403/ 410	416/ 416Se	440A	440C
Yield Strength (10^3 psi)	42	40	35	35	40	40	30	40	40	45	50	40	40	60	65
Tensile Strength (10^3 psi) 70°F	105	105	90	85	90	95	80	95	65	65	80	75	75	105	110
-320°F	200	275	220	220	—	152	185	200	—	90	—	158	—	—	—
Impact Strength, Izod (ft-lb)70°F	115	100	110	110	110	110	110	110	20-35	35	5-15	85	20-64	—	—
-300°F	42-120	110	110	110	—	85	—	95	—	2	—	5	—	—	—
Elongation in 2 in. (%)	55	55	55	55	45	45	55	45	25	30	25	30	25	20	14
Modulus of Elasticity (10^6 psi)	29	28	28	28	29	30	28	28	29	29	29	29	29	29	29

Typical Properties of Cast Iron

	Ductile				White	Gray					
	0-55-06	60-40-18	100-70-03	120-90-02		20	25	30	40	50	60
Yield Strength (10^3 psi)	60-75	45-60	75-90	90-125	—	*	*	*	*	*	*
Tensile Strength (10^3 psi)	90-110	60-80	100-120	120-150	—	20-25[1]	25-30[1]	30-35[1]	40-48[1]	50-57[1]	60-66[1]
Impact Strength Charpy unnotched (ft-lb)	15-65	60-115	35-50	25-40	3.5-10.0	—	—	—	—	—	—
Elongation in 2 in. (%)	3-10	10-25	6-10	2-7	—	=1%	=1%	=1%	=0.8%	=0.5%	=0.5%
Modulus of Elasticity (10^6 psi)	22-25	22-25	22-25	22-25	—	12	13	15	17	19	20

[1] Yield strength for gray iron is considered to be strength required to produce 0.1% permanent strain, usually about 65-80% of tensile strength.

Typical Properties of Carbon Steel[1]

	AISI									
	1018		1020		1035		1040		1045	
	HR	CR	HR	CR	HR	CR	HR	CR	HR	CR
Yield Strength (10^3 psi)	32	45	30	51	39	—	42	71	45	77
Tensile Strength (10^3 psi)	58	64	55	61	72	—	76	85	82	91
Hardness BHN	116	126	111	121	143	—	149	170	163	179
Impact Strength Izod (ft-lb)	85	—	85	—	36	—	36	39	23	—
Elongation in 2 in. (%)	25	15	25	15	18	—	18	12	16	12
Modulus of Elasticity (10^6) psi)	30	30	30	30	30	30	30	30	30	30
Machinability Rating (C1212)	52	70	52	65	65	—	60	—	56	65

Typical Properties of Wrought Copper Alloys

	Alloy							
	Tellurium-copper	Leaded beryllium copper	Med. Leaded brass (65Cu-34Zn)	Free-Cutting brass	Leaded phos.-bronze (88Cu-4Zn)	Aluminum-silicon-bronze (91Cu-7Al-2Si)	Silicon-bronze (97Cu-3Si)	Manganese bronze
	CDA Designation							
	145	173	340	360	544	642	655	675
Yield strength (10^3 psi)	10-44	25-178	19-42	18-45	na-57	55-68	22-60	30-60
Tensile strength (10^3 psi)	32-48	68-195	50-55	49-58	na-68	90-102	58-108	65-84
Elongation in 2 in. (%)	50-20	48-5	60-40	53-25	na-20	28-22	60-13	33-19
Modulus of Elasticity (10^6) psi)	17	18.5	15	14	15	16	15	15
Coeff. of therm. expansion (in./in.-°F x 10^{-7})	95	99	113	114	96	100	100	118
Density (lb/in^3)	0.323	0.298	0.306	0.307	0.321	0.278	0.308	0.302

Typical Properties of Alloy Steel[1]

	AISI							
	4130		4140		4340		8620	
	HR	CR	HR	CR	HR	CR	HR	CR
Yield strength (10^3 psi)	56	87	62	90	69	99	63	85
Tensile Strength (10^3 psi)	86	98	89	102	101	111	85	101
Hardness BHN	183	201	187	223	207	223	183	212
Impact Strength Izod (ft-lb)	53	—	25	—	25	—	82	—
Elongation in 2 in. (%)	29	21	26	18	21	16	28	22
Modulus of Elasticity (10^6) psi	30	30	30	30	30	30	30	30
Machinability Rating (C1212)	65	70	57	66	45	55	58	63

Typical Properties of Structural Steel & Pressure Vessel Plate[1]

	Structural Steel		Pressure Vessel Plate		
	A36	A283 Gr D	A285 Gr C	A515 Gr 55	A516 Gr 70
Yield strength (10^3 psi)	36	33	30	30	38
Tensile Strength (10^3 psi)	58	60	55	55	70
Hardness BHN	—	—	—	—	—
Impact Strength Izod (ft-lb)	—	—	—	—	—
Elongation in 2 in. (%)	—	—	—	—	—
Modulus of Elasticity (10^6) psi)	30	30	30	30	30
Machinability Rating	—	—	—	—	—

[1] Typical properties based on 1" diameter bar. Larger sections have lower mechanical properties.

Approximate Hardness Conversion Numbers For Steel

Based on Brinell Hardness Numbers

Brinell Indentation Dia.	Hardness No. 10 mm Tungsten Carbide Ball, 3,000 Kg Load	Rockwell Hardness No. C-Scale 150 Kg Load, Brale Penetrator	Rockwell Hardness No. B-Scale 100 Kg Load, 1/16 in. Diam. Ball	Rockwell Hardness No. A-Scale 60 Kg Load, Brale Penetrator	Diamond Pyramid Hardness No. Vickers	Shore Scleroscope Hardness No.	Tensile Strength (Approximate) in 1,000 psi
2.25	745	65.3		84.1	840	91	
2.30	712						
2.35	682	61.7		82.2	737	84	
2.40	653	60.0		81.2	697	81	
2.45	627	58.7		80.5	667	79	
2.50	601	57.3		79.8	640	77	
2.55	578	56.0		79.1	615	75	
2.60	555	54.7		78.4	591	73	298
2.65	534	53.5		77.8	569	71	288
2.70	514	52.1		76.9	547	70	274
2.75	495	51.0		76.3	528	68	264
2.80	477	49.6		75.6	508	66	252
2.85	461	48.5		74.9	491	65	242
2.90	444	47.1		74.2	472	63	230
2.95	429	45.7		73.4	455	61	219
3.00	415	44.5		72.8	440	59	212
3.05	401	43.1		72.0	425	58	202
3.10	388	41.8		71.4	410	56	193
3.15	375	40.4		70.6	396	54	184
3.20	363	39.1		70.0	383	52	177
3.25	352	37.9	(110.0)	69.3	372	51	171
3.30	341	36.6	(109.0)	68.7	360	50	164
3.35	331	35.5	(108.5)	68.1	350	48	159
3.40	321	34.3	(108.0)	67.5	339	47	154
3.45	311	33.1	(107.5)	66.9	328	46	149
3.50	302	32.1	(107.0)	66.3	319	45	146
3.55	293	30.9	(106.0)	65.7	309	43	141
3.60	285	29.9	(105.5)	65.3	301		138
3.65	277	28.8	(104.5)	64.6	292	41	134
3.70	269	27.6	(104.0)	64.1	284	40	130
3.75	262	26.6	(103.0)	63.6	276	39	127
3.80	255	25.4	(102.0)	63.0	269	38	123
3.85	248	24.2	(101.0)	62.5	261	37	120
3.90	241	22.8	100.0	61.8	253	36	116
3.95	235	21.7	99.0	61.4	247	35	114
4.00	229	20.5	98.2	60.8	241	34	111
4.05	223	(18.8)	97.3		234		
4.10	217	(17.5)	96.4		228	33	105
4.15	212	(16.0)	95.5		222		102
4.20	207	(15.2)	94.6		218	32	100
4.25	201	(13.8)	93.8		212	31	98
4.30	197	(12.7)	92.8		207	30	95
4.35	192	(11.5)	91.9		202	29	93
4.40	187	(10.0)	90.7		196		90
4.45	183	(9.0)	90.0		192	28	89
4.50	179	(8.0)	89.0		188	27	87
4.55	174	(6.4)	87.8		182		85
4.60	170	(5.4)	86.8		178	26	83
4.65	167	(4.4)	86.0		175		81
4.70	163	(3.3)	85.0		171	25	79
4.80	156	(0.9)	82.9		163		76
4.90	149		80.8		156	23	73
5.00	143		78.7		150	22	71
5.10	137		76.4		143	21	67
5.20	131		74.0		137		65

American Welding Society Symbols

The following information on welding symbols has been reproduced with permission of the "American Welding Society".

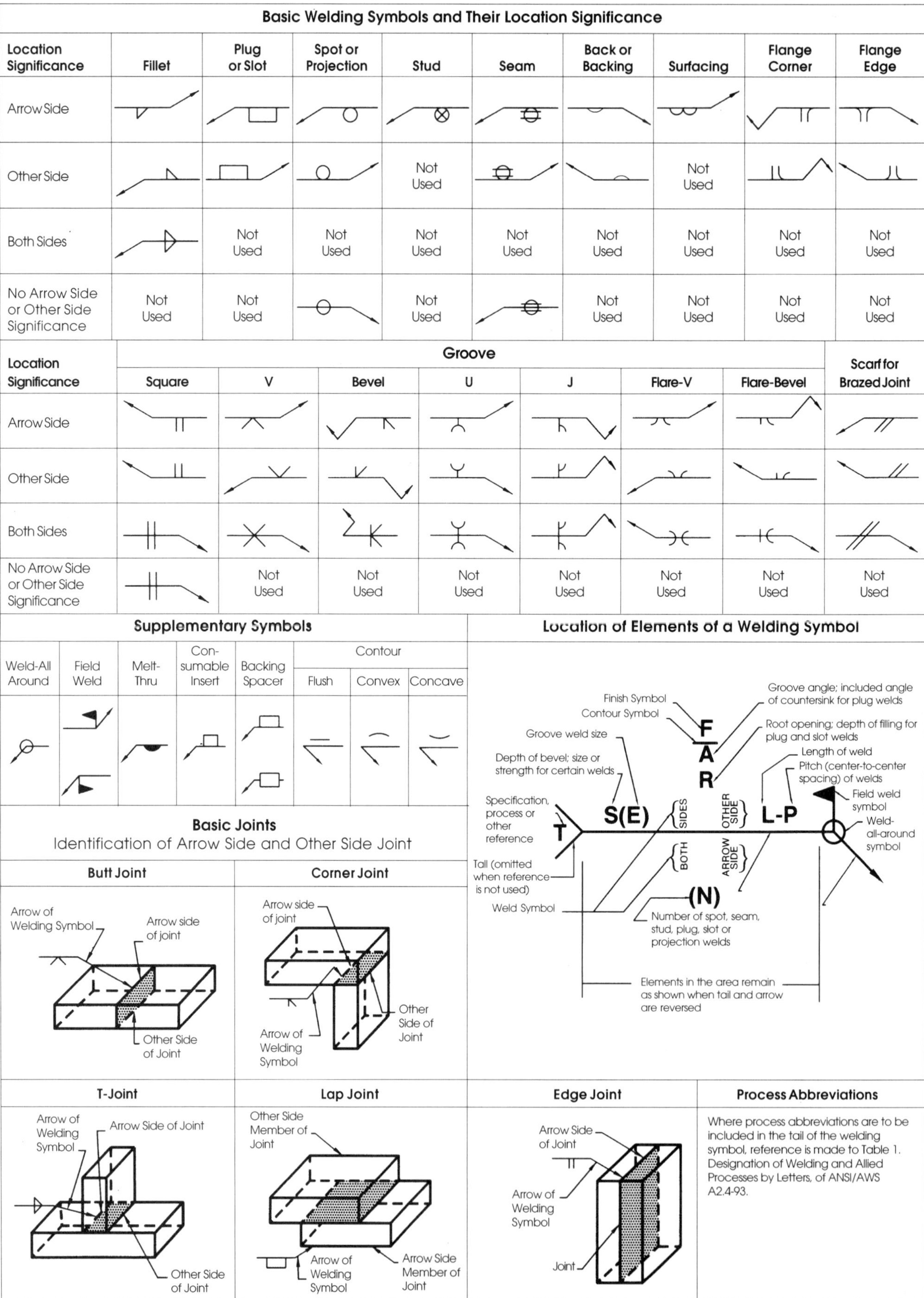

Basic Welding Symbols and Their Location Significance

Location Significance	Fillet	Plug or Slot	Spot or Projection	Stud	Seam	Back or Backing	Surfacing	Flange Corner	Flange Edge
Arrow Side									
Other Side				Not Used			Not Used		
Both Sides		Not Used	Not Used	Not Used	Not Used	Not Used	Not Used	Not Used	Not Used
No Arrow Side or Other Side Significance	Not Used	Not Used		Not Used		Not Used	Not Used	Not Used	Not Used

Location Significance	Groove							Scarf for Brazed Joint
	Square	V	Bevel	U	J	Flare-V	Flare-Bevel	
Arrow Side								
Other Side								
Both Sides								
No Arrow Side or Other Side Significance		Not Used	Not Used	Not Used	Not Used	Not Used	Not Used	Not Used

American Welding Society Symbols

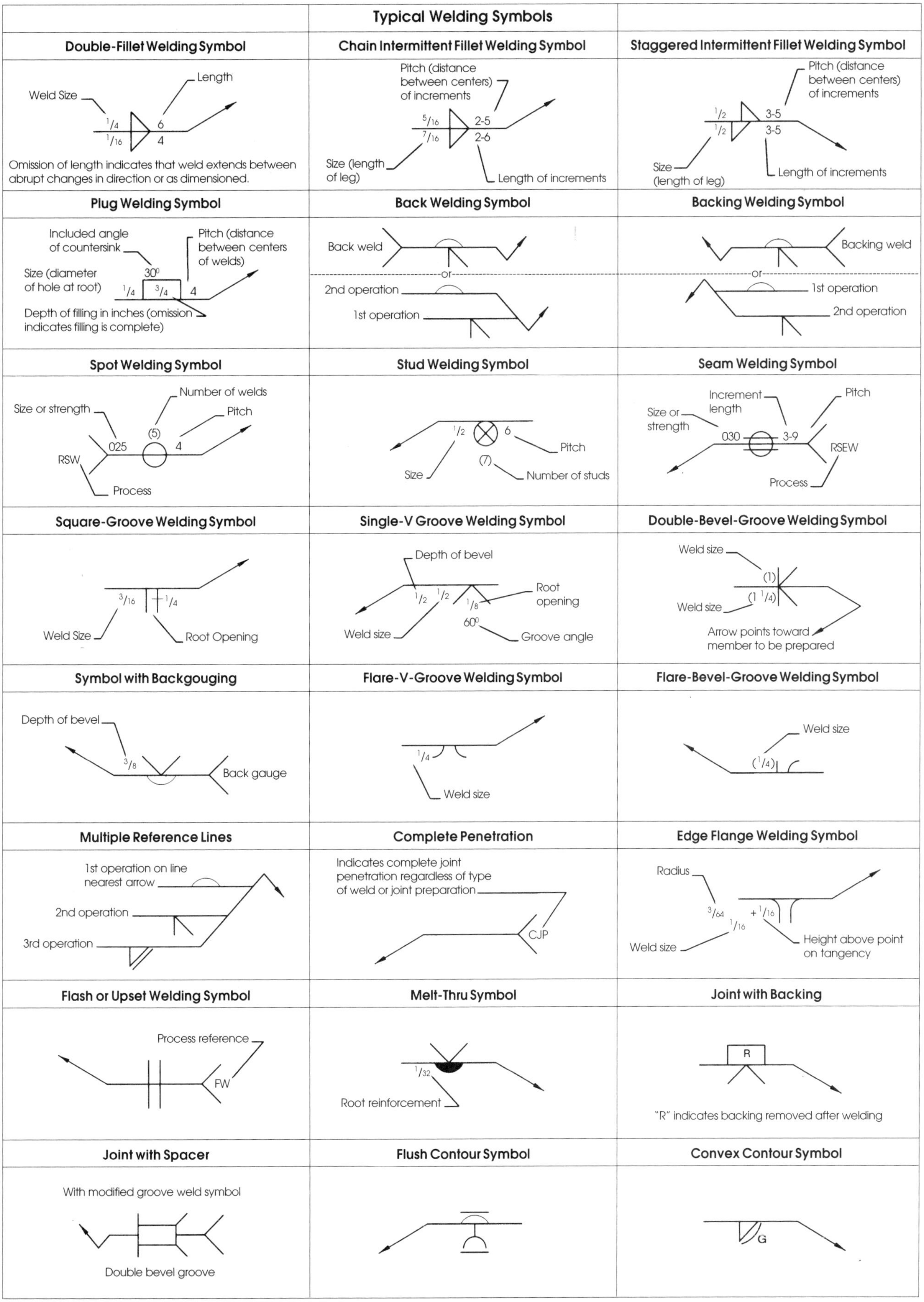

Typical Welding Symbols

Double-Fillet Welding Symbol

Weld Size
Length
$^1/_4$ 6
$^1/_{16}$ 4

Omission of length indicates that weld extends between abrupt changes in direction or as dimensioned.

Chain Intermittent Fillet Welding Symbol

Pitch (distance between centers) of increments
$^5/_{16}$ 2-5
$^7/_{16}$ 2-6
Size (length of leg)
Length of increments

Staggered Intermittent Fillet Welding Symbol

Pitch (distance between centers) of increments
$^1/_2$ 3-5
$^1/_2$ 3-5
Size (length of leg)
Length of increments

Plug Welding Symbol

Included angle of countersink
Pitch (distance between centers of welds)
Size (diameter of hole at root)
30°
$^1/_4$ $^3/_4$ 4
Depth of filling in inches (omission indicates filling is complete)

Back Welding Symbol

Back weld
or
2nd operation
1st operation

Backing Welding Symbol

Backing weld
or
1st operation
2nd operation

Spot Welding Symbol

Size or strength
Number of welds
Pitch
(5)
025 4
RSW
Process

Stud Welding Symbol

$^1/_2$ 6
Pitch
(7)
Size
Number of studs

Seam Welding Symbol

Increment length
Pitch
Size or strength
030 3-9
RSEW
Process

Square-Groove Welding Symbol

$^3/_{16}$ $^1/_4$
Weld Size
Root Opening

Single-V Groove Welding Symbol

Depth of bevel
$^1/_2$ $^1/_2$ $^1/_8$
Root opening
60°
Weld size
Groove angle

Double-Bevel-Groove Welding Symbol

Weld size
(1)
(1 $^1/_4$)
Weld size
Arrow points toward member to be prepared

Symbol with Backgouging

Depth of bevel
$^3/_8$
Back gauge

Flare-V-Groove Welding Symbol

$^1/_4$
Weld size

Flare-Bevel-Groove Welding Symbol

Weld size
($^1/_4$)

Multiple Reference Lines

1st operation on line nearest arrow
2nd operation
3rd operation

Complete Penetration

Indicates complete joint penetration regardless of type of weld or joint preparation
CJP

Edge Flange Welding Symbol

Radius
$^3/_{64}$ + $^1/_{16}$
$^1/_{16}$
Weld size
Height above point on tangency

Flash or Upset Welding Symbol

Process reference
FW

Melt-Thru Symbol

$^1/_{32}$
Root reinforcement

Joint with Backing

R

"R" indicates backing removed after welding

Joint with Spacer

With modified groove weld symbol

Double bevel groove

Flush Contour Symbol

Convex Contour Symbol

G

Standard Thread Relief

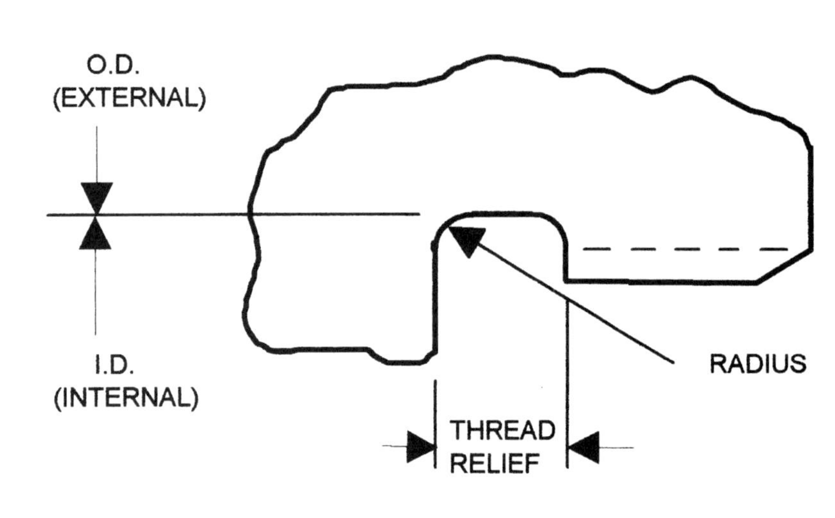

Thread Relief	Radius
.109	.030
.125 to .328	.062
.375 to .453	.094

Pitch	Thread Relief ± .015	I.D. +.005 -.000	O.D. +.000 -.005
1/4-28	.108	.260	.195
1/4-20	.150	.260	.178
5/16-24	.125	.323	.250
5/16-18	.167	.323	.233
3/8-24	.125	.386	.313
3/8-16	.188	.386	.287
7/16-20	.150	.448	.365
7/16-14	.215	.448	.339
1/2-20	.150	.510	.427
1/2-13	.230	.510	.394
9/16-18	.167	.573	.483
9/16-12	.250	.573	.449
5/8-18	.167	.635	.545
5/8-11	.273	.635	.502

Pitch	Thread Relief ± .015	I.D. +.005 -.000	O.D. +.000 -.005
3/4-16	.188	.760	.662
3/4-10	.300	.760	.616
7/8-14	.215	.885	.776
7/8-9	.333	.885	.727
1-12	.250	1.010	.886
1-8	.375	1.010	.835
1 1/8-12	.250	1.135	1.011
1 1/8-7	.428	1.135	.938
1 1/4-12	.250	1.260	1.136
1 1/4-7	.428	1.260	1.063
1 3/8-12	.250	1.385	1.261
1 3/8-6	.500	1.385	1.158
1 1/2-12	.250	1.510	1.386
1 1/2-6	.500	1.510	1.283

Drill Sizes

Drill Sizes									
Drill Size	Dec. Equiv.	Drill Size	Dec. Equiv.	Drill Size	Dec. Equiv.	Drill Size	Dec. Equiv.	Drill Size	Dec. Equiv.
80	0.0135	43	0.089	8	0.199	25/64	0.3906	1	1.000
79	0.0145	42	0.0935	7	0.201	X	0.397	1 1/64	1.0156
1/64	0.0156	3/32	0.0938	13/64	0.2031	Y	0.404	1 1/32	1.0312
78	0.016	41	0.096	6	0.204	13/32	0.4062	1 3/64	1.0469
77	0.018	40	0.098	5	0.2055	Z	0.413	1 1/16	1.0625
76	0.020	39	0.0995	4	0.209	27/64	0.4219	1 5/64	1.0781
75	0.021	38	0.1015	3	0.213	7/16	0.4375	1 3/32	1.0938
74	0.0225	37	0.104	7/32	0.2188	29/64	0.4531	1 7/64	1.1094
73	0.024	36	0.1065	2	0.221	15/32	0.4688	1 1/8	1.125
72	0.025	7/64	0.1094	1	0.228	31/64	0.4844	1 9/64	1.1406
71	0.026	35	0.110	A	0.234	1/2	0.500	1 5/32	1.1562
70	0.028	34	0.111	15/64	0.2344	33/64	0.5156	1 11/64	1.1719
69	0.0292	33	0.113	B	0.238	17/32	0.5312	1 3/16	1.1875
68	0.031	32	0.116	C	0.242	35/64	0.5469	1 13/64	1.2031
1/32	0.0312	31	0.120	D	0.246	9/16	0.5625	1 7/32	1.2188
67	0.032	1/8	0.125	1/4	0.250	37/64	0.5781	1 15/64	1.2344
66	0.033	30	0.1285	F	0.257	19/32	0.5938	1 1/4	1.250
65	0.035	29	0.136	G	0.261	39/64	0.6094	1 9/32	1.2812
64	0.036	28	0.1405	17/64	0.2656	5/8	0.625	1 5/16	1.3125
63	0.037	9/64	0.1406	H	0.266	41/64	0.6406	1 11/32	1.3438
62	0.038	27	0.144	I	0.272	21/32	0.6562	1 3/8	1.375
61	0.039	26	0.147	J	0.277	43/64	0.6719	1 13/32	1.4062
60	0.040	25	0.1495	K	0.281	11/16	0.6875	1 7/16	1.4375
59	0.041	24	0.152	9/32	0.2812	45/64	0.7031	1 15/32	1.4688
58	0.042	23	0.154	L	0.290	23/32	0.7188	1 1/2	1.500
57	0.043	5/32	0.1562	M	0.295	47/64	0.7344	1 9/16	1.5625
56	0.0465	22	0.157	19/64	0.2969	3/4	0.750	1 5/8	1.625
3/64	0.0469	21	0.159	N	0.302	49/64	0.7656	1 11/16	1.6875
55	0.052	20	0.161	5/16	0.3125	25/32	0.7812	1 3/4	1.750
54	0.055	19	0.166	O	0.316	51/64	0.7969	1 13/16	1.8125
53	0.0595	18	0.1695	P	0.323	13/16	0.8125	1 7/8	1.875
1/16	0.0625	11/64	0.1719	21/64	0.3281	53/64	0.8281	1 15/16	1.9375
52	0.0635	17	0.173	Q	0.332	27/32	0.8438	2	2.000
51	0.067	16	0.177	R	0.339	55/64	0.8594		
50	0.070	15	0.180	11/32	0.3438	7/8	0.875		
49	0.073	14	0.182	S	0.348	57/64	0.8906		
48	0.076	13	0.185	T	0.358	29/32	0.9062		
5/64	0.0781	3/16	0.1875	23/64	0.3594	59/64	0.9219		
47	0.0785	12	0.189	U	0.368	15/16	0.9375		
46	0.081	11	0.191	3/8	0.375	61/64	0.9531		
45	0.082	10	0.1935	V	0.377	31/32	0.9688		
44	0.086	9	0.196	W	0.386	63/64	0.9844		

Fractions With Decimal Equivalent

8THS	16THS	32NDS		64THS			
1/8-.125	1/16-.0625	1/32-.03125	17/32-.53125	1/64-.015625	17/64-.265625	33/64-.515625	49/64-765625
1/4-.250	3/16-.1875	3/32-.09375	19/32-.59375	3/64-.046875	19/64-.296875	35/64-.546875	51/64-.796875
3/8-.375	5/16-.3125	5/32-.15625	21/32-.65625	5/64-.078125	21/64-.328125	37/64-.578125	53/64-.828125
1/2-.500	7/16-.4375	7/32-.21875	23/32-.71875	7/64-.109375	23/64-.359375	39/64-.609375	55/64-.859375
5/8-.625	9/16-.5625	9/32-.28125	25/32-.78125	9/64-.140625	25/64-.390625	41/64-.640625	57/64-.890625
3/4-.750	11/16-.6875	11/32-.34375	27/32-.84375	11/64-.171875	27/64-.421875	43/64-.671875	59/64-.921875
7/8-.875	13/16-.8125	13/32-.40625	29/32-.90625	13/64-.203125	29/64-.453125	45/64-.703125	61/64-.953125
	15/16-.9375	15/32-.46875	31/32-.96875	15/64-.234375	31/64-.484375	47/64-.734375	63/64-.984375

Identification of SAE Bolt Grades Head Markings

Grades 0, 1 and 2: no marking

Grade 3: 2 radial dashes 180^0 apart

Grade 5: 3 radial dashes 120^0 apart

Grade 6: 4 radial dashes 90^0 apart

Grade 7: 5 radial dashes 72^0 apart

Grade 8: 6 radial dashes 60^0 apart

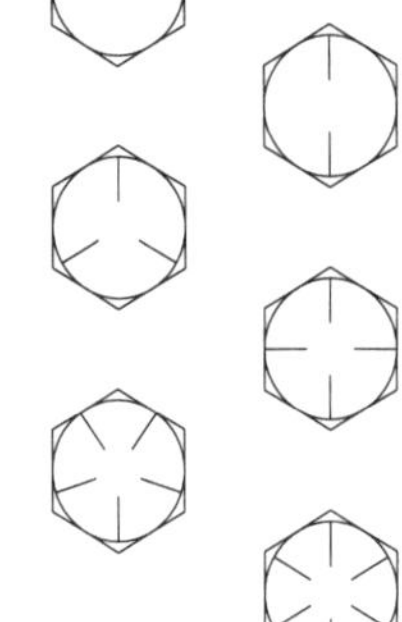

Bolt, Cap Screw & Stud

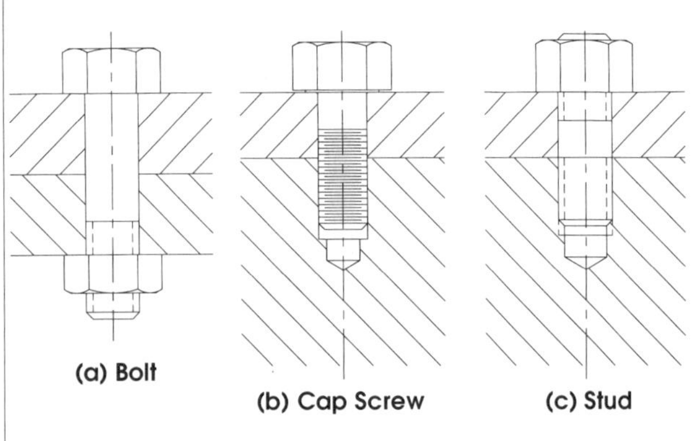

(a) Bolt (b) Cap Screw (c) Stud

Hexagon Head Bolts and Screws

Lengths 3/4" to 8", increment = 1/4"
Lengths 8" to 20", increment = 1/2"
Lengths 20" to 30", increment = 1"

For bolts or screws up to 6" in length:
thread length = 2D + 1/4"
For bolts or screws over 6" in length:
thread length = 2D + 1/2"

Socket set screw lengths are standardized as follows[1]:

Lengths 1/16" through 3/16", increment = 1/32"
Lengths 1/8" through 1/2", increment = 1/16"
Lengths 1/2" through 1", increment = 1/8"
Lengths 1" through 2", increment = 1/4"
Lengths 2" through 6", increment = 1/2"
Lengths over 6", increment = 1"

[1]Applicable only to sizes 0 (.060") to 3 (.099") inclusive

Taps and Tap Drill Sizes

Course Thread	
Thread Size and Designation	Tap Drill Size
(#2) .086 - 56 UNC	.070 (#50)
(#4) .112 - 40 UNC	.089 (#43)
(#6) .138 - 32 UNC	.106 (#36)
(#8) .164 - 32 UNC	.136 (#29)
(#10) .190 - 24 UNC	.149 (#25)
(#12) .216 - 24 UNC	.177 (#16)
.250 - 20 UNC	.201 (#7)
.3125 - 18 UNC	.257 (F)
.375 - 16 UNC	5/16
.4375 - 14 UNC	23/64
.500 - 13 UNC	27/64
.5625 - 12 UNC	31/64
.625 - 11 UNC	17/32
.750 - 10 UNC	21/32
.875 - 9 UNC	49/64
1.000 - 8 UNC	7/8
1.125 - 7 UNC	63/64
1.250 - 7 UNC	1 3/32
1.500 - 6 UNC	1 21/64

Fine Thread	
Thread Size and Designation	Tap Drill Size
(#2) .086 - 64 UNF	.070 (#50)
(#4) .112 - 48 UNF	.093 (#42)
(#6) .138 - 40 UNF	.113 (#33)
(#8) .164 - 36 UNF	.136 (#29)
(#10) .190 - 32 UNF	.159 (#21)
(#12) .216 - 28 UNF	.182 (#14)
.250 - 28 UNF	.213 (#3)
.3125 - 24 UNF	.272 (I)
.375 - 24 UNF	.332 (Q)
.4375 - 20 UNF	25/64
.500 - 20 UNF	29/64
.5625 - 18 UNF	1/2
.625 - 18 UNF	9/16
.750 - 16 UNF	11/16
.875 - 14 UNF	13/16
1.000 - 12 UNF	29/32
1.125 - 12 UNF	1 3/64
1.250 - 12 UNF	1 11/64
1.500 - 12 UNF	1 27/64

Mechanical Requirements For Screws

Grade Designation		Nominal Size of Product In.	Proof Stress ksi	Proof Stress MPA	Yield Strength ksi Min	Yield Strength MPa Min
Inch system	Metric System					
SAE Grade 1		1/4 to 1 1/2	33	(228)	36	(248)
SAE Grade 2		1/4 to 3/4	55	(379)	57	(393)
		3/4 to 1 1/2	33	(228)	36	(248)
	5.8	M5 — M24	(55)	380	(61)	420
SAE Grade 5		1/4 to 1	85	(586)	92	(634)
		1 1/8 to 1 1/2	74	(510)	81	(559)
	8.8	M16 — M72	(87)	600	(96)	660
SAE Grade 8		1/4 to 1 1/2	120	(828)	130	(897)
	10.9	M5 — M100	(120)	830	(136)	940

NOTE: SAE Grade 2 properties apply to screws up to 6" long. See Grade 1 for screws over 6" long.

Inch Series ▪ Hex Cap Screws

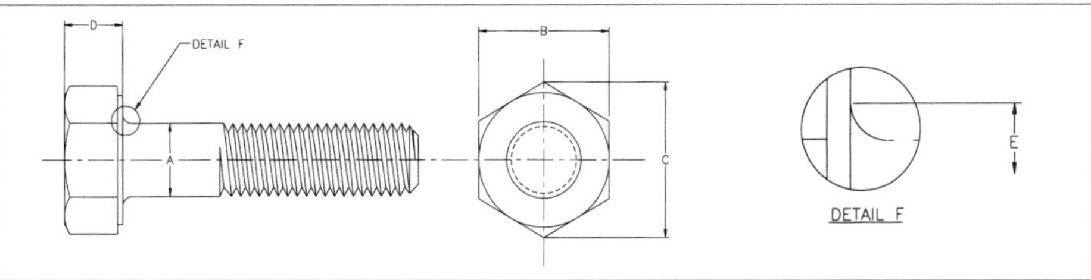

Nominal Size or Basic Product Dia			A		B			C		D			E
			Body Body Dia		Width Across Flats			Width Across Corners		Height			Fillet Transition Dia[2]
	Inch	mm[1]	Max	Min	Basic	Max	Min	Max	Min	Basic	Max	Min	Max
1/4	0.2500	6.35	0.2500	0.2450	7/16	0.438	0.428	0.505	0.488	5/32	0.163	0.150	.300
5/16	0.3125	7.94	0.3125	0.3065	1/2	0.500	0.489	0.577	0.557	13/64	0.211	0.195	.362
3/8	0.3750	9.52	0.3750	0.3690	9/16	0.562	0.551	0.650	0.628	15/64	0.243	0.226	.425
7/16	0.4375	11.11	0.4375	0.4305	5/8	0.625	0.612	0.722	0.698	9/32	0.291	0.272	.488
1/2	0.5000	12.70	0.5000	0.4930	3/4	0.750	0.736	0.866	0.840	5/16	0.323	0.302	.550
9/16	0.5625	14.29	0.5625	0.5545	13/16	0.812	0.798	0.938	0.910	23/64	0.371	0.348	.652
5/8	0.6250	15.88	0.6250	0.6170	15/16	0.938	0.922	1.083	1.051	25/64	0.403	0.378	.715
3/4	0.7500	19.05	0.7500	0.7410	1 1/8	1.125	1.100	1.299	1.254	15/32	0.483	0.455	.840
7/8	0.8750	22.22	0.8750	0.8660	1 5/16	1.312	1.285	1.516	1.465	35/64	0.563	0.531	1.005
1	1.0000	25.40	1.0000	0.9900	1 1/2	1.500	1.469	1.732	1.675	39/64	0.627	0.591	1.190
1 1/8	1.1250	28.58	1.1250	1.1140	1 11/16	1.688	1.631	1.949	1.859	11/16	0.718	0.658	1.315
1 1/4	1.2500	31.75	1.2500	1.2390	1 7/8	1.875	1.812	2.165	2.066	25/32	0.813	0.749	1.440
1 3/8	1.3750	34.92	1.3750	1.3630	2 1/16	2.062	1.994	2.382	2.273	27/32	0.878	0.810	1.565
1 1/2	1.5000	38.10	1.5000	1.4880	2 1/4	2.250	2.175	2.598	2.480	1 5/16	0.974	0.902	1.690
1 3/4	1.7500	44.45	1.7500	1.7380	2 5/8	2.625	2.538	3.031	2.893	1 3/32	1.134	1.054	1.940
2	2.0000	50.80	2.0000	1.9880	3	3.000	2.900	3.464	3.306	1 7/32	1.263	1.175	2.190
2 1/4	2.2500	57.15	2.2500	2.2380	3 3/8	3.375	3.262	3.897	3.719	1 3/8	1.423	1.327	2.440
2 1/2	2.5000	63.50	2.5000	2.4880	3 3/4	3.750	3.625	4.330	4.133	1 17/32	1.583	1.479	2.690
2 3/4	2.7500	69.85	2.7500	2.7380	4 1/8	4.125	3.988	4.763	4.546	1 11/16	1.744	1.632	2.940
3	3.0000	76.20	3.0000	2.9880	4 1/2	4.500	4.350	5.196	4.959	1 7/8	1.935	1.815	3.190

[1] For reference only
[2] For both "short" and "long" bolts

Inch Series ▪ Hex Nuts and Hex Jam Nuts

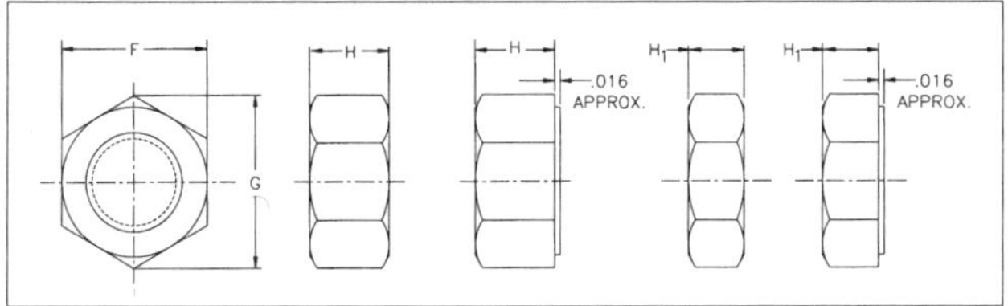

Nominal Size or Basic Major Dia of Thread		F			G		H			H_1		
		Width Across Flats			Width Across Corners		Thickness Hex Nuts			Thickness Hex Jam Nuts		
		Basic	Max	Min	Max	Min	Basic	Max	Min	Basic	Max	Min
1/4	0.2500	7/16	0.438	0.428	0.505	0.488	7/32	0.226	0.212	5/32	0.163	0.150
5/16	0.3125	1/2	0.500	0.489	0.577	0.557	17/64	0.273	0.258	3/16	0.195	0.180
3/8	0.3750	9/16	0.562	0.551	0.650	0.628	21/64	0.337	0.320	7/32	0.227	0.210
7/16	0.4375	11/16	0.688	0.675	0.794	0.768	3/8	0.385	0.365	1/4	0.260	0.240
1/2	0.5000	3/4	0.750	0.736	0.866	0.840	7/16	0.448	0.427	5/16	0.323	0.302
9/16	0.5625	7/8	0.875	0.861	1.010	0.982	31/64	0.496	0.473	5/16	0.324	0.301
5/8	0.6250	15/16	0.938	0.922	1.083	1.051	35/64	0.559	0.535	3/8	0.387	0.363
3/4	0.7500	1 1/8	1.125	1.088	1.299	1.240	41/64	0.665	0.617	27/64	0.446	0.398
7/8	0.8750	1 5/16	1.312	1.269	1.516	1.447	3/4	0.776	0.724	31/64	0.510	0.458
1	1.0000	1 1/2	1.500	1.450	1.732	1.653	55/64	0.887	0.831	35/64	0.575	0.519
1 1/8	1.1250	1 11/16	1.688	1.631	1.949	1.859	31/32	0.999	0.939	39/64	0.639	0.579
1 1/4	1.2500	1 7/8	1.875	1.812	2.165	2.066	1 1/16	1.094	1.030	23/32	0.751	0.687
1 3/8	1.3750	2 1/16	2.062	1.994	2.382	2.273	1 11/64	1.206	1.138	25/32	0.815	0.747
1 1/2	1.5000	2 1/4	2.250	2.175	2.598	2.480	1 9/32	1.317	1.245	27/32	0.880	0.808

Inch Series ▪ Plain Washers

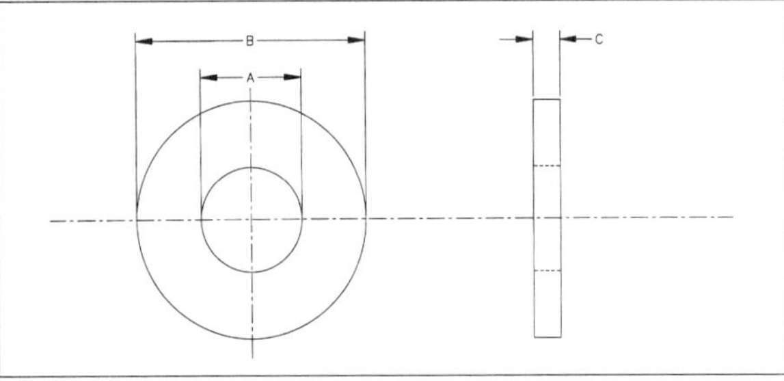

Nominal Washer Size			A Inside Diameter			B Outside Diameter			C Thickness		
			Basic	Tolerance Plus	Tolerance Minus	Basic	Tolerance Plus	Tolerance Minus	Basic	Max	Min
			0.078	0.000	0.005	0.188	0.000	0.005	0.020	0.025	0.016
			0.094	0.000	0.005	0.250	0.000	0.005	0.020	0.025	0.016
			0.125	0.008	0.005	0.312	0.008	0.005	0.032	0.040	0.025
No. 6	0.138		0.156	0.008	0.005	0.375	0.015	0.005	0.049	0.065	0.036
8	0.164		0.188	0.008	0.005	0.438	0.015	0.005	0.049	0.065	0.036
10	0.190		0.219	0.008	0.005	0.500	0.015	0.005	0.049	0.065	0.036
3/16	0.188		0.250	0.015	0.005	0.562	0.015	0.005	0.049	0.065	0.036
12	0.216		0.250	0.015	0.005	0.562	0.015	0.005	0.065	0.080	0.051
1/4	0.250	N	0.281	0.015	0.005	0.625	0.015	0.005	0.065	0.080	0.051
1/4	0.250	W	0.312	0.015	0.005	0.734	0.015	0.007	0.065	0.080	0.051
5/16	0.312	N	0.344	0.015	0.005	0.688	0.015	0.007	0.065	0.080	0.051
5/16	0.312	W	0.375	0.015	0.005	0.875	0.030	0.007	0.083	0.104	0.064
3/8	0.375	N	0.406	0.015	0.005	0.812	0.015	0.007	0.065	0.080	0.051
3/8	0.375	W	0.438	0.015	0.005	1.000	0.030	0.007	0.083	0.104	0.064
7/16	0.438	N	0.469	0.015	0.005	0.922	0.015	0.007	0.065	0.080	0.051
7/16	0.438	W	0.500	0.015	0.005	1.250	0.030	0.007	0.083	0.104	0.064
1/2	0.500	N	0.531	0.015	0.005	1.062	0.030	0.007	0.095	0.121	0.074
1/2	0.500	W	0.562	0.015	0.005	1.375	0.030	0.007	0.109	0.132	0.086
9/16	0.562	N	0.594	0.015	0.005	1.156	0.030	0.007	0.095	0.121	0.074
9/16	0.562	W	0.625	0.015	0.005	1.469	0.030	0.007	0.109	0.132	0.086
5/8	0.625	N	0.656	0.030	0.007	1.312	0.030	0.007	0.095	0.121	0.074
5/8	0.625	W	0.688	0.030	0.007	1.750	0.030	0.007	0.134	0.160	0.108
3/4	0.750	N	0.812	0.030	0.007	1.469	0.030	0.007	0.134	0.160	0.108
3/4	0.750	W	0.812	0.030	0.007	2.000	0.030	0.007	0.148	0.177	0.122
7/8	0.875	N	0.938	0.030	0.007	1.750	0.030	0.007	0.134	0.160	0.108
7/8	0.875	W	0.938	0.030	0.007	2.250	0.030	0.007	0.165	0.192	0.136
1	1.000	N	1.062	0.030	0.007	2.000	0.030	0.007	0.134	0.160	0.108
1	1.000	W	1.062	0.030	0.007	2.500	0.030	0.007	0.165	0.192	0.136
1 1/8	1.125	N	1.250	0.030	0.007	2.250	0.030	0.007	0.134	0.160	0.108
1 1/8	1.125	W	1.250	0.030	0.007	2.750	0.030	0.007	0.165	0.192	0.136
1 1/4	1.250	N	1.375	0.030	0.007	2.500	0.030	0.007	0.165	0.192	0.136
1 1/4	1.250	W	1.375	0.030	0.007	3.000	0.030	0.007	0.165	0.192	0.136
1 3/8	1.375	N	1.500	0.030	0.007	2.750	0.030	0.007	0.165	0.192	0.136
1 3/8	1.375	W	1.500	0.045	0.010	3.250	0.045	0.010	0.180	0.213	0.153
1 1/2	1.500	N	1.625	0.030	0.007	3.000	0.030	0.007	0.165	0.192	0.136
1 1/2	1.500	W	1.625	0.045	0.010	3.500	0.045	0.010	0.180	0.213	0.153
1 5/8	1.625		1.750	0.045	0.010	3.750	0.045	0.010	0.180	0.213	0.153
1 3/4	1.750		1.875	0.045	0.010	4.000	0.045	0.010	0.180	0.213	0.153
1 7/8	1.875		2.000	0.045	0.010	4.250	0.045	0.010	0.180	0.213	0.153
2	2.000		2.125	0.045	0.010	4.500	0.045	0.010	0.180	0.213	0.153
2 1/4	2.250		2.375	0.045	0.010	4.750	0.045	0.010	0.220	0.248	0.193
2 1/2	2.500		2.625	0.045	0.010	5.000	0.045	0.010	0.238	0.280	0.210
2 3/4	2.750		2.875	0.065	0.010	5.250	0.065	0.010	0.259	0.310	0.228
3	3.000		3.125	0.065	0.010	5.500	0.065	0.010	0.284	0.327	0.249

NOTES:

1. Preferred sizes are for the most part from series previously designated "Standard Plate" and "SAE". Where common sizes existed in the two series, the SAE size is designated "N" (narrow) and the Standard Plate "W" (wide). These sizes as well as all other sizes of Type A Plain Washers are to be ordered by ID, OD and thickness dimensions.
2. Nominal washer sizes are intended for use with comparable nominal screw or bolt sizes.
3. The 0.734 in., 1.156 in. and 1.469 in. outside diameters avoid washers which could be used in coin operated devices.

Inch Series • Regular Helical Spring Lock Washers[1]

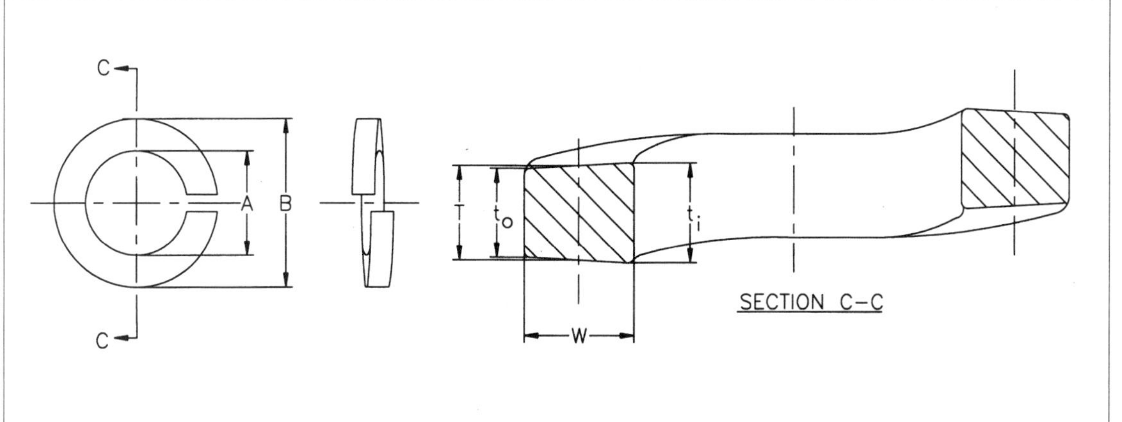

Nominal Washer Size		A Inside Diameter		B Outside Diameter	T Mean Section Thickness $\left(\frac{t_i + t_o}{2}\right)$	W Section Width
		Max	Min	Max[2]	Min	Min
No. 2	0.086	0.094	0.088	0.172	0.020	0.035
No. 3	0.099	0.107	0.101	0.195	0.025	0.040
No. 4	0.112	0.120	0.114	0.209	0.025	0.040
No. 5	0.125	0.133	0.127	0.236	0.031	0.047
No. 6	0.138	0.148	0.141	0.250	0.031	0.047
No. 8	0.164	0.174	0.167	0.293	0.040	0.055
No. 10	0.190	0.200	0.193	0.334	0.047	0.062
No. 12	0.216	0.227	0.220	0.377	0.056	0.070
1/4	0.250	0.262	0.254	0.489	0.062	0.109
5/16	0.312	0.326	0.317	0.586	0.078	0.125
3/8	0.375	0.390	0.380	0.683	0.094	0.141
7/16	0.438	0.455	0.443	0.779	0.109	0.156
1/2	0.500	0.518	0.506	0.873	0.125	0.171
9/16	0.562	0.582	0.570	0.971	0.141	0.188
5/8	0.625	0.650	0.635	1.079	0.156	0.203
11/16	0.688	0.713	0.698	1.176	0.172	0.219
3/4	0.750	0.775	0.760	1.271	0.188	0.234
13/16	0.812	0.843	0.824	1.367	0.203	0.250
7/8	0.875	0.905	0.887	1.464	0.219	0.266
15/16	0.938	0.970	0.950	1.560	0.234	0.281
1	1.000	1.042	1.017	1.661	0.250	0.297
1 1/16	1.062	1.107	1.080	1.756	0.266	0.312
1 1/8	1.125	1.172	1.144	1.853	0.281	0.328
1 3/16	1.188	1.237	1.208	1.950	0.297	0.344
1 1/4	1.250	1.302	1.271	2.045	0.312	0.359
1 5/16	1.312	1.366	1.334	2.141	0.328	0.375
1 3/8	1.375	1.432	1.398	2.239	0.344	0.391
1 7/16	1.438	1.497	1.462	2.334	0.359	0.406
1 1/2	1.500	1.561	1.525	2.430	0.375	0.422

[1] Formerly designated medium helical spring lock washers.
[2] The maximum outside diameters specified allow for the commercial tolerances on cold drawn wire.

Weights of Hex Cap Screws — lbs Per 100 Screws

Screw Length in. \ Nom. Length in.	1/4	5/16	3/8	7/16	1/2	9/16	5/8	3/4	7/8	1	1 1/8	1 1/4	1 3/8	1 1/2	1 3/4	2
3/8	1.05															
1/2	1.18	1.99	2.93													
5/8	1.31	2.20	3.24													
3/4	1.44	2.41	3.55	5.08	8.10											
7/8	1.57	2.62	3.86	5.50	8.66											
1	1.74	2.83	4.17	5.92	9.22	11.28	15.16	24.46	36.78	51.42	70.4	94.3	121.9	156.9	239.7	340.4
1 1/4	2.08	3.37	4.86	6.76	10.34	12.71	16.93	27.06	40.35	56.10	76.3	101.7	130.8	167.6	254.3	359.5
1 1/2	2.43	3.91	5.64	7.82	11.59	14.14	18.70	29.66	43.92	60.78	82.2	109.1	139.7	178.4	268.9	378.6
1 3/4	2.77	4.45	6.42	8.88	12.97	15.89	20.66	32.36	47.49	65.46	88.1	116.6	148.6	189.1	283.5	397.7
2	3.12	4.99	7.19	9.94	14.35	17.64	22.82	35.12	51.06	70.14	94.0	124.0	157.5	199.8	298.0	416.8
2 1/4	3.46	5.52	7.97	11.00	15.73	19.39	24.97	38.23	54.96	74.82	99.9	131.5	166.4	210.6	312.6	435.9
2 1/2	3.81	6.06	8.74	12.06	17.11	21.14	27.13	41.34	59.19	79.92	105.8	138.9	175.3	221.3	327.2	455.0
2 3/4	4.15	6.60	9.52	13.12	18.49	22.90	29.28	44.45	63.42	85.44	112.3	146.3	184.2	232.0	341.8	474.1
3	4.50	7.14	10.30	14.18	19.87	24.65	31.44	47.56	67.65	90.96	119.3	154.4	193.1	242.8	356.4	439.2
3 1/4	4.84	7.68	11.07	15.24	21.25	26.40	33.59	50.67	71.88	96.48	126.3	163.0	202.8	253.5	370.9	512.3
3 1/2	5.19	8.21	11.85	16.30	22.63	28.15	35.75	53.78	76.11	102.00	133.3	171.7	213.3	265.1	385.5	531.4
3 3/4	5.53	8.75	12.62	17.36	24.01	29.90	37.90	56.89	80.34	107.52	140.3	180.3	223.8	277.6	400.1	550.5
4	5.88	9.29	13.40	18.41	25.39	31.65	40.06	60.00	84.57	113.04	147.3	188.9	234.2	290.0	415.9	569.6
4 1/2	6.57	10.37	14.95	20.53	28.15	35.15	44.37	66.22	93.03	124.08	161.3	206.2	255.2	314.9	449.7	609.3
5	7.26	11.44	16.50	22.65	30.91	38.65	48.68	72.44	101.49	135.12	175.4	223.5	276.2	339.9	483.6	653.6
5 1/2	7.95	12.52	18.06	24.77	33.67	42.16	52.99	78.66	109.95	146.16	189.4	240.8	297.1	364.8	517.5	697.9
6	8.64	13.59	19.61	26.89	36.43	45.66	57.30	84.88	118.41	157.20	203.4	258.1	318.1	389.7	551.4	742.2
For each additional inch of screw length ADD	1.38	2.15	3.10	4.24	5.52	7.00	8.62	12.44	16.90	22.1	28.0	34.6	42.0	49.9	67.8	88.6

[1] Weights given in the table are those of carbon steel, coarse thread Hex Cap Screws. Carbon steel, fine thread Hex Cap Screws weigh approximately 2 percent more. For screws of longer lengths add the weight differential.

[2] All weights are approximate

Weights of Nuts — lbs Per 1000 Nuts

Screw Length in. \ Nom. Length in.	Square	Hex Flat	Hex Flat Jam	Hex	Hex Jam	Hex Slotted	Hex Thick	Hex Thick Slotted	Heavy Square	Heavy Hex	Heavy Hex Jam	Heavy Hex Slotted
1/4	8.93			7.35	5.15	6.03	9.53	8.21	13.7	11.6	8.52	9.90
5/16	18.1			11.0	7.67	9.40	13.7	12.1	20.6	17.2	11.7	15.1
3/8	26.3			16.0	10.5	13.1	20.0	17.1	37.4	31.4	20.2	26.9
7/16	43.6			28.4	18.6	23.6	34.5	29.7	50.1	41.6	26.1	35.9
1/2	57.8			37.5	26.2	31.6	48.4	42.5	78.7	65.4	40.0	57.0
9/16	—			58.3	36.8	49.4	74.1	65.2	—	81.5	49.1	71.3
5/8	108			73.3	49.3	60.7	97.4	84.8	143	119	69.6	102
3/4	154			119	77.0	103	153	137	235	193	110	171
7/8	245			190	120	171	232	213	362	297	167	272
1	363			283	176	250	331	298	515	425	235	385
1 1/8	525	403	247	403	247	358	489	444	724	592	324	537
1 1/4	706	543	361	543	361	475	642	574	955	786	458	706
1 3/8	945	730	479	730	479	655	860	785	1250	1020	593	933
1 1/2	1220	943	609	943	609	829	1110	996	1610	1310	748	1180
1 5/8										1620	916	—
1 3/4										2040	1140	1840
1 7/8										2410	1340	—
2										2990	1650	2740
2 1/4										4190	2270	3910
2 1/2										5640	3320	5160
2 3/4										7380	4290	6860
3										9500	5450	8820
3 1/4										11940	6510	11210
3 1/2										15260	8510	14490
3 3/4										18120	10050	17300
4										21800	12000	20930

[1] Weights given in the table are those of carbon steel, coarse thread Hex Cap Screws. Carbon steel, fine thread Hex Cap Screws weigh approximately 2 percent more. For screws of longer lengths add the weight differential.

[2] All weights are approximate

Recommended Torque Values
SAE Grade 2 Bolt

SAE Grade 2 Bolt							
					Assembly Torque		
Size	Basic Major Dia	Stress Area (Sq In)	Proof Load (psi)	Clamp Load (pounds)	Lubricated Max	Units	Dry Max
4-40	.1120	.00604	55,000	240	4.3	inch-lb	5
4-48	.1120	.00661	▲	280	4.8	▲	6
6-32	.1380	.00909		380	8.1		10
6-40	.1380	.01015		420	9.1		12
8-32	.1640	.01400		580	10.6		19
8-36	.1640	.01474		600	15.7		20
10-24	.1900	.0175		720	21.6		27
10-32	.1900	.0200		820	24.6		31
1/4-20	.2500	.0318		1,320	51.6	▼	66
1/4-28	.2500	.0364		1,500	59	inch-lb	76
5/16-18	.3125	.0524		2,160	8.8	ft-lb	11
5/16-24	.3125	.0580		2,400	9.8	▲	12
3/8-16	.3750	.0775		3,200	15.7		20
3/8-24	.3750	.0878		3,620	17.8		23
7/16-14	.4375	.1063		4,380	25		30
7/16-20	.4375	.1187		4,900	28		35
1/2-13	.5000	.1419	▼	5,840	38		50
1/2-20	.5000	.1599	55,000	6,600	43		55
9/16-12	.5625	.182	52,000	1,100	52		70
9/16-18	.5625	.203	▲	7,900	58		80
5/8-11	.6250	.226		8,800	72		100
5/8-18	.6250	.256		10,000	81		110
3/4-10	.7500	.334	▼	13,000	128		175
3/4-16	.7500	.373	52,000	14,550	143		195
7/8-9	.8750	.462	28,000	9,700	111		165
7/8-14	.8750	.509	▲	10,700	122		185
1-8	1.0000	.606		12,700	167		250
1-12	1.0000	.663		13,900	182		270
1 1/8-7	1.1250	.763		16,000	236		350
1 1/8-12	1.1250	.856		18,000	265		400
1 1/4-7	1.2500	.969		20,350	333		500
1 1/4-12	1.2500	1.073		22,550	369		550
1 3/8-6	1.3750	1.155		24,300	432		660
1 3/8-12	1.375	1.315		27,600	498		740
1 1/2-6	1.5000	1.405		29,800	580	▼	670
1 1/2-12	1.5000	1.580	▼	33,200	653	ft-lb	980

Recommended Torque Grades
SAE Grade 5 Bolt

SAE Grade 5 Bolt							
					Assembly Torque		
Size	Basic Major Dia	Stress Area (Sq In)	Proof Load (psi)	Clamp Load (pounds)	Lubricated Max	Units	Dry Max
4-40	.1120	.00604	85,000	380	6.7	inch-lb	8
4-48	.1120	.00661	▲	420	7.4	▲	9
6-32	.1380	.00909		580	12.5		16
6-40	.1380	.01015		640	14		18
8-32	.1640	.01400		900	23		30
8-36	.1640	.01474		940	24		31
10-24	.1900	.0175		1,120	33		43
10-32	.1900	.0200		1,285	38		49
1/4-20	.2500	.0316		2,020	79	▼	96
1/4-28	.2500	.0364		2,320	91	inch-lb	120
5/16-18	.3125	.0524		3,340	13.7	ft-lb	17
5/16-24	.3125	.0580		3,700	15.1	▲	19
3/8-16	.3750	.0775		4,940	24		30
3/8-24	.3750	.0878		5,600	27		35
7/16-14	.4375	.1063		6,800	38		50
7/16-20	.4375	.1187		7,550	43		55
1/2-13	.5000	.1419		9,050	59		75
1/2-20	.5000	.1599		10,700	66		90
9/16-12	.5625	.182		11,600	85		110
9/16-18	.5625	.203		12,950	95		120
5/8-11	.6250	.226		14,400	118		150
5/8-18	.6250	.256		16,950	133		170
3/4-10	.7500	.334	▼	21,300	209		260
3/4-16	.7500	.373	85,000	23,800	234		300
7/8-9	.8750	.462	78,000	27,000	310		430
7/8-14	.8750	.509		29,800	341		470
1-8	1.0000	.606	▼	35,500	465		640
1-12	1.0000	.663	78,000	38,800	509		700
1 1/8-7	1.1250	.763	74,000	42,300	625		800
1 1/8-12	1.1250	.856	▲	47,500	701		880
1 1/4-7	1.2500	.969		53,800	882		1120
1 1/4-12	1.2500	1.073		59,600	977		1240
1 3/8-6	1.3750	1.155		64,100	1156		1460
1 3/8-12	1.3750	1.315		73,000	1317		1680
1 1/2-6	1.5000	1.405		78,000	1535	▼	1940
1 1/2-12	1.5000	1.580	▼	87,700	1726	ft-lb	2200

Recommended Torque Values
SAE Grade 8 Bolt

SAE Grade 8 Bolt					Assembly Torque		
Size	Basic Major Dia	Stress Area (Sq In)	Proof Load (psi)	Clamp Load (pounds)	Lubricated Max	Units	Dry Max
4-40	.1120	.00604	120,000	540	9.5	inch-lb	12
4-48	.1120	.00661	▲	600	10.5	▲	13
6-32	.1380	.00909		820	17.7		23
6-40	.1380	.01015		920	19.8		25
8-32	.1640	.01400		1,260	32		41
8-36	.1640	.01474		1,320	34		43
10-24	.1900	.0175		1,580	47	▼	60
10-32	.1900	.0200		1,800	53	inch-lb	66
1/4-20	.2500	.0316		2,860	9.3	ft-lb	12
1/4-28	.2500	.0364		3,260	10.7		14
5/16-18	.3125	.0524		4,720	19.3		25
5/16-24	.3125	.0580		5,220	21.4		28
3/8-16	.3750	.0775		7,000	34		45
3/8-24	.3750	.0878		7,900	38		50
7/16-14	.4375	.1063		9,550	54		70
7/16-20	.4375	.1187		10,700	61		80
1/2-13	.5000	.1419		12,750	83		110
1/2-20	.5000	.1599		14,400	94		120
9/16-12	.5625	.182		16,400	120		150
9/16-18	.5625	.203		18,250	134		170
5/8-11	.6250	.226		20,350	166		220
5/8-18	.6250	.256		23,000	189		240
3/4-10	.7500	.334		30,100	295		360
3/4-16	.7500	.373		33,600	330		420
7/8-9	.8750	.462		41,600	477		600
7/8-14	.8750	.509		45,800	526		660
1-8	1.0000	.606		54,500	715		900
1-12	1.0000	.663		59,700	783		1000
1 1/8-7	1.1250	.763		68,700	1013		1280
1 1/8-12	1.1250	.856		77,000	1137		1440
1 1/4-7	1.2500	.969		87,200	1430		1820
1 1/4-12	1.2500	1.073		96,600	1584		2000
1 3/8-6	1.3750	1.155		104,000	1875		2380
1 3/8-12	1.3750	1.315		118,400	2135		2720
1 1/2-6	1.5000	1.405		126,500	2489	▼	3160
1 1/2-12	1.5000	1.580	▼	142,200	2799	ft-lb	3550

Socket and Socket Wrenches

The minimum ratchet wrench handle swing provided shall be 35°.

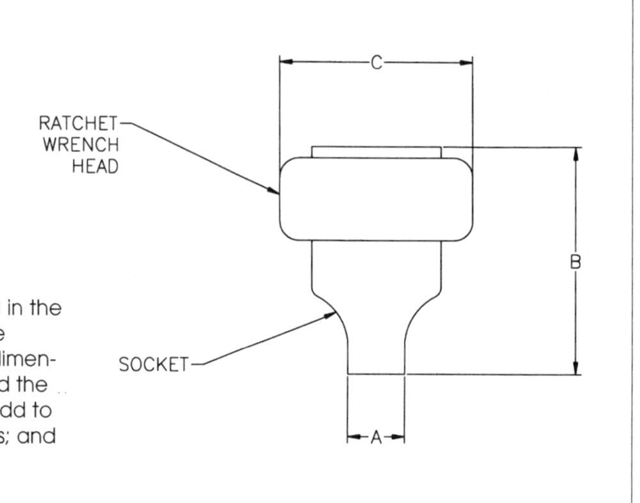

NOTE:
In addition to the dimensions indicated in the sketch, the following considerations are necessary: **FOR BOLTS:** Add to the "B" dimension: the length of the tapped hole; and the thickness of the bolt head. **FOR NUTS:** Add to the "B" dimension: the washer thickness; and the length of three threads of the stud.

	Standard Length Sockets (6 and 12 Point)											
	1/4 Inch drive				3/8 Inch Drive				1/2 Inch Drive			
	C — 1.125				C — 1.875				C — 1.938			
Hex Size	Drive End Dia	A Hex End Dia	Socket Length	B	Drive End Dia	A Hex End Dia	Socket Length	B	Drive End Dia	A Hex End Dia	Socket Length	B
1/16	0.51	0.32	1.00	1.57								
1/32	0.51	0.36	1.00	1.57								
1/4	0.51	0.40	1.00	1.57	0.69	0.40	1.25	2.13				
7/32	0.51	0.41	1.00	1.57	—	—	—	—				
5/16	0.51	0.48	1.00	1.57	0.69	0.47	1.25	2.13				
11/32	0.52	0.52	1.00	1.57	0.69	0.50	1.25	2.13				
3/8	0.58	0.56	1.00	1.57	0.69	0.56	1.25	2.13	0.88	0.57	1.50	2.50
7/16	0.69	0.69	1.00	1.57	0.88	0.69	1.25	2.13	0.94	0.69	1.50	2.50
1/2					0.88	0.75	1.25	2.13	0.94	0.76	1.50	2.50
9/16					0.94	0.82	1.25	2.13	0.94	0.82	1.50	2.50
19/32					0.94	0.86	1.25	2.13	0.97	0.86	1.57	2.57
5/8					1.00	0.90	1.25	2.13	1.00	0.90	1.57	2.57
11/16					1.06	0.97	1.25	2.13	1.07	0.97	1.57	2.57
3/4					1.12	1.06	1.25	2.13	1.13	1.06	1.57	2.57
25/32									1.13	1.09	1.63	2.63
13/16									1.23	1.15	1.63	2.63
7/8									1.29	1.22	1.75	2.75
15/16									1.41	1.30	1.75	2.75
1									1.41	1.40	1.75	2.75
1 1/16									1.51	1.48	1.85	2.85
1 1/8									1.57	1.54	1.94	2.94
1 3/16									—	—	—	—
1 1/4									1.73	1.82	2.00	3.00

TOOL CLEARANCES
Open End Wrenches

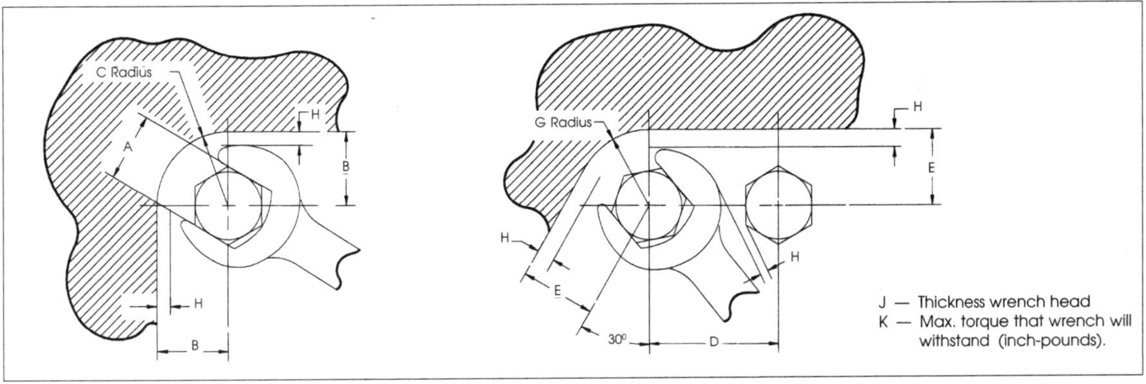

Open End Wrench 15° Dim. in In.											
Bolt Size	A	B Min	C Max	D Min	E Min	F Min	G Max	H Ref	J Max	K Max	Bolt Size
1/4	.438	.470	.590	.890	.420	.520	.640	.050	.250	375	1/4
5/16	.500	.520	.640	1.000	.470	.580	.660	.050	.266	490	5/16
3/8	.562	.590	.770	1.130	.520	.660	.700	.050	.297	700	3/8
7/16	.625	.640	.830	1.230	.550	.700	.700	.050	.344	935	7/16
1/2	.750	.770	.920	1.510	.670	.880	.800	.060	.375	1500	1/2
9/16	.813	.910	1.120	1.660	.720	.970	.860	.060	.406	1710	9/16
1/2[1]	.875	.970	1.150	1.810	.800	1.060	.910	.060	.438	2250	1/2[1]
5/8	.938	.970	1.150	1.850	.810	1.060	.950	.060	.438	2750	5/8
5/8[1]	1.062	1.090	1.250	2.100	.970	1.200	1.200	.080	.500	3500	5/8[1]
3/4	1.125	1.140	1.370	2.210	1.000	1.270	1.230	.080	.500	4000	3/4
3/4[1]	1.250	1.270	1.420	2.440	1.080	1.390	1.310	.080	.562	5250	3/4[1]
7/8	1.312	1.390	1.690	2.630	1.170	1.520	1.340	.080	.562	6000	7/8
7/8[1]	1.438	1.470	1.720	2.800	1.250	1.590	1.340	.090	.641	7500	7/8[1]
1	1.500	1.470	1.720	2.840	1.270	1.590	1.450	.090	.641	8250	1
1[1]	1.625	1.560	1.880	3.100	1.380	1.750	1.560	.090	.641	9000	1[1]

12 Point Box Wrench

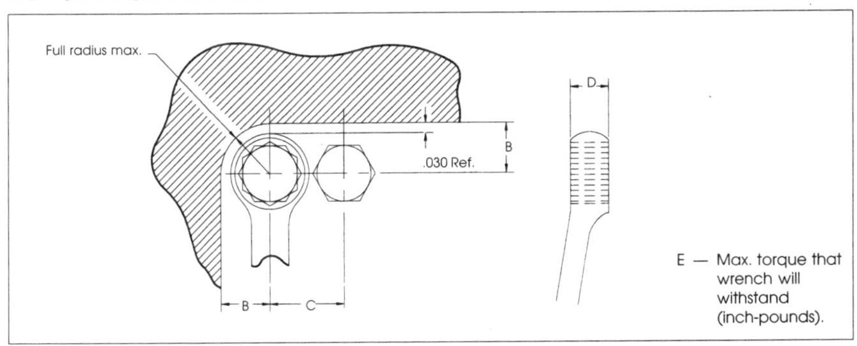

12 Point Box Wrench, Dim. in In.											
Bolt Size	A	B Min	C Min	D Max	E Max	Bolt Size	A	B Min	C Min	D Max	E Max
1/4	.438	.400	.650	.359	650	5/8[1]	1.062	.840	1.450	.781	5400
5/16	.500	.450	.740	.375	1020	3/4	1.125	.950	1.600	.844	5900
3/8	.562	.500	.830	.406	1200	3/4[1]	1.250	.980	1.700	.875	7200
7/16	.625	.560	.920	.469	2000	7/8	1.312	1.090	1.850	.906	8000
1/2	.750	.660	1.090	.594	2600	7/8[1]	1.438	1.220	2.050	1.000	8400
9/16	.813	.720	1.190	.594	3000	1	1.500	1.270	2.140	1.062	10450
1/2[1]	.875	.750	1.260	.594	3300	1[1]	1.625	1.340	2.280	1.156	11750
5/8	.938	.780	1.320	.656	4100						

[1] Heavy Structural Bolt

Metric Hex Cap Screws

Dimensions in mm

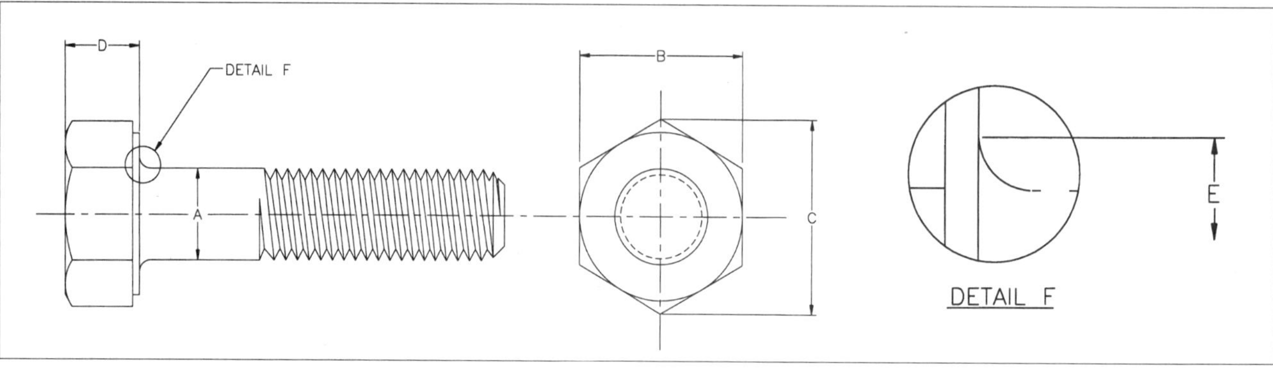

Nom Screw Dia and Thread Pitch	A Body Diameter		B Width Across Flats		C Width Across Corners		D Head Height		E Fillet Trans Dia For Short & Long Screws
	Max	Min	Max	Min	Max	Min	Max	Min	Max
M5 x 0.8	5.00	4.82	8.00	7.78	9.24	8.79	3.65	3.35	5.7
M6 x 1	6.00	5.82	10.00	9.78	11.55	11.05	4.15	3.85	6.8
M8 x 1.25	8.00	7.78	13.00	12.73	15.01	14.38	5.50	5.10	9.2
M10 x 1.5	10.00	9.78	16.00	15.73	18.48	17.77	6.63	6.17	11.2
M12 x 1.75	12.00	11.73	18.00	17.73	20.78	20.03	7.76	7.24	13.7
M14 x 2	14.00	13.73	21.00	20.67	24.25	23.35	9.09	8.51	15.7
M16 x 2	16.00	15.73	24.00	23.67	27.71	26.75	10.32	9.68	17.7
M20 x 2.5	20.00	19.67	30.00	29.16	34.64	32.95	12.88	12.12	22.4
M24 x 3	24.00	23.67	36.00	35.00	41.57	39.55	15.44	14.56	26.4
M30 x 3.5	30.00	29.67	46.00	45.00	53.12	50.85	19.48	17.92	33.4
M36 x 4	36.00	35.61	55.00	53.80	63.51	60.79	23.38	21.62	39.4
M42 x 4.5	42.00	41.38	65.00	62.90	75.06	71.71	26.97	25.03	45.4
M48 x 5	48.00	47.38	75.00	72.60	86.60	82.76	31.07	28.93	52.0
M56 x 5.5	56.00	55.26	85.00	82.20	98.15	93.71	36.20	33.80	62.0
M64 x 6	64.00	63.26	95.00	91.80	109.70	104.65	41.32	38.68	70.0
M72 x 6	72.00	71.26	105.00	101.40	121.24	115.60	46.45	43.55	78.0
M80 x 6	80.00	79.26	115.00	111.00	132.72	126.54	51.58	48.42	86.0
M90 x 6	90.00	89.13	130.00	125.50	150.11	143.07	57.74	54.26	96.0
M100 x 6	100.0	99.13	145.00	140.00	167.43	159.60	63.90	60.10	107.0

Weights of Metric Hex Nuts

Approximate Weight of 1000 Tapped Steel Nuts In Kilograms

Nom Nut Size	Type of Nut Hex, Style 1 ANSI B18.2.4.1M	Hex, Style 2 ANSI B18.2.4.2M	Heavy Hex ANSI B18.2.4.6M
M1.6 x 0.35	0.053		
M2 x 0.4	0.109		
M2.5 x 0.45	0.224		
M3 x 0.5	0.317	0.392	
M3.5 x 0.6	0.429	0.515	
M4 x 0.7	0.677	0.820	
M5 x 0.8	1.22	1.34	
M6 x 1	2.27	2.51	
M8 x 1.25	4.94	5.50	
M10 x 1.5(16)	9.26	10.3	
M10 x 1.5(15)	7.58	8.46	
M12 x 1.75	14.5	16.2	24.7
M14 x 2	22.8	25.3	36.9
M16 x 2	34.4	38.4	55.4

Nom Nut Size	Type of Nut Hex, Style 1 ANSI B18.2.4.1M	Hex, Style 2 ANSI B18.2.4.2M	Heavy Hex ANSI B18.2.4.6
M20 x 2.5	63.6	72.3	106
M22 x 2.5			132
M24 x 3	108	122	183
M27 x 3			263
M30 x 3.5	219	246	340
M36 x 4	377	426	586
M42 x 4.5			900
M48 x 5			1350
M56 x 5.5			1910
M64 x 6			2620
M72 x 6			3460
M80 x 6			4450
M90 x 6			6330
M100 x 6			8610

Carbon and Alloy Steel Metric Nuts
Nut Proof Load Values, kN

Nom Dia & Thread Pitch	Thread Stress Area, mm²	Property Class of Nut								
		5	5 (Over-tapped)	9	10	12	12 (Over-tapped)	8S and 8S3	10S and 10S3	10S (Over-tapped)
M1.6 x 0.35	1.27	0.66			1.32					
M2 x 0.4	2.07	1.08			2.15					
M2.5 x 0.45	3.39	1.76			3.53					
M3 x 0.5	5.03	2.62		4.53	5.23					
M3.5 x 0.6	6.78	3.53		6.10	7.05					
M4 x 0.7	8.78	4.57		7.90	9.13					
M5 x 0.8	14.2	8.23	6.60	13.0	14.8	16.3	13.1			
M6 x 1	20.1	11.7	9.35	18.4	20.9	23.1	18.5			
M8 x 1.25	36.6	21.6	17.2	34.4	38.1	42.5	34.0			
M10 x 1.5	58.0	34.2	27.3	54.5	60.3	67.3	53.9			
M12 x 1.75	84.3	51.4	41.3	80.1	88.5	100	80.1	90.6	105	98.2
M14 x 2	115	70.2	56.4	109	121	137	109	124	143	134
M16 x 2	157	95.8	76.9	149	165	187	149	169	195	183
M20 x 2.5	245	154	123	225	260	294	235	263	305	285
M22 x 2.5	303	—	—	—	—	—	291	326	377	353
M24 x 3	353	222	177	325	374	424	339	379	439	411
M27 x 3	459	—	—	—	—	—	—	493	571	535
M30 x 3.5	561	353	281	516	595	673	539	603	698	654
M36 x 4	817	515	409	752	866	980	784	878	1020	952
M42 x 4.5	1120	706	560	1030		1340	1080			
M48 x 5	1470	920	735	1350		1760	1410			
M56 x 5.5	2030	1280	1020	1870		2440	1950			
M64 x 6	2680	1690	1340	2470		3220	2570			
M72 x 6	3460	2180	1730	3180		4150	3320			
M80 x 6	4340	2730	2170	3990		5210	4170			
M90 x 6	5590	3520	2800	5140		6710	5370			
M100 x 6	6990	4400	3500	6430		8390	6710			

Metric Hardened Steel Washers
Dimensions in mm

Nom Washer Size[1]	A		B		C		E
	Inside Dia		Outside Dia		Thickness		Clipped Width[2]
	Max	Min	Max	Min	Max	Min	Min
12	14.4	14.0	27.0	25.7	4.6	3.1	10.5
14	16.4	16.0	30.0	28.7	4.6	3.1	12.2
16	18.4	18.0	34.0	32.4	4.6	3.1	14.0
20	22.5	22.0	42.0	40.4	4.6	3.1	17.5
22	24.5	24.0	44.0	42.4	4.6	3.4	19.2
24	26.5	26.0	50.0	48.4	4.6	3.4	21.0
27	30.5	30.0	56.0	54.1	4.6	3.4	23.6
30	33.6	33.0	60.0	58.1	4.6	3.4	26.2
36	39.6	39.0	72.0	70.1	4.6	3.4	31.5
42	45.6	45.0	84.0	81.8	7.2	4.6	36.7
48	52.7	52.0	95.0	92.8	7.2	4.6	42.0
56	62.7	62.0	107.0	104.8	8.7	6.1	49.0
64	70.7	70.0	118.0	115.8	8.7	6.1	56.0
72	78.7	78.0	130.0	127.5	8.7	6.1	63.0
80	86.9	86.0	142.0	139.5	8.7	6.1	70.0
90	96.9	96.0	159.0	156.5	8.7	6.1	78.7
100	107.9	107.0	176.0	173.5	8.7	6.1	87.5

[1] Nominal washer sizes are intended for use with fasteners of the same nominal thread diameter
[2] Washers may be clipped on one side not closer to the center of the washer than width E

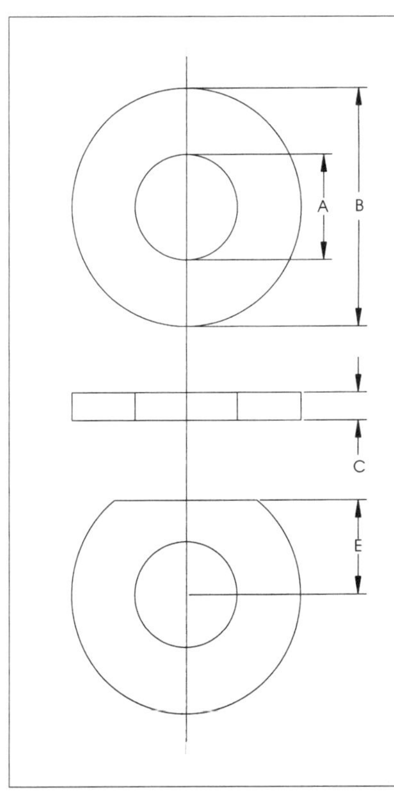

Metric Hex Nuts, Style 1 ▪ Dimensions in mm

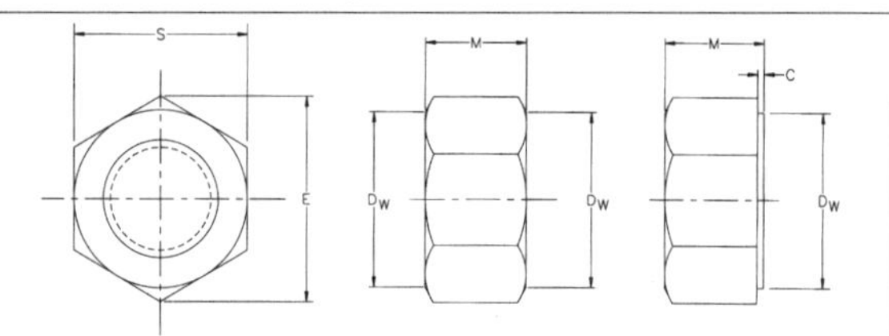

Nominal Nut Dia & Thread Pitch	S		E		M		D_W	C		Total Runout of Bearing Surface FIM
	Width Across Flats		Width Across Corners		Thickness		Bearing Face Dia	Washer Face Thickness		
	Max	Min	Max	Min	Max	Min	Min	Max	Min	Max
M1.6 x 0.35	3.20	3.02	3.70	3.41	1.30	1.05	2.4	—	—	—
M2 x 0.4	4.00	3.82	4.62	4.32	1.60	1.35	3.1	—	—	—
M2.5 x 0.45	5.00	4.82	5.77	5.45	2.00	1.75	4.1	—	—	—
M3 x 0.5	5.50	5.32	6.35	6.01	2.40	2.15	4.6	—	—	—
M3.5 x 0.6	6.00	5.82	6.93	6.58	2.80	2.55	5.1	—	—	—
M4 x 0.7	7.00	6.78	8.08	7.66	3.20	2.90	5.9	—	—	—
M5 x 0.8	8.00	7.78	9.24	8.79	4.70	4.40	6.9	—	—	0.30
M6 x 1	10.00	9.78	11.55	11.05	5.20	4.90	8.9	—	—	0.33
M8 x 1.25	13.00	12.73	15.01	14.38	6.80	6.44	11.6	—	—	0.36
M10 x 1.5	16.00	15.73	18.48	17.77	8.40	8.04	14.6	—	—	0.39
M12 x 1.75	18.00	17.73	20.78	20.03	10.80	10.37	16.6	—	—	0.42
M14 x 2	21.00	20.67	24.25	23.35	12.80	12.10	19.6	—	—	0.45
M16 x 2	24.00	23.67	27.71	26.75	14.80	14.10	22.5	—	—	0.48
M20 x 2.5	30.00	29.16	34.64	32.95	18.00	16.90	27.7	0.8	0.4	0.56
M24 x 3	36.00	35.00	41.57	39.55	21.50	20.20	33.2	0.8	0.4	0.64
M30 x 3.5	46.00	45.00	53.12	50.85	25.60	24.30	42.7	0.8	0.4	0.76
M36 x 4	55.00	53.80	63.51	60.79	31.00	29.40	51.1	0.8	0.4	0.89

Metric Hex Nuts, Style 2 ▪ Dimensions in mm

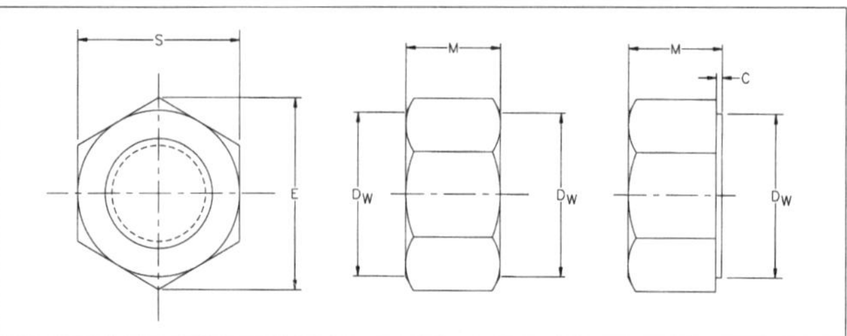

Nominal Nut Dia & Thread Pitch	S		E		M		D_W	C		Total Runout of Bearing Surface FIM
	Width Across Flats		Width Across Corners		Thickness		Bearing Face Dia	Washer Face Thickness		
	Max	Min	Max	Min	Max	Min	Min	Max	Min	Max
M3 x 0.5	5.50	5.32	6.35	6.01	2.90	2.65	4.6	—	—	—
M3.5 x 0.6	6.00	5.82	6.93	6.58	3.30	3.00	5.1	—	—	—
M4 x 0.7	7.00	6.78	8.08	7.66	3.80	3.50	5.9	—	—	—
M5 x 0.8	8.00	7.78	9.24	8.79	5.10	4.80	6.9	—	—	0.30
M6 x 1	10.00	9.78	11.55	11.05	5.70	5.40	8.9	—	—	0.33
M8 x 1.25	13.00	12.73	15.01	14.38	7.50	7.14	11.6	—	—	0.36
M10 x 1.5	16.00	15.73	18.48	17.77	9.30	8.94	14.6	—	—	0.39
M12 x 1.75	18.00	17.73	20.78	20.03	12.00	11.57	16.6	—	—	0.42
M14 x 2	21.00	20.67	24.25	23.35	14.10	13.40	19.6	—	—	0.45
M16 x 2	24.00	23.67	27.71	26.75	16.40	15.70	22.5	—	—	0.48
M20 x 2.5	30.00	29.16	34.64	32.95	20.30	19.00	27.7	0.8	0.4	0.56
M24 x 3	36.00	35.00	41.57	39.55	23.90	22.60	33.2	0.8	0.4	0.64
M30 x 3.5	46.00	45.00	53.12	50.85	28.60	27.30	42.7	0.8	0.4	0.76
M36 x 4	55.00	53.80	63.51	60.79	34.70	33.10	51.1	0.8	0.4	0.89

Metric Plain Washers

Dimensions in mm

Nom Washer Size	Washer Series	A Inside Dia		B Outside Dia		C Thickness	
		Max	Min	Max	Min	Max	Min
1.6	Narrow	2.09	1.95	4.00	3.70	0.70	0.50
	Regular	2.09	1.95	5.00	4.70	0.70	0.50
	Wide	2.09	1.95	6.00	5.70	0.90	0.60
2	Narrow	2.64	2.50	5.00	4.70	0.90	0.60
	Regular	2.64	2.50	6.00	5.70	0.90	0.60
	Wide	2.64	2.50	8.00	7.64	0.90	0.60
2.5	Narrow	3.14	3.00	6.00	5.70	0.90	0.60
	Regular	3.14	3.00	8.00	7.64	0.90	0.60
	Wide	3.14	3.00	10.00	9.64	1.20	0.80
3	Narrow	3.68	3.50	7.00	6.64	0.90	0.60
	Regular	3.68	3.50	10.00	9.64	1.20	0.80
	Wide	3.68	3.50	12.00	11.57	1.40	1.00
3.5	Narrow	4.18	4.00	9.00	8.64	1.20	0.80
	Regular	4.18	4.00	10.00	9.64	1.40	1.00
	Wide	4.18	4.00	15.00	14.57	1.75	1.20
4	Narrow	4.88	4.70	10.00	9.64	1.20	0.80
	Regular	4.88	4.70	12.00	11.57	1.40	1.00
	Wide	4.88	4.70	16.00	15.57	2.30	1.60
5	Narrow	5.78	5.50	11.00	10.57	1.40	1.00
	Regular	5.78	5.50	15.00	14.57	1.75	1.20
	Wide	5.78	5.50	20.00	19.48	2.30	1.60
6	Narrow	6.87	6.65	13.00	12.57	1.75	1.20
	Regular	6.87	6.65	18.80	18.37	1.75	1.20
	Wide	6.87	6.65	25.40	24.88	2.30	1.60
8	Narrow	9.12	8.90	18.80	18.37	2.30	1.60
	Regular	9.12	8.90	25.40	24.48	2.30	1.60
	Wide	9.12	8.90	32.00	31.38	2.80	2.00
10	Narrow	11.12	10.85	20.00	19.48	2.30	1.60
	Regular	11.12	10.85	28.00	27.48	2.80	2.00
	Wide	11.12	10.85	39.00	38.38	3.50	2.50
12	Narrow	13.57	13.30	25.40	24.88	2.80	2.00
	Regular	13.57	13.30	34.00	33.38	3.50	2.50
	Wide	13.57	13.30	44.00	43.38	3.50	2.50
14	Narrow	15.52	15.25	28.00	27.48	2.80	2.00
	Regular	15.52	15.25	39.00	38.38	3.50	2.50
	Wide	15.52	15.25	50.00	49.38	4.00	3.00
16	Narrow	17.52	17.25	32.00	31.38	3.50	2.50
	Regular	17.52	17.25	44.00	43.38	4.00	3.00
	Wide	17.52	17.25	56.00	54.80	4.60	3.50
20	Narrow	22.32	21.80	39.00	38.38	4.00	3.00
	Regular	22.32	21.80	50.00	49.38	4.60	3.50
	Wide	22.32	21.80	66.00	64.80	5.10	4.00
24	Narrow	26.12	25.60	44.00	43.38	4.60	3.50
	Regular	26.12	25.60	56.00	54.80	5.10	4.00
	Wide	26.12	25.60	72.00	70.80	5.60	4.50
30	Narrow	33.02	32.40	56.00	54.80	5.10	4.00
	Regular	33.02	32.40	72.00	70.80	5.60	4.50
	Wide	33.02	32.40	90.00	88.60	6.40	5.00
36	Narrow	38.92	38.30	66.00	64.80	5.60	4.50
	Regular	38.92	38.30	90.00	88.60	6.40	5.00
	Wide	38.92	38.30	110.00	108.60	8.50	7.00

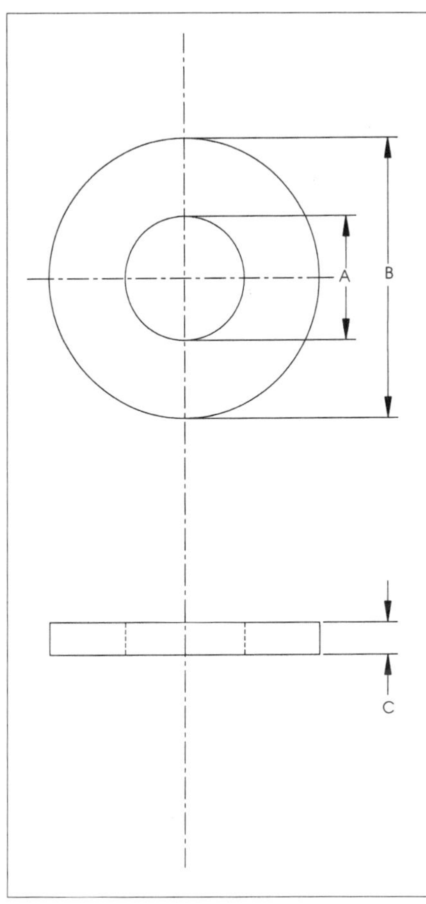

GENERAL ENGINEERING

Weights of Metric Hex Cap Screws — Approximate Weight of 100 Steel Screws In Kilograms

Screw Length \ Screw Size	M5 x 0.8	M6 x 1	M8 x 1.25	M10 x 1.5 (16)	M10 x 1.5 (15)	M12 x 1.75 (15)	M14 x 2	M16 x 2	M20 x 2.5	M24 x 3
8	0.237									
10	0.261	0.425								
12	0.285	0.459	0.939							
14	0.308	0.493	1.00	1.74	1.61					
16	0.332	0.527	1.06	1.84	1.71	2.71	4.07			
20	0.385	0.595	1.19	2.03	1.90	2.99	4.46			
25	0.459	0.699	1.34	2.28	2.15	3.35	4.94	7.00		
30	0.533	0.806	1.53	2.52	2.39	3.70	5.44	7.66	13.3	
35	0.608	0.914	1.72	2.82	2.69	4.06	5.92	8.31	14.4	22.7
40	0.681	1.02	1.91	3.12	2.99	4.90	6.41	8.96	15.4	24.2
45	0.756	1.13	2.10	3.42	3.29	5.33	7.00	9.64	16.4	25.7
50	0.830	1.24	2.29	3.72	3.59	5.77	7.59	10.4	17.4	27.1
55		1.34	2.49	4.02	3.89	6.20	8.18	11.2	18.5	28.6
60		1.45	2.68	4.32	4.19	6.63	8.78	12.0	19.7	30.1
65			2.85	4.63	4.50	7.07	9.37	12.7	20.9	31.7
70			3.06	4.93	4.80	7.50	9.96	13.5	22.1	33.4
75			3.25	5.23	5.10	7.93	10.5	14.3	23.4	35.2
80			3.44	5.53	5.40	8.36	11.1	15.1	24.6	36.9
85				5.83	5.70	8.80	11.7	15.8	25.8	38.7
90				6.13	6.00	9.23	12.3	16.6	27.0	40.4
100				6.73	6.60	10.1	13.5	18.2	29.4	43.9
110						11.0	14.7	19.7	31.8	47.4
120						11.8	15.9	21.3	34.2	50.9
130							16.9	22.7	36.4	54.1
140							18.1	24.2	38.9	57.6
150								25.8	41.3	61.1
160								27.3	43.7	64.6
170									46.1	68.1
180									48.5	71.6
190									51.0	75.1
200									52.9	77.8
220										84.8
240										91.8

Screw Length \ Screw Size	M30 x 3.5	M36 x 4	M42 x 4.5	M48 x 5	M56 x 5.5	M64 x 6	M72 x 6	M80 x 6	M90 x 6	M100 x 6
40	44.2									
45	46.5	74.5								
50	48.9	77.9	117							
55	51.2	81.3	121							
60	53.6	84.7	126	181						
65	55.9	88.1	131	187						
70	58.2	91.5	135	193	246					
75	60.6	94.8	140	200	263					
80	63.2	98.2	145	206	279	408				
85	65.9	102	149	212	296	419				
90	68.7	105	154	218	313	430	572			
100	74.1	113	163	230	330	452	601	776		
110	79.6	121	173	242	347	474	629	812		
120	85.1	129	182	255	363	496	657	847	1150	1510
130	90.1	136	194	268	380	518	686	883	1190	1570
140	95.6	144	205	282	397	540	714	918	1240	1620
150	101	152	215	296	416	562	742	953	1290	1680
160	107	160	226	310	435	584	771	989	1330	1740
170	112	167	237	324	454	609	799	1020	1380	1790
180	117	175	247	338	473	634	827	1060	1420	1850
190	123	183	258	352	492	659	856	1090	1470	1910
200	127	190	267	364	508	680	888	1130	1510	1960
220	138	206	288	392	546	730	951	1210	1600	2080
240	149	221	310	420	584	780	1010	1290	1700	2190
260	160	237	331	448	622	830	1080	1360	1800	2310
280	171	253	353	476	660	880	1140	1440	1900	2430
300	182	269	374	504	699	930	1200	1520	2000	2560
340		300	417	560	775	1029	1330	1680	2190	2800
380		332	460	616	851	1129	1460	1830	2390	3040
420			502	672	927	1229	1580	1990	2590	3290
460				727	1000	1328	1710	2140	2790	3530
500				783	1080	1428	1840	2300	2980	3780
600					1270	1677	2150	2690	3480	4390
700						1926	2470	3080	3970	5000
800								3470	4460	5610
900									4960	6220
1000										6820

Torque, Metric Grades

Torque, Metric Grade 5.8											
						Assembly Torque		Calculated inch information provided for reference only			
										Assembly Torque, lb-ft	
Size	Basic Major Dia. mm	Stress Area mm ²	Proof Load MPa	Clamp Load kN	Tap Drill Dia. mm	Lubricated (0.15) Max. N-mm	Dry (0.2) Max. N-mm	Basic Major Dia. inch	Clamp Load lb	Lubricated (0.15) Max	Dry (0.2) Max
M5 x 0.8	5	14.2	380	4.05	4.2	3.04	4.05	0.197	910	26.86	35.82[1]
M6 x 1	6	20.1	380	5.73	5	5.16	6.87	0.236	1288	45.63	60.84[1]
M8 x 1.25	8	36.6	380	10.43	6.8	12.52	16.69	0.315	2345	9.23	12.31
M10 x 1.5	10	58	380	16.53	8.5	24.80	33.06	0.394	3716	18.29	24.38
M12 x 1.75	12	84.3	380	24.03	10.2	43.25	57.66	0.472	5401	31.90	42.53
M14 x 2	14	115	380	32.78	12	68.83	91.77	0.551	7368	50.76	67.69
M16 x 2	16	157	380	44.75	14	107.39	143.18	0.630	10059	79.21	105.61
M20 x 2.5	20	245	380	69.83	17.5	209.48	279.30	0.787	15697	154.50	206.00
M24 x 3	24	353	380	100.61	21	362.18	482.90	0.945	22617	267.13	356.17

Torque, Metric Grade 8.8											
						Assembly Torque		Calculated inch information provided for reference only			
										Assembly Torque, lb-ft	
Size	Basic Major Dia. mm	Stress Area mm ²	Proof Load MPa	Clamp Load kN	Tap Drill Dia. mm	Lubricated (0.15) Max. N-mm	Dry (0.2) Max. N-mm	Basic Major Dia. inch	Clamp Load lb	Lubricated (0.15) Max	Dry (0.2) Max
M16 x 2	16	157	600	70.65	14	169.56	226.08	0.630	15883	125.06	166.75
M20 x 2.5	20	245	600	110.25	17.5	330.75	441.00	0.787	24785	243.95	325.26
M22 x 2.5	22	303	600	136.35	19.5	449.96	599.94	0.866	30653	331.87	442.49
M24 x 3	24	353	600	158.85	21	571.86	762.48	0.945	35711	421.78	562.38
M27 x 3	27	459	600	206.55	24	836.53	1115.37	1.063	46434	616.99	822.65
M30 x 3.5	30	561	600	252.45	26.5	1136.03	1514.70	1.181	56753	837.89	1117.19
M36 x 4	36	817	600	367.65	32	1985.31	2647.08	1.417	82651	1464.29	1952.39

Torque, Metric Grade 10.9											
						Assembly Torque		Calculated inch information provided for reference only			
										Assembly Torque, lb-ft	
Size	Basic Major Dia. mm	Stress Area mm ²	Proof Load MPa	Clamp Load kN	Tap Drill Dia. mm	Lubricated (0.15) Max. N-mm	Dry (0.2) Max. N-mm	Basic Major Dia. inch	Clamp Load lb	Lubricated (0.15) Max	Dry (0.2) Max
M5 x 0.8	5	14.2	830	8.84	4.2	6.63	8.84	0.197	1987	4.89	6.52
M6 x 1	6	20.1	830	12.51	5	11.26	15.01	0.236	2813	8.31	11.07
M8 x 1.25	8	36.6	830	22.78	6.8	27.34	36.45	0.315	5122	20.17	26.89
M10 x 1.5	10	58	830	36.11	8.5	54.16	72.21	0.394	8117	39.94	53.26
M12 x 1.75	12	84.3	830	52.48	10.2	94.46	125.94	0.472	11797	69.67	92.89
M14 x 2	14	115	830	71.59	12	150.33	200.45	0.551	16094	110.88	147.84
M16 x 2	16	157	830	97.73	14	234.56	312.74	0.630	21971	173.00	230.67
M20 x 2.5	20	245	830	152.51	17.5	457.54	610.05	0.787	34286	337.46	449.95
M22 x 2.5	22	303	830	188.62	19.5	622.44	829.92	0.866	42403	459.09	612.12
M24 x 3	24	353	830	219.74	21	791.07	1054.76	0.945	49400	583.47	777.95
M27 x 3	27	459	830	285.73	24	1157.20	1542.93	1.063	64234	853.50	1138.01
M30 x 3.5	30	561	830	349.22	26.5	1571.50	2095.34	1.181	78508	1159.08	1545.44
M36 x 4	36	817	830	508.58	32	2746.35	3661.79	1.417	114334	2025.60	2700.80
M42 x 4.5	42	1120	830	697.20	37.5	4392.36	5856.48	1.654	156737	3239.64	4319.52
M48 x 5	48	1470	830	915.08	43	6588.54	8784.72	1.890	205717	4859.46	6479.28
M56 x 5.5	56	2030	830	1263.68	50.5	10614.87	14153.16	2.205	284085	7829.12	10438.83

[1] Lb-in.

Wrench Size

Nom Size of Wrench Also Basic Width Across Flats of Fastener	Nom Thread Dia of Hex Fastener for Which Wrench Size is Suited
3	—
3.2	M1.6, M2
4	M2, M2.5
5	M2.5, M3
5.5	M3, M3.5
6	M3.5
7	M4
8	M5
10	M6
13	M8
15	M10
16	M10
18	M12
21	M14
24	M16
27	—
30	M20
34	—
36	M24
41	—
46	M30
50	—
55	M36
60	—
65	M42
70	—
75	M48
80	—
85	M56
90	—
95	M64
100	—
105	M72
110	—
115	M80
120	—
130	M90
135	—
145	M100
150	—

Clearance Holes For Bolts, Screws and Studs

Nom Fastener Size	D_h — Clearance Hole Diameter, Basic		
	Close Clearance	Normal Clearance (Preferred)	Loose Clearance
1.6	1.7	1.8	2.0
2	2.2	2.4	2.6
2.5	2.7	2.9	3.1
3	3.2	3.4	3.6
3.5	3.7	3.9	4.2
4	4.3	4.5	4.8
5	5.3	5.5	5.8
6	6.4	6.6	7.0
8	8.4	9.0	10.0
10	10.5	11.0	12.0
12	13.0	13.5	14.5
14	15.0	15.5	16.5
16	17.0	17.5	18.5
20	21.0	22.0	24.0
22	—	24.0	—
24	25.0	26.0	28.0
27	—	30.0	—
30	31.0	33.0	35.0
36	37.0	39.0	42.0
42	43.0	45.0	48.0
48	50.0	52.0	56.0
56	58.0	62.0	66.0
64	66.0	70.0	74.0
72	74.0	78.0	82.0
80	82.0	86.0	91.0
90	93.0	96.0	101.0
100	104.0	107.0	112.0

Hydraulic Symbols In Accordance With DIN-ISO 1219

PUMP

One Flow Direction
Fixed Displacement — Variable Displacement

Two Flow Directions
Fixed Displacement — Variable Displacement

HYDRAULIC MOTOR

One Flow Direction
Fixed displacement — Variable Displacement

Two Flow Directions
Fixed Displacement — Variable Displacement

PUMP/MOTOR

Reversible Flow Direction
Fixed Displacement — Variable Displacement

One Flow Direction
Fixed Displacement — Variable Displacement

Two Flow Directions
Fixed Displacement — Variable Displacement

HYDROSTATIC TRANSMISSION

Remote Drive Transmission
Simplified representation, without auxiliary equipment

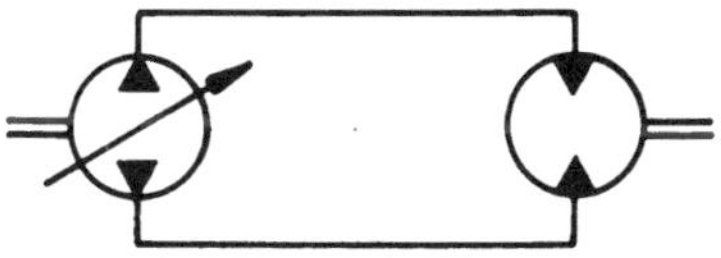

HYDROSTATIC TRANSMISSION

Variable-Speed Drive Unit

OSCILLATING MOTOR

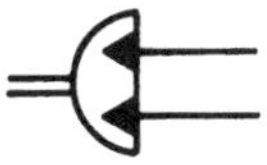

CYLINDER

Single-Acting

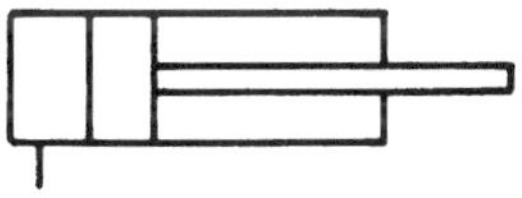

Single-Acting With Spring Return

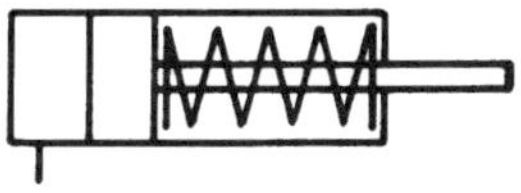

Double-Acting Differential

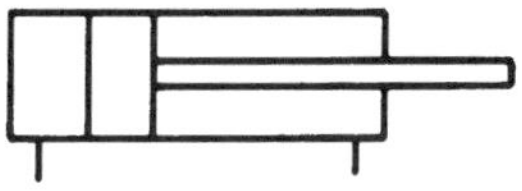

Double-Acting With Piston Rod On Both Ends

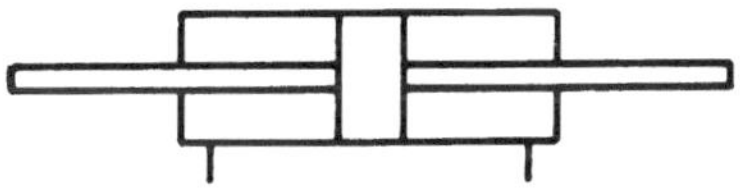

With Cushion at End Position

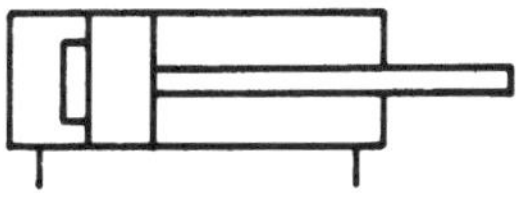

Cushion Adjustable at Both Ends

Telescope Cylinder

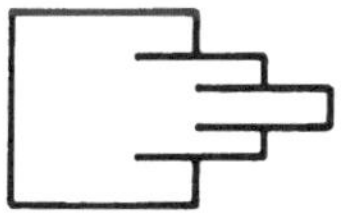

CYLINDER

Pressure Intensifier

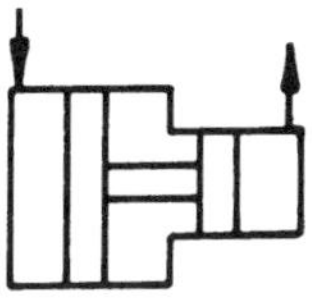

DIRECTIONAL CONTROL VALVES

Valves which are used to open or close various through-flow paths. Essentially, these valves are characterized by the following features:

- the number of switched (distinct) positions. This is shown by a corresponding number of squares, and identification is by means of letters such as 0, a, b (*).
- the number of ports and connections within the different switched positions. These are shown by lines and arrows.

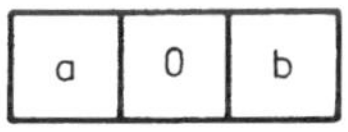

a | 0 | b

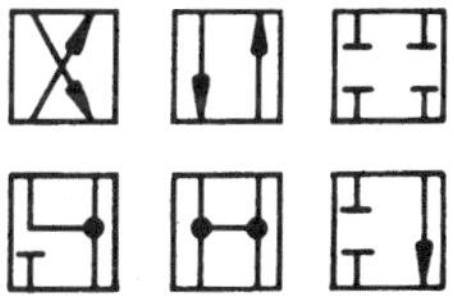

Marking of ports by means of letter (at the basic switched position "0")[1]

P	Pump, Pressure
T	Tank, Return
A, B	Load
X, Y, Z	Auxiliary Pilot Ports
L	Leakage Oil

Designation:
Example: 4/3 directional-control valve
4 = Number of Ports
3 = Number of Switched Positions

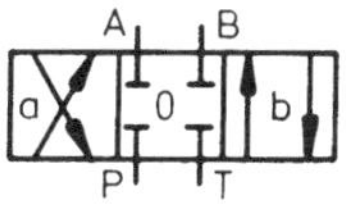

A B
a 0 b
P T

2/2 Directional-Control Valve

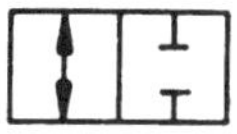

3/2 Directional-Control Valve

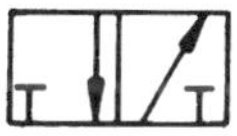

[1] The designation of the switch positions and the ports is not (yet) part of the DIN ISO Standard 1219.

Hydraulic Symbols In Accordance With DIN-ISO 1219

DIRECTIONAL CONTROL VALVES

4/3 Directional-Control Valve

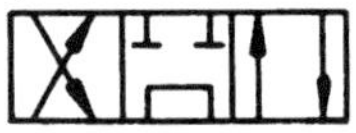

6/3 Directional-Control Valve

Direct Operated
By Manual Control Lever With Detent — By Pushbutton

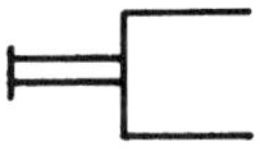

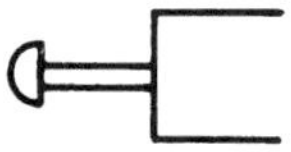

Direct Operated
By Pedal — By Plunger

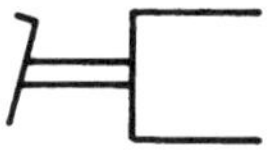

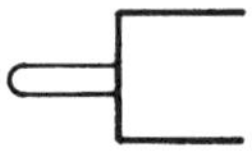

Direct Operated
By Roller — With Spring Return

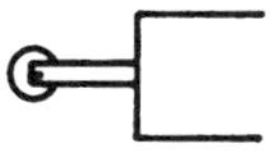

Direct Operated
With Spring Centering

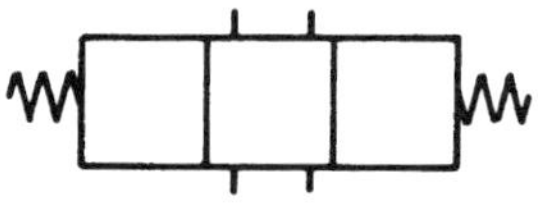

Direct Operated
By Solenoid, Spring Return

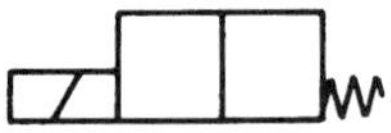

Direct Operated
Double Solenoid, Spring Centered

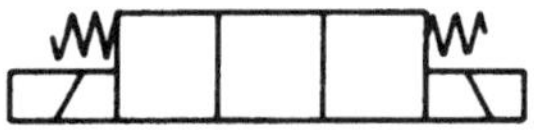

Direct Operated
Hydraulic Actuation — Pneumatic Actuation

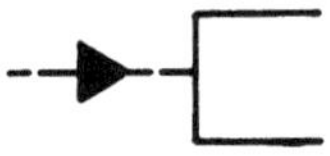

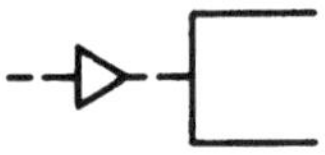

DIRECTIONAL-CONTROL VALVES

Pilot Operated
Hydraulically Actuated,
Electromagentically Controlled

Detailed

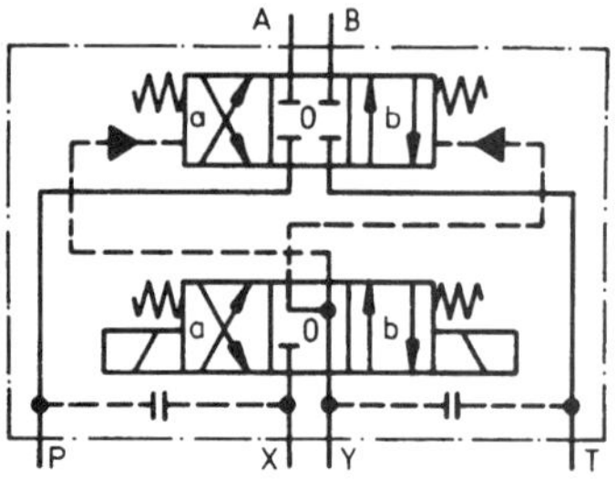

A B a 0 b P X Y T

Simplified

A B a 0 b a b X P T Y

Pilot Operated
Hydraulically Actuated,
Pneumatically Controlled

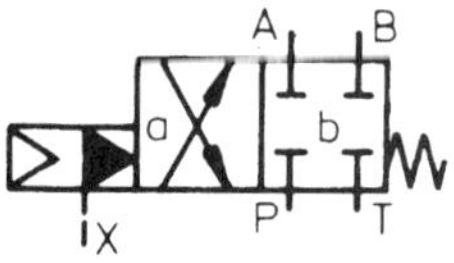

A B a b X P T

DIRECTIONAL-CONTROL VALVES WITHOUT FIXED POSITIONS

These are valves with gradual transition between the individual switched positions and with variable throttling effect. Represented by parallel lines running along the length of the symbol.

Tracer Valve With Plunger Operated Against a Return Spring

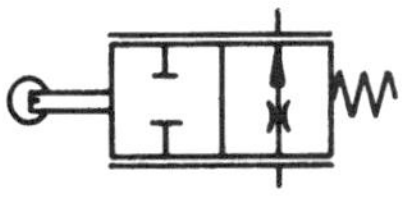

Solenoid-Operated Proportional Valve

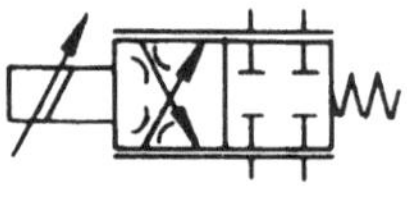

Electro-Hydraulic Servo Valve

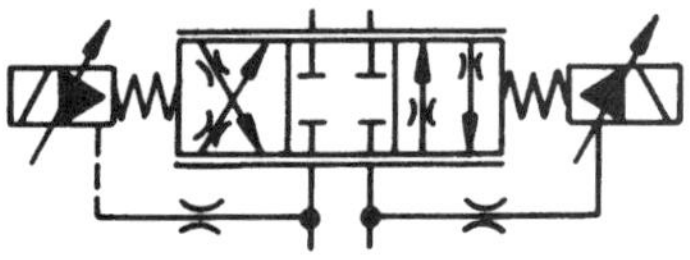

PRESSURE CONTROL VALVES

Valves which influence the pressure which are represented by a single square with an arrow. The throttling area is continuously variable.

Throttling Area
Normally Open — Normally Closed

Pressure Relief Valve, Directly Operated

Fixed Setting — Adjustable

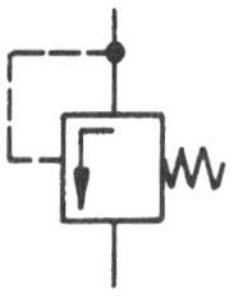

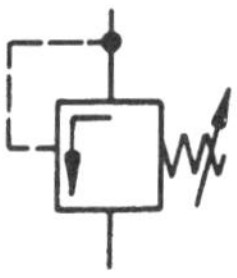

Pressure Relief Valve, Pilot Operated

Detailed

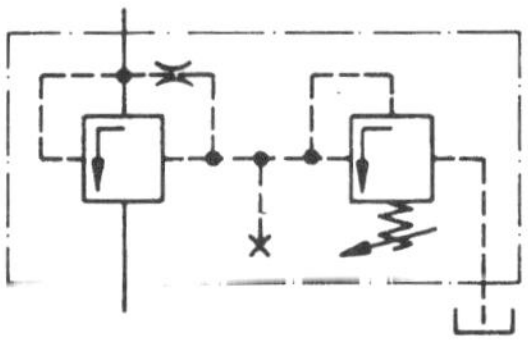

Simplified

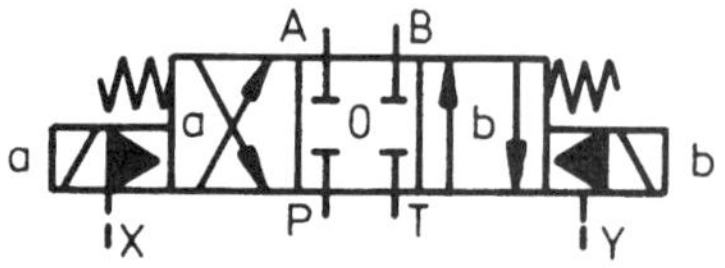

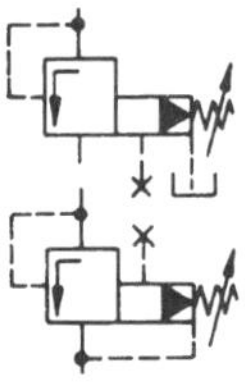

Pressure-Reducing Valve
(directly operated pressure regulator)

Fixed Setting — Adjustable

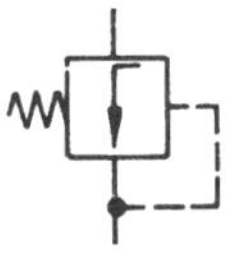

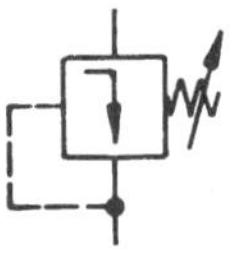

Pressure-Reducing Valve, Pilot Operated

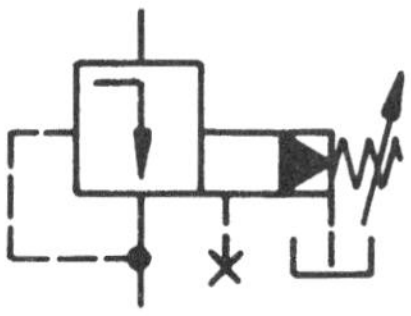

Hydraulic Symbols In Accordance With DIN-ISO 1219

PRESSURE CONTROL VALVES

Pressure-Reducing Valve With Discharge
(3-Way pressure reducing valve)

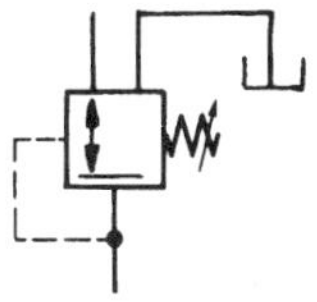

Sequence Valve With External Signal Input

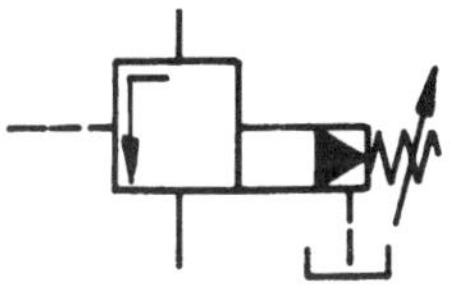

Discharge Valve, Pilot Operated

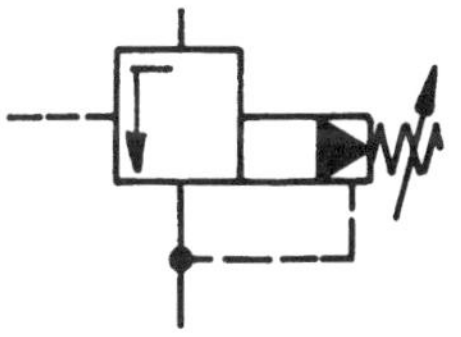

Electrical Pressure Switch

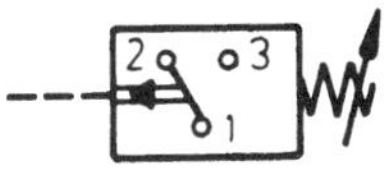

2 3 1

FLOW CONTROL VALVES

Valves which influence the volumetric flow are represented by a restriction of the line area.

Throttle, Fixed Setting
Not Compensated

Throttle, Adjustable
Not Compensated

2-Way Flow Regulator
Pressure Compensated

3-Way Flow Regulator
Pressure Compensated

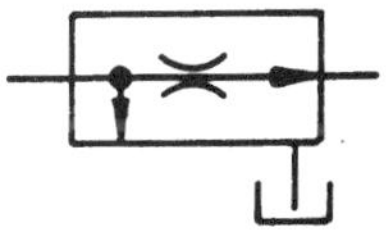

FLOW CONTROL VALVES

Flow Divider

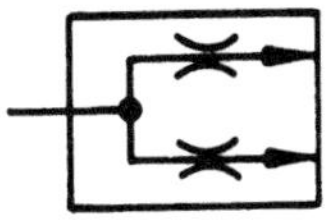

CHECK VALVES

Non-Return Valves, Free

Non-Return Valve, Spring Loaded

Pilot-Controlled
Internal control-oil drain

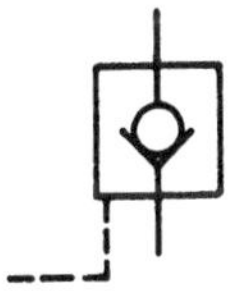

External control-oil drain

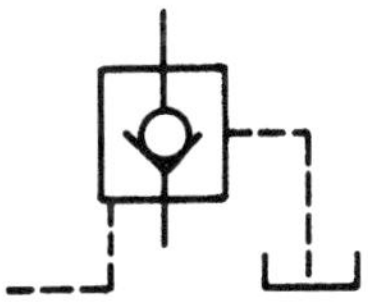

Throttle Valve With Non-Return Valve

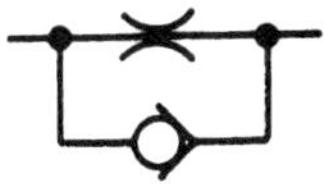

Shuttle Valve

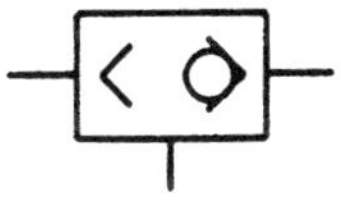

ENERGY TRANSMISSION & ACCESSORIES

Pressure Source

Electric Motor

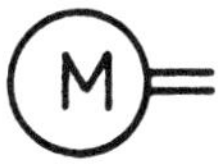

M

Heat Engine

M

Input & Output Shafts

Shaft Coupling

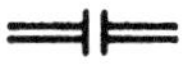

Hydraulic Main Line

Pilot Line

Drain Line

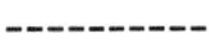

ENERGY TRANSMISSION & ACCESSORIES

Flexible Pipe

Line Junction

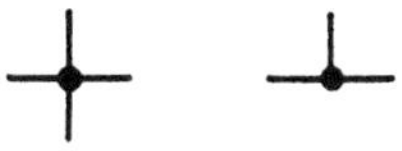

Crossing Lines

Air Bleed

Power Take-Off

Quick-Acting Couplings

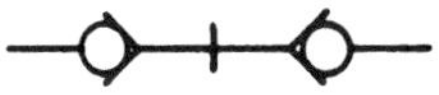

Rotary Connection

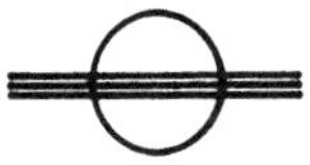

Reservoir With Lines Below Oil Level

Accumulator

Filter

Cooler

Heater

Pressure Gauge

Flow Meter

Pneumatic Symbols

PNEUMATIC SYMBOLS

Direction of Flow
In Pneumatic Systems

Manual Drain Filter

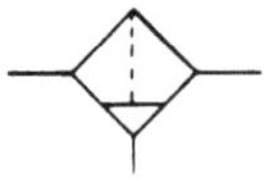

Automatic Drain Filter

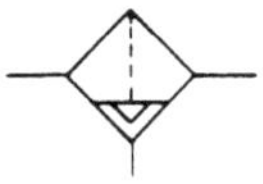

Lubricator

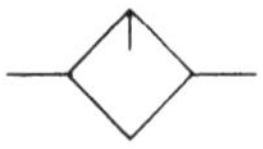

Airline Pressure Regulator
Adjustable, Relieving

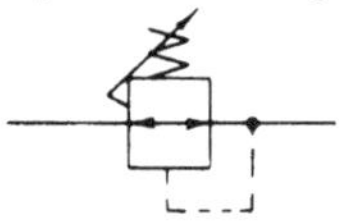

Airline Filter, Regulator & Lubricator With Gauge

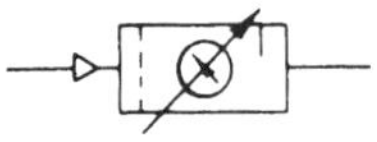

Pressure Gauge

Muffler

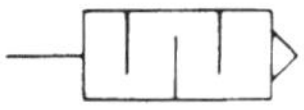

Fixed Displacement Compressor

Unidirectional Motor

Bidirectional Motor

Adjustable Flow Control

Check Valve

VALVE OPERATORS

Spring

Push Button

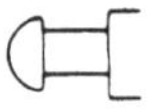

Manual Actuator

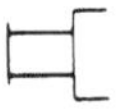

Pedal or Treadle

Mechanical Actuator

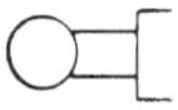

Push-Pull Lever

Pressure Compensated Actuator

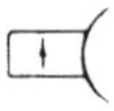

Solenoid Actuator

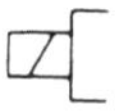

Solenoid and Pilot Actuator

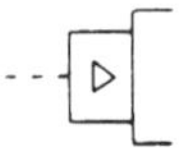

2-WAY VALVES

Normally Closed

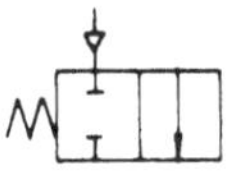

Normally Open

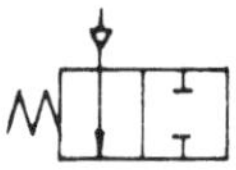

3-WAY 2-POSITION VALVES

Normally Closed

Normally Open

Distributor

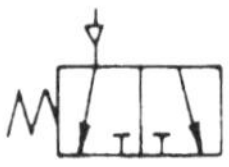

Selector

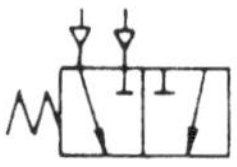

4-WAY 2-POSITION VALVES

Single Pressure

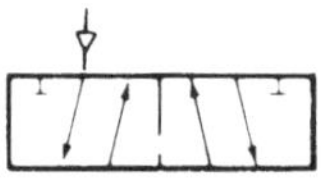

Dual Pressure

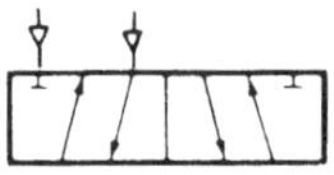

4-WAY 3-POSITION VALVES

Single Pressure, All Ports Blocked Center

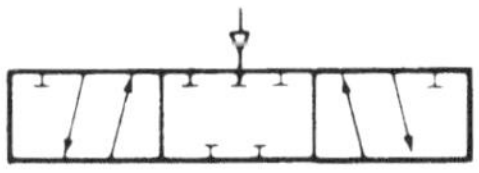

Dual Pressure, All Ports Blocked Center

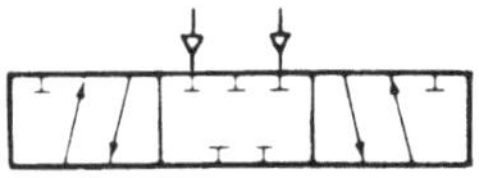

Single Pressure, Inlet to Cylinder Center

Dual Pressure, Inlet to Cylinder Center

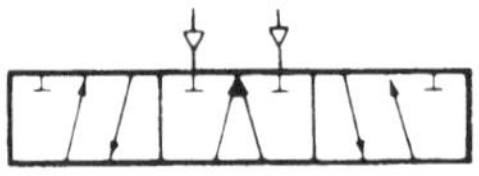

Single Pressure, Cylinder to Exhaust Center

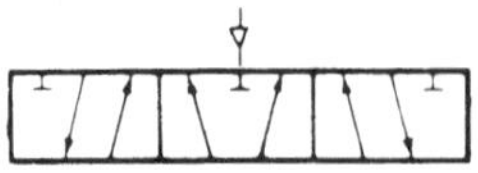

Dual Pressure, Cylinder to Exhaust Center

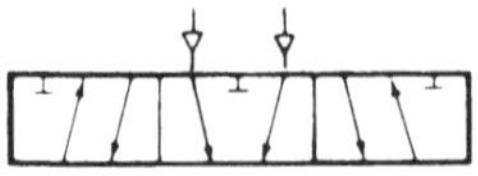

Electronic/Electrical Symbols

RESISTOR

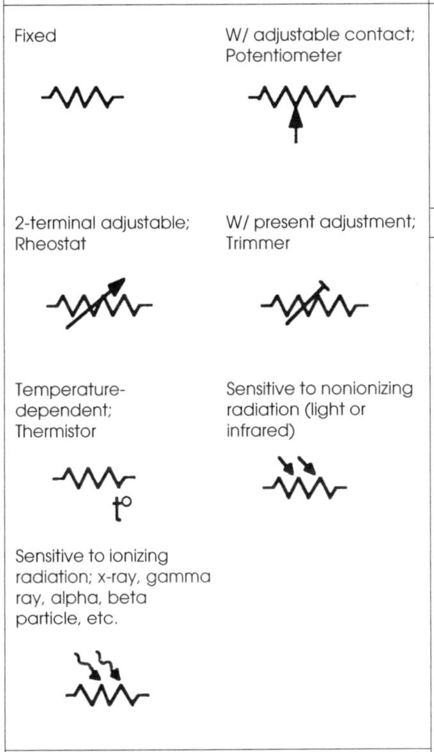

Fixed

W/ adjustable contact; Potentiometer

2-terminal adjustable; Rheostat

W/ present adjustment; Trimmer

Temperature-dependent; Thermistor

t°

Sensitive to nonionizing radiation (light or infrared)

Sensitive to ionizing radiation; x-ray, gamma ray, alpha, beta particle, etc.

CAPACITOR

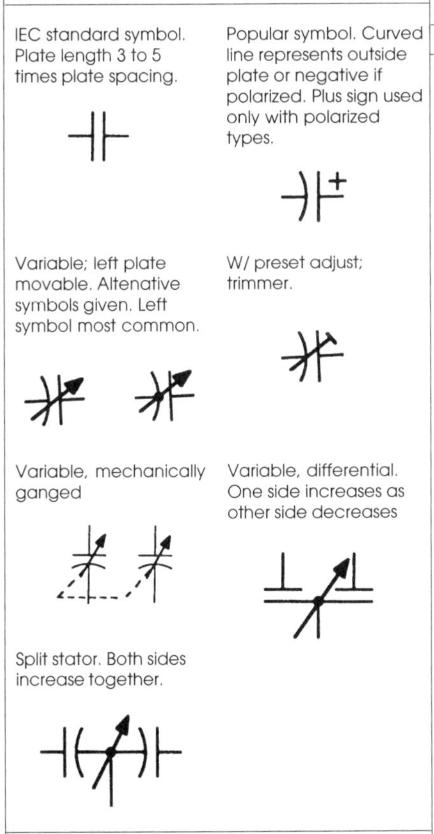

IEC standard symbol. Plate length 3 to 5 times plate spacing.

Popular symbol. Curved line represents outside plate or negative if polarized. Plus sign used only with polarized types.

Variable; left plate movable. Altenative symbols given. Left symbol most common.

W/ preset adjust; trimmer.

Variable, mechanically ganged

Variable, differential. One side increases as other side decreases

Split stator. Both sides increase together.

INDUCTOR

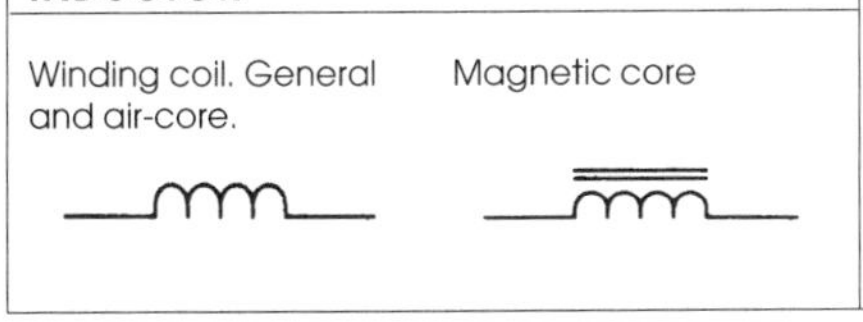

Winding coil. General and air-core.

Magnetic core

INDUCTOR

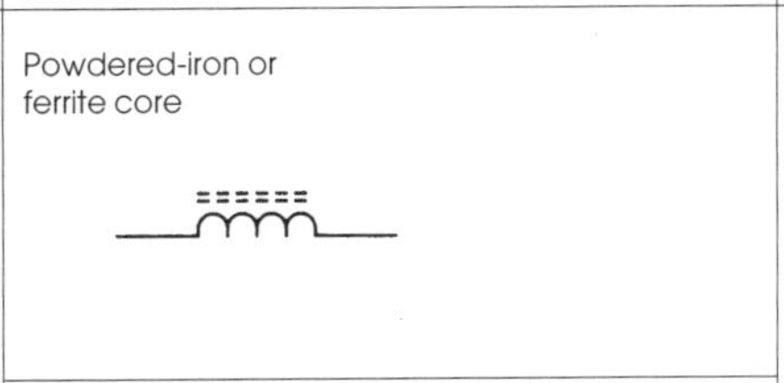

Powdered-iron or ferrite core

TRANSFORMER

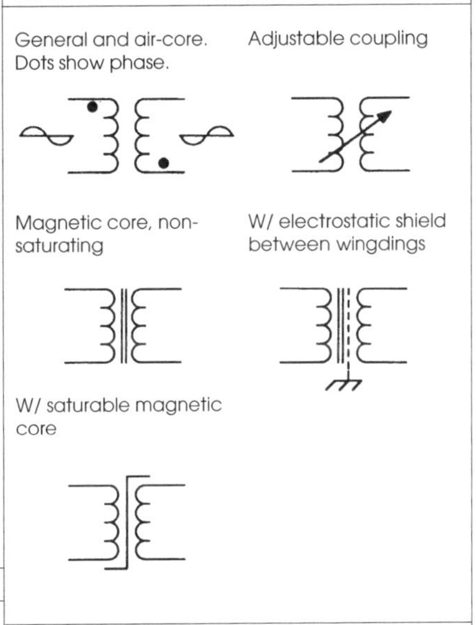

General and air-core. Dots show phase.

Adjustable coupling

Magnetic core, non-saturating

W/ electrostatic shield between wingdings

W/ saturable magnetic core

MISCELLANEOUS

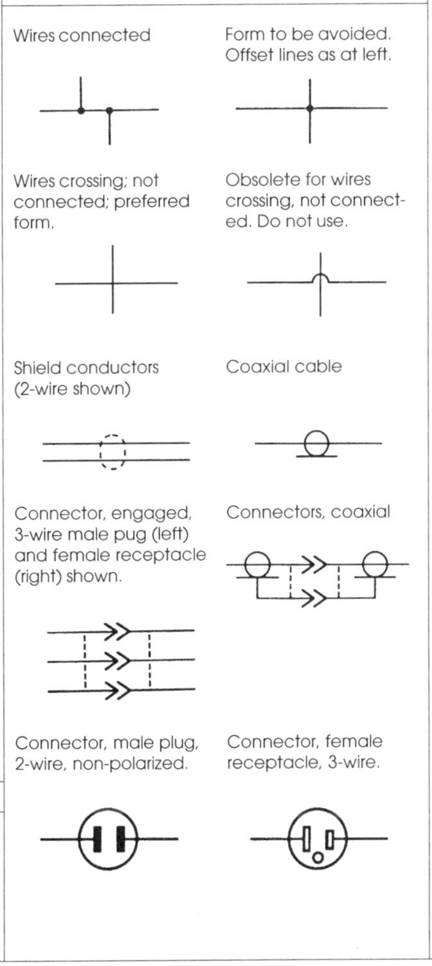

Wires connected

Form to be avoided. Offset lines as at left.

Wires crossing; not connected; preferred form.

Obsolete for wires crossing, not connect-ed. Do not use.

Shield conductors (2-wire shown)

Coaxial cable

Connector, engaged, 3-wire male pug (left) and female receptacle (right) shown.

Connectors, coaxial

Connector, male plug, 2-wire, non-polarized.

Connector, female receptacle, 3-wire.

MISCELLANEOUS

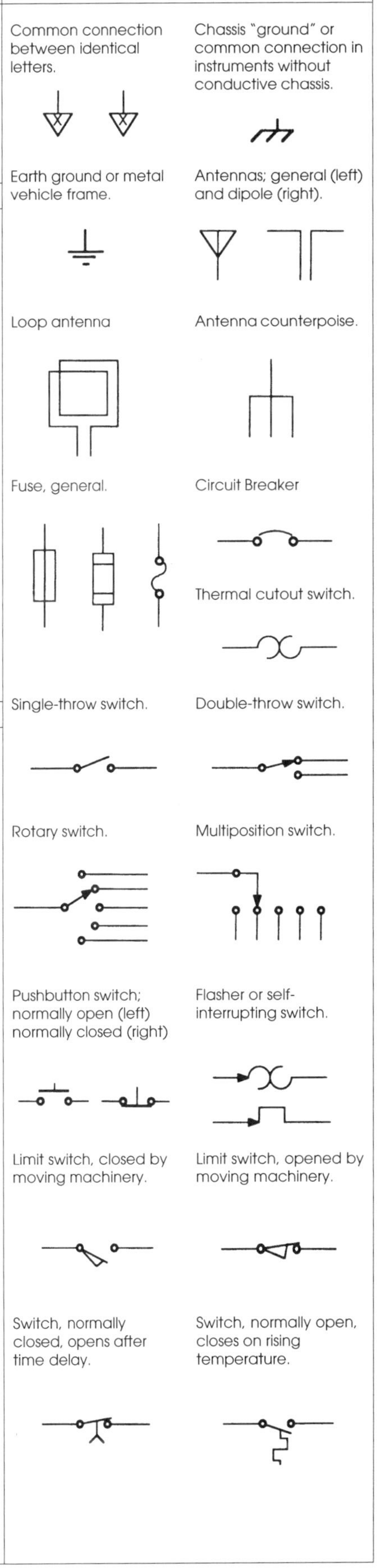

Common connection between identical letters.

Chassis "ground" or common connection in instruments without conductive chassis.

Earth ground or metal vehicle frame.

Antennas; general (left) and dipole (right).

Loop antenna

Antenna counterpoise.

Fuse, general.

Circuit Breaker

Thermal cutout switch.

Single-throw switch.

Double-throw switch.

Rotary switch.

Multiposition switch.

Pushbutton switch; normally open (left) normally closed (right)

Flasher or self-interrupting switch.

Limit switch, closed by moving machinery.

Limit switch, opened by moving machinery.

Switch, normally closed, opens after time delay.

Switch, normally open, closes on rising temperature.

Electronic/Electrical Symbols

MISCELLANEOUS

Switch, normally closed, opens on rising liquid level.

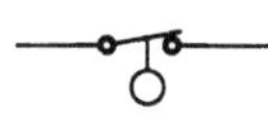

Switch, normally open closes on increased fluid flow.

Switch, normally closed, opens on increased pressure.

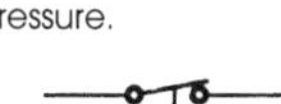

Centrifugal switch, opens on increased angular speed.

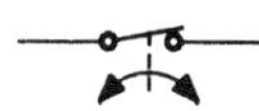

Relay or solenoid coil.

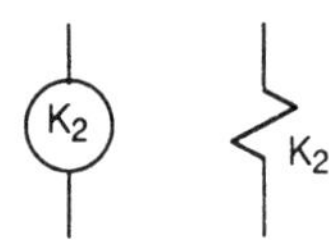

Relay contacts; normally open (left) and closed (right). Make length no more than 1.5 times spacing to avoid confusion with capacitor symbol.

Alternative contacts.

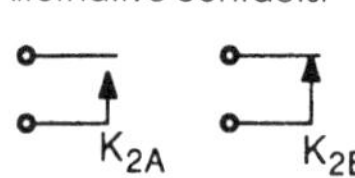

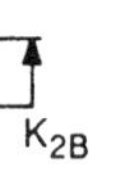

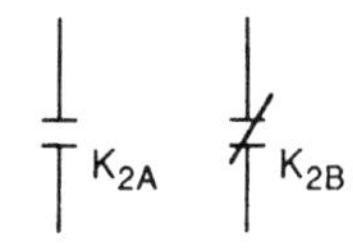

Battery; single cell (left) and multicell (right). Polarity sign optional.

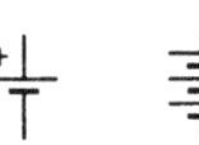

AC source, general. Oscillator.

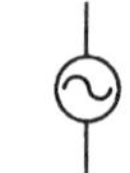

Thermocouple

Piezoelectric crystal

Magnetic head; erase (x), record (right arrow), and read (left arrow) functions.

Microphone

Loudspeaker

Headphone; earphone

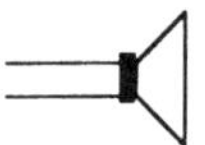

Incandescent lamp; pilot or indicating.

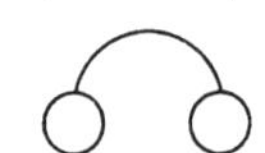

Incandescent lamp; illuminating.

Lamp, fluorescent, with heaters

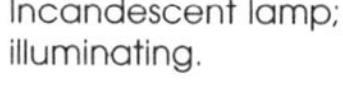

Cold-cathode glow amp; neon lamp

Meter, general. Label defines type.

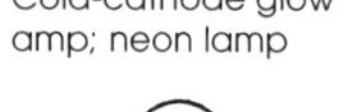

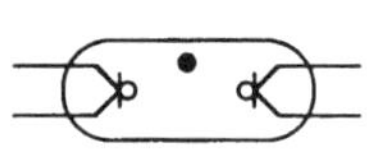

Rotating machine Label defines type.

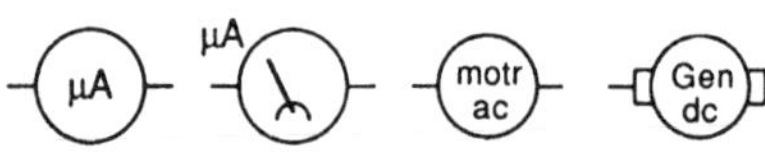

MISCELLANEOUS

Semiconductor diode.

Regulator, avalanche, or zener diode.

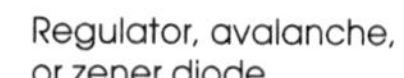

Varactor diode; varicap

Esaki (tunnel) diode

Transistor, NPN

Transistor, PNP

Unijunction transistor

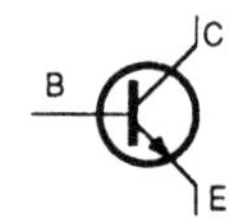

Junction FET, N-channel. Reverse arrow for P-channel

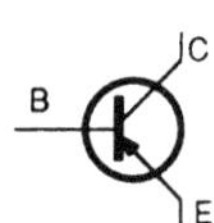

Insulated-gate or MOS FET, N-channel, depletion type.

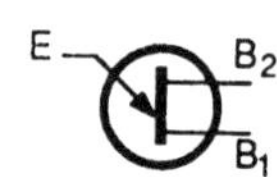

Insulated-gate FET P-channel, enhancement type.

Silicon controlled rectifier (SCR)

Gate turn-off SCR (GTO)

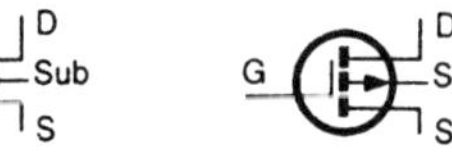

Triac

Diac

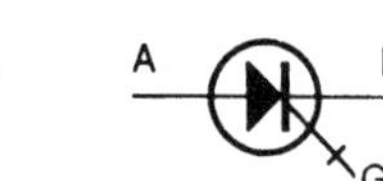

Silicon bilateral switch (SBS)

Programmable UJT (PUT). Pins compared to UJT.

Amplifier, general. Signs indicate inverting and noninverting inputs.

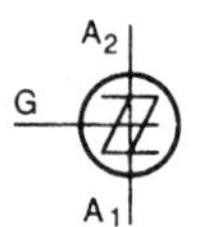

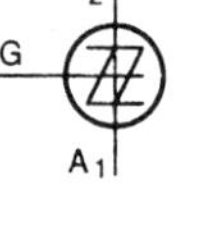

Integrator, general.

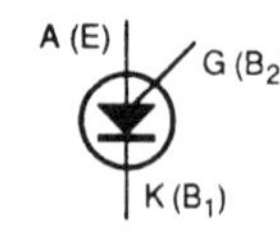

Analog multiplier

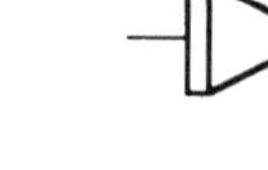

Analog divider

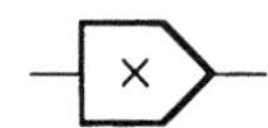

Definitions of Technical Terms

ABSOLUTE PRESSURE
The indicated value of the weight of the earth's atmosphere. At sea level, this value is approximately 14.65 psi (pounds per square inch).

ACCUMULATOR
A vessel, normally cylindrical, which is used to store fluid and gas for future release of the energy in the compressed fluid and gas. Normally contains a diaphragm or piston between the fluid (liquid) and gas chambers. Fluid is normally introduced at one end and the gas at the opposite end.

ACCURACY
The ability of the servo system to achieve the desired output.

ACTUATOR
A device for converting hydraulic energy into mechanical energy, i.e., a motor or cylinder.

ADAPTER
A mechanical device used to align the shaft of an electric motor (or other rotary device) with the shaft of a hydraulic pump to maintain radial and parallel shaft alignment.

AERATION
Air trapped in the hydraulic fluid. Excessive aeration causes the fluid to appear milky and components to operate erratically.

AIR BLEEDER AUTOMATIC
A valve that is fit into a hydraulic pipe line to facilitate automatic release of air trapped in the pipeline.
See also AIR BLEEDER, MANUAL.

AIR BLEEDER MANUAL
A valve that is fit into a hydraulic pipe line to facilitate manually initiated release of air trapped in the pipeline.
See also AIR BLEEDER, AUTOMATIC.

AIR BREATHER
A mechanical device which contains a fine mesh filter element. Normally attached to the top of a reservoir or tank to allow air to pass in and out of the reservoir or tank.

AMBIENT
The current condition of temperature, humidity and atmospheric pressure.

AMBIENT NOISE LEVEL (BACKGROUND NOISE)
The noise level in the area surrounding the machine or component to be tested with machine being tested not operating.

AMPLIFIER
An electronic device that receives an input voltage or current signal and modifies the signal into a driving voltage or current at a different level.

ANALOG
An infinitely variable device.

ANALOG DEVICE
An electronic device that requires or produces an infinitely variable signal, usually voltage or current, in response to a state change within the device.

ANNULAR AREA
A ring shaped area — often refers to the net effective area of the rod side of a cylinder piston, i.e., the piston area of the rod.

ANSI FLANGE
A mechanical device that is used to connect two pieces of pipe together to form a pressure tight joint. ANSI flanges are round, use through bolts and/or nuts to attach two matched flanges together or to a valve or other mechanical device. See ANSI standards for pressure and temperature ratings.

ATMOSPHERE (ONE)
A measure of pressure equal to about 14.7 psi.

ATMOSPHERIC PRESSURE
See absolute pressure.

ATTENUATION
Opposite of *gain* (see gain).

BACK PRESSURE
The level of pressure on the return or downstream side of a device or system.

BACKUP BOTTLE
A vessel, normally cylindrical, which is used to store gas for future release of the energy in the compressed gas to an accumulator.

BACK-UP RING
A fabric or plastic device that is used with an o-ring or other gasket to prevent extrusion of the o-ring or gasket into an adjacent space or crevice.

BAFFLE
A separator found in a reservoir, tank or other chamber to divert fluid flow in specific direction(s) for de-aeration of moving fluid.

BALL VALVE
A valve that may be used to divert the flow of fluid in a passage. Most normally configured in a two way pattern which is either open or closed.

BAR
The measure of pressure in the metric system. One (1) bar = 14.5 psig.

BETA RATIO (β)
The amount, expressed as a ratio, of particles in a fluid stream upstream of a filter, after the fluid passes through a filter, divided by the amount of particles downstream, for a particular size particle. See PARTICLE COUNT.

BLADDER
A separator or diaphragm, usually found in a chamber to facilitate separation of two (2) fluids or gases.
See ACCUMULATOR.

BLEED-OFF
To divert a specific, controllable portion of pump delivery directly to reservoir.

BOLT KIT
A set of bolts or screws that are selected to suit a particular application, i.e. pre-selected length, threads and strength to match the mounted component.

BREATHER
A device which permits air to move in and out of a container or component to maintain atmospheric pressure.

BURST PRESSURE
The level of pressure at which a component, pipe, tube, hose or other fluid passage will burst during application of internal pressure. Normally 2.5 - 4.0 times working pressure. See also WORKING PRESSURE and PROOF PRESSURE.

BYPASS
A secondary passage for fluid flow.

C

CASE DRAIN LINE
The line or passage from the internal cavity of a pump or other component that will carry fluid leakage from the device to a low pressure reservoir or tank.

CAVITATION
A localized gaseous condition within a liquid stream which occurs where the pressure is reduced to the vapor pressure.

CHARGING ASSEMBLY
A system of valves and passages that allow addition or deletion of gas to the gas chamber of an accumulator without discharging any existing gas.

CHECK VALVE
A valve that allows fluid flow in one direction, yet stops flow in the opposite direction.

CIRCUIT
A combination of passages, components and devices that form a working set of logic for a particular application.

CLEAN OUT
A hole in a reservoir or tank that is normally covered with a plate that may be removed to allow cleaning of the interior of the reservoir or tank.

CLOSED CENTER VALVE
A condition where pump output is not unloaded to sump when the valve is in its center or neutral operating position.

CLOSED CIRCUIT
A piping arrangement in which pump output, after passing through other hydraulic components, returns directly to pump inlet.

CLOSED LOOP
In a control system, a type of control that has an input signal and a feedback of the result of the input signal which is used to modulate the input signal automatically.
See OPEN LOOP.

COMMAND SIGNAL
An external signal to which the servo must respond.

COMPENSATOR CONTROL
A displacement control for variable pumps and motors which alters displacement in response to pressure changes in the system as related to its adjusted pressure setting.

COMPOUND GAUGE
A visual indicator of pressure that is set for 'zero' psi at atmospheric pressure and includes a dial which will continue to indicate the level of pressure above or below atmospheric pressure.

COMPRESSIBILITY
The change in volume of a unit of fluid when it is subjected to a unit change in pressure (in^2/lb).

COMPRESSION
The name used to describe the change in pressure in a hydraulic system from low pressure to an elevated pressure. Normally the change in pressure is made in a controlled amount of time to cause an even application of energy into the system. See DECOMPRESSION.

CONNECTOR
A mechanical device used to attach two pieces of tubing together or to attach a piece of tubing to a component.

CONTROLLABILITY
The finest adjustable increment of a system.

COOLER
A mechanical device used to transfer heat from a fluid to air or another fluid. Normally constructed of finned tubes with one fluid on the inside and the other fluid or air on the outside of the tubes. See HEAT EXCHANGER.

COUNTERBALANCE VALVE
A valve used to balance the weight of a machine or dead load by causing a back pressure in the system cylinders of sufficient magnitude to support the weight. Normally closed, opened by internal pressure in the counterbalance valve or from a separate source of fluid, that is connected to the opposite end of the balanced cylinder.

COUPLING
A mechanical device used to attach the shaft of an electric motor or other motive power device to a hydraulic pump.

CRACKING PRESSURE
The pressure at which a pressure operated valve begins to pass fluid.

CUSHION
A mechanical device fitted into a hydraulic cylinder that closes off the flow path of fluid to effect a smooth deceleration and stop of the cylinder at the end of the stroke.

CYCLE
The time of activation of a device or system that is one complete movement from the start position to an extreme position and back to the original position.

CYCLING
A rhythmic change of the factor under control.

CYLINDER, DOUBLE ACTING
A hydro-mechanical device, usually a cylindrical chamber with one closed end and a movable shaft at the other end. When fluid flow is applied to a port in the closed end, the shaft extends until the collar or piston reaches the shaft end. When fluid is applied to the shaft end port, the shaft will retract into the chamber until the piston or collar reaches the closed end. The cylinder will produce an output force at the shaft end in proportion to its internal area multiplied times the pressure potential of the fluid power system.

CYLINDER, SINGLE ACTING
A hydro-mechanical device, usually a cylindrical chamber with one closed end and a movable shaft at the other end. When fluid flow is applied to a port in the closed end, the shaft extends until the collar or piston reaches the shaft end. When an external weight or load is placed on the shaft, the shaft will retract into the chamber until the piston or collar reaches the closed end. The cylinder will produce an output force at the shaft end in proportion to its internal closed end area multiplied times the pressure potential of the fluid power system.

CYLINDER, TELESCOPING, SINGLE ACTING
A hydro-mechanical device, usually a cylindrical chamber with one closed end and a telescoping movable shaft at the other end. When fluid flow is applied to a port in the closed end, the telescoping shaft extends until all collars or pistons reach their limit. When an external weight is applied to the shaft end, the shaft will retract into the chamber until all the pistons or collars reach their closed end limits. The cylinder will produce an output force at the shaft end in proportion to its internal closed end area multiplied times the pressure potential of the fluid power system.

CYLINDER, DOUBLE ROD
A hydro-mechanical device, usually a cylindrical chamber with movable shafts at the both ends. When fluid flow is applied to a port in either end, the shaft extends until the collar or piston reaches the shaft end. When fluid is applied to the other port, the shaft will retract into the chamber until the piston or collar reaches the opposite end. The cylinder will produce an output force at the shaft end in proportion to its internal area multiplied times the pressure potential of the fluid power system.

CYLINDER, TELESCOPING, DOUBLE ACTING
A hydro-mechanical device, usually a cylindrical chamber with one closed end and a telescoping movable shaft at the other end. When fluid flow is applied to a port in the closed end, the telescoping shaft extends until all collars or pistons reach their limit. When fluid is applied to the shaft end port, the shaft will retract into the chamber until all the pistons or collars reach their closed end limits. The cylinder will produce an output force at the shaft end in proportion to its internal area multiplied times the pressure potential of the fluid power system.

DEADBAND
The region or band of no response where an error signal will not cause a corresponding actuation of the controlled variable.

DEADTIME
Any definite delay between two related actions. Measured in units of time.

DECIBEL (dB)
A non-dimensional number used to express sound pressure and sound power. It is logarithmic expression of the ratio of a measure quantity to a reference quantity.

dB (A) & (C)
A sound level reading in decibels made on the A- & C-weighted network, respectively of a sound level meter.

DECOMPRESSION
The name used to describe the change in pressure in a hydraulic system from elevated pressure to a lower pressure. Normally the change in pressure is made in a controlled amount of time to cause an even release of energy in the system. See COMPRESSION.

DELIVERY
The volume of fluid discharged by a pump in a given time, usually expressed in gallons per minute (gpm).

DE-VENT
To close the vent connection of a pressure control valve, permitting the valve to function at its adjusted pressure setting.

DEVICE
A combination of individual components that are arranged to form a unit with a specific set of operating parameters.

DIAGRAM
A formal drawing showing the arrangement of components or devices.

DIFFERENTIAL CURRENT
The algebraic summation of the current in the torque motor; measures in MA (milliamperes).

DIFFERENTIAL CYLINDER
Any cylinder in which the two opposed pistons are not equal.

DIFFERENTIAL PRESSURE
The value or magnitude of pressure measured as the absolute difference of the inlet pressure and outlet pressure.

DIGITAL
The production of a discrete signal based on a change in state. See ANALOG.

DIGITAL DEVICE
A device or component that responds to or produces a discrete function based on a change in state. See ANALOG.

DIRECTIONAL VALVE
A valve whose primary function is to direct or prevent flow through selected channels.

DIRT CAPACITY
The measure of volume (or weight) of particles that a filter or strainer will hold at the limit of operation.

DISPLACEMENT
The volume for one revolution or stroke or for one radiant when so stated.

DITHER
A cyclic application of voltage across a solenoid or coil. Most often used to assure that the device driven by the coil or solenoid remains in a state of constant motion, thus reducing breakaway friction.

DOWNSTREAM
The passage beyond a device, normally at the outlet of direction of flow.

DRAIN LINE
A passage in a hydraulic system that is sized and assigned to components that require a connection to a low pressure passage to the system reservoir.

DRIFT
The measure of movement of a device after a preset condition is applied. Normally drift is measured with varying temperature, although drift may be plotted against any variable, such as humidity, etc.

DUROMETER
The measure of hardness of a rubber or other synthetic compound.

DYNAMIC BEHAVIOR
Describes how a control system or an individual unit reacts with time when subjected to an input signal.

DYNAMIC ERROR
The error that results during the transient state, that is, the state when the system is moving from one steady state condition to another.

EFFICIENCY
The ratio of output to input. Volumetric efficiency of a pump is the actual output, in gpm, divided by the theoretical or design output. The overall efficiency of a hydraulic system is the output power divided by the input power. Efficiency is usually expressed as a percent.

ELECTRIC MOTOR
An electro-mechanical device that converts electrical power into rotary motion. The resultant power output is measured in horsepower.

ELECTROHYDRAULIC SERVO-VALVE
A servo-valve which is capable of continuously controlling hydraulic output as a function of an electrical input.

ENERGY
See JOULE.

ELEMENT
See FILTER ELEMENT.

EROSION
Degradation of a surface which is the result of mixtures of fluid and air or fluid and dirt particles passing over the surface at the same time as a change in pressure occurs.

ERROR (SIGNAL)
The signal which is the algebraic summation of an input signal and feedback signal.

EXHAUST LINE
A passage that is open to atmosphere. Normally used in systems using pressurized air or gas which may be dispersed into the atmosphere.

FEEDBACK
Part of a closed loop system which monitors back information about the condition under control for comparison.

FEEDBACK LOOP
Any closed circuit consisting of one or more forward elements and one or more feedback elements.

FILLER CAP
A mechanical device which provides an access for filling a reservoir or tank. Normally equipped with a fine screen to strain out dirt particles.

FILTER
A mechanical device used to house a filter element. See FILTER ELEMENT.

FILTER ELEMENT
A series of wire or fabric meshes which are bonded together by caps or perforated cylinders and are fitted into hydraulic system passages to strain fine particles and silt from fluid passed through the passage.

FITTING
A mechanical device used to attach two pieces of tubing/piping together or to attach a piece of tubing/pipe to a component.

FLOW CONTROL VALVE, PRESSURE COMPENSATED
A valve used to cause a variable pressure drop in a fluid passage, thus reducing the amount of fluid that may pass through the passage regardless of the pressure level at the inlet of the valve. Often fitted with a check valve that permits free flow of fluid in the opposite direction.

FLOW CONTROL VALVE, NON-PRESSURE COMPENSATED
A valve used to cause a variable pressure drop in a fluid passage, thus potentially reducing the amount of fluid that may pass through the passage regardless of the pressure level at the inlet of the valve. Varying pressures at the inlet of the valve will change the flow capacity. Often fitted with a check valve that permits free flow of fluid in the opposite direction.

FLOW DIVIDER
A mechanical device used to divide the fluid in a passage into two or more separate fluid streams.

FLOW RATE
The volume mass, or weight of a fluid passing through any conductor per unit of time.

FLOW SWITCH
A digital device that opens or closes a contact when a preset flow passes over the sensing element. Normally mounted in a fluid flow passage with a paddle or wand perpendicular to the fluid stream.

FLOWMETER
An analog device which indicates the volume of fluid passing through its interior passage. The output signal may be a visual one or a low level electrical signal.

FLUID
A media used in a fluid power system for transfer of energy (work). See FLUID POWER SYSTEM.

FLUID FRICTION
The measure of the resistance of flow of fluid in a passage, measured in psi (pounds per square inch) or other measures of pressure. Fluid friction results in increased fluid temperature and loss of work potential in the fluid power system.

FLUID MOTOR
A mechanical device that transforms the flow of pressurized fluid into rotary motion.

FLUID POWER SYSTEM
The term used to describe a system of components that use a pressurized fluid to transfer energy (do work).

FORCE
The measure of the result of pressurized fluid acting upon a chamber in a fluid power system. Normally the measure is in pounds and is most often used to state the force in pounds that will be available at the rod of a cylinder when acted upon by pressurized fluid from a fluid power system. The system of units normally used are square inches, pounds per square inch, and pounds.

FOUR WAY
A term used to describe a valve that has four ports, normally a pressure (inlet) port, a return (tank) port, an 'A' ('1') work port and a 'B' ('2') work port. Used to change direction of a cylinder or other output device.

FOUR WAY VALVE, MANUALLY & DIRECT OPERATED
A valve having a four way functional capability that may be manually activated to directly control the operating spool. Movement of the spool from extreme end to extreme end reverses the flow paths of the ports. See FOUR WAY.

FOUR WAY VALVE, PROPORTIONAL CONTROL & DIRECT OPERATED
A valve having a four way functional capability that may be proportionately actuated by a solenoid to control the operating spool in infinite resolution. Movement of the spool from extreme end to extreme end completely reverses the flow paths of the ports. See FOUR WAY.

FOUR WAY VALVE, SOLENOID & DIRECT OPERATED
A valve having a four way functional capability that may be solenoid activated to directly control the operating spool. Movement of the spool from extreme end to extreme end reverses the flow paths of the ports. See FOUR WAY.

FOUR WAY VALVE, SOLENOID & PILOT OPERATED
A valve having a four way functional capability that may be solenoid activated to directly control the operating spool which then controls a secondary, larger spool. Movement of the secondary spool from extreme end to extreme end reverses the flow paths of the ports. See FOUR WAY.

FREQUENCY BANDS
A division of the audible range of frequencies into sub-groups for detailed analysis of sound.

FREQUENCY RESPONSE ANALYSIS
A control system analysis which by introducing a varying rhythmic change (like alternating current) into a process or control unit observes what effect these changes have on the output. Since the information determines how a system or control unit will react, it is possible to use this method of analysis to predict what the addition of new equipment will mean to an operation.

FULL FLOW
A filter in which all the fluid must pass through the filter element or medium.

GAIN
Ratio of increase in a signal (or measurement) as it passes through a control system or a specific control element. If a signal gets smaller, it is said to be attenuated.

GAS BOTTLE
See BACK-UP BOTTLE.

GASKET
A seal, made from rubber or other synthetic material in the shape of a circle and of polygonal cross-section.
See O-RING.

GATE VALVE
A two-way valve that may be opened or closed to block the flow of fluid in a passage. Normally manually operated, but may be automated, especially for larger sizes. Normally designed so that when open, the opening of the passage is not restricted, but there will be some small pressure loss.
See GLOBE VALVE and NEEDLE VALVE.

GAUGE PRESSURE
A term used to state that any pressure stated is corrected for atmospheric pressure. Normally abbreviated psig (pounds per square inch gauge).

GLAND
A mechanical device that is used to contain a seal, o-ring or gasket in a specified space to result in a leak-proof connection between two or more mechanical components.

GLOBE VALVE
A two-way valve that may be opened or closed to block the flow of fluid in a passage. Normally manually operated, but may be automated, especially for larger sizes. Normally designed so that the flow of fluid must make a non-straight turn inside the valve body which results in a loss of pressure across the valve when open, which is greater than the loss across a gate valve. See GATE VALVE and NEEDLE VALVE.

HEAD
The measure of pressure at the base or other reference point of a column of fluid. Normally measured in feet of water.

HEAT
The form of energy that has the capacity to create warmth or to increase the temperature of a substance. Any energy that is wasted or used to overcome friction is converted to heat. Heat is measured in calories or British Thermal Units (BTU's). One BTU is the amount of heat required to raise the temperature of one pound of water one degree Fahrenheit. In the metric system one calorie is the amount of heat required to raise the temperature of one gram of water from $3.5^{\circ}C$ to $4.5^{\circ}C$ (called a small calorie). If the temperature change is from $14.5^{\circ}C$ to $15.5^{\circ}C$, the unit is the normal calorie.

HEATER
An electro-mechanical device that converts electricity into heat, normally for use in raising the temperature of fluid stored in a reservoir or tank.

HEAT EXCHANGER
See COOLER.

HORSEPOWER
The measure of energy used in description of the normal power level in a system. 1 horsepower = 550 lb.-ft./minute of work.

HOSE
A passage used to transport fluid between components in a fluid power system. Normally constructed from multiple layers of rubber or other synthetic materials interlaced and bonded with wire mesh to form a flexible passage. Normally fitted with metal end connections to permit connection to pipe threads or other joints.

HUNTING
Tendency for a system to oscillate continuously.

HYDRAULIC BALANCE
A condition of equal opposed hydraulic forces acting on a part in a hydraulic component.

HYDRAULIC CONTROL
A control which is actuated by hydraulically induced forces.

HYDRAULIC POWER
See FLUID POWER.

HYDRAULIC MOTOR
See FLUID MOTOR.

HYDRODYNAMICS
Engineering science pertaining to the energy of liquid flow and pressure.

HYDROKINETICS
Engineering science pertaining to the energy of liquids in motion.

HYDROPNEUMATICS
Pertaining to the combination of hydraulic and pneumatic fluid power.

HYDROSTATICS
Engineering science pertaining to the energy of liquids at rest.

HYSTERESIS
The difference between the response of a unit or system to an increasing signal and the response to a decreasing signal.

HZ (HERTZ)
A measure of the number of cycles that occur in a specific period of time. Usually the time base is the second, but the time base may be any acceptable measure of time. Synonymous term for "cycles per second."

i

INDICATOR
A mechanical device with points to a scale to provide a visual perspective of the state of a component. See NEEDLE.

INPUT
Incoming signal to a control unit or system.

INTAKE LINE
A passage at the inlet port of a component, normally at the inlet port of a pump.

j

JOULE
A unit of work, energy, or heat. 1 J (joule) = 1 Nm (Newton meter).

k

KINETIC ENERGY
Energy that a substance or body has by virtue of its mass (weight) and velocity.

l

LAG
Preferred engineering term for delay in response (usually in degrees).

LAMINAR FLOW
A condition of flow in a passage that is typified by slow movement of fluid in a relatively straight path along the centerline of a passage. See TURBULENT FLOW.

LEVEL TRANSMITTER
An electro-mechanical device which senses the level of fluid in a chamber and produces an analog signal that corresponds with the change of state in the chamber. See LEVEL SWITCH.

LEVEL SWITCH
An electro-mechanical device which senses the level of fluid in a chamber and opens or closes a digital switch to indicate a change of state. See LEVEL TRANSMITTER.

LEVERAGE
A gain in output force over input force by sacrificing the distance moved. Mechanical advantage or force multiplication.

LIFT
The measure of the capability of a pump to raise fluid from a lower to higher level at its inlet port without damage to the pump. Normally expressed in feet of water.

LINE
A connection between components, a passage for fluid or gas transfer. See PIPE, and TUBE and HOSE.

LINEAR ACTUATOR
A device for converting hydraulic energy into linear motion, i.e. a cylinder or ram.

LINEAR VARIABLE TRANSFORMER (LVT)
An electro-mechanical linear device that produces an analog signal in proportion to the difference in velocity between a magnet and a separate fixed coil.

LINEAR VARIABLE DIFFERENTIAL TRANSFORMER (LVDT)
An electro-mechanical linear device that produces an analog signal in proportion to the difference in distance between a magnet and separate fixed coil.

LINEARITY (SERVOVALVE)
The degree of straightness of the hysteresis plot.

LIQUID LEVEL GAUGE
Gauge to visually indicate the fluid level in a reservoir or tank.

LITER
A metric measure of volume. One (1) liter = 0.2642 gallons.

LUBRICATOR
A mechanical device which is used to inject drops or mist of oil into an air line for lubrication purposes.

m

MANIFOLD
A fabricated system of passages to which various components are attached to form a working assembly or sub-assembly.

MANUAL CONTROL
A control actuated by the operator.

MANUAL OVERRIDE
A means of manually actuating an automatically-controlled device.

MECHANICAL CONTROL
A control actuated by linkages, gears, screws, cams or other mechanical elements.

METER
39.37 inches. The measure of distance in the metric system.

METER
To regulate the amount or rate of fluid flow.

METER-IN
To regulate the amount of fluid flow into an actuator or system.

METER-OUT
To regulate the flow of the discharge fluid out of an actuator or system.

MICRON
1/1000th of a millimeter or 0.00003937 inches. The measure used to determine the particle size of contaminants in a fluid system.

MICRON RATING
The size, in microns, of the particles a filter will remove.

MUFFLER
A mechanical device which provides a complex path for exhaust of air from a pressurized chamber, thus reducing the noise level of the exhausting air.

n

NEEDLE
See INDICATOR.

NEEDLE VALVE
A two-way valve that may be opened or closed to block the flow of fluid in a passage. Normally manually operated, but may be automated, especially for larger sizes. Normally designed so that the flow of fluid must make a non-straight turn inside the valve body which results in a desired loss of pressure across the valve when open which is greater than the loss across a gate valve.
See GATE VALVE and GLOBE VALVE.

NEWTON
A unit of force based on the unit of mass, Kg (kilogram), multiplied by the acceleration, m/s^2 (meters per second per second) which produces Kgm/s^2, called the Newton. 1 N = 1 Kgm/s^2 = 0.1225 lbs. (f) - (pounds force).

NIPPLE
A short length of pipe. May be threaded or plain end.

NITROGEN
An inert gas used to serve as an energy source for accumulators or to be used as a cleaning agent when pure, non-explosive gases are required.

NULL
The position of a device that is its normal or otherwise preset 'zero' condition.

O-RING
A seal, made from rubber or other synthetic material in the shape of a circle and of circular or other polygonal cross-section. See GASKET.

OPEN LOOP
In a control system, a type of control that has an input signal, but no feedback of the result of the input signal.
See CLOSED LOOP.

OPERATING PRESSURE
The level of pressure at which a component, pipe, tube, hose or other fluid passage will experience during application of maximum expected fluid pressure.
See also BURST PRESSURE and PROOF PRESSURE.

ORIFICE
A narrowing of the passage size. Normally constructed in a connector or fitting of a sharp edged metallic component.

OUTPUT STAGE
A spool or other device that is controlled by a smaller spool or torque motor.

OVERLAP
The condition of a spool and body in a servo valve or other spool valve wherein the spool must move a specified amount (the overlap) before exposing two adjacent cavities to each other.

OVERSHOOT
Occurs when the process exceeds the target value as operating conditions change.

OXIDATION
The absorption of oxygen into fluid and the subsequent plating of the oxygen/fluid mixture onto metal surfaces.

PACKING
A seal or gasket. See SEAL, O-RING and GASKET.

PARTICLE
A piece of debris (sand, dirt, metal, fabric, etc.) found in a fluid.

PARTICLE COUNT
The visual or electronic summation of the quantity of particles, grouped by size, in a fluid sample of specified size.

PASSAGE
A hole through which fluid is passed in a fluid power system.
See TUBE, PIPE, HOSE and MANIFOLD.

PETROLEUM FLUID
A hydraulic oil (fluid) that is made from a petroleum base. Normally will support combustion if heated to a specific temperature.

PH (PHASE)
A term used to describe the quantity of cyclic electrical power sources in a high voltage system. Most commonly 1-phase or 3-phase.

PHASE SHIFT
A time difference between the input and output signal of a control unit or system, usually measured in degrees.

PHOSPHATE ESTER FLUID
A hydraulic oil (fluid) that is made from an ester base. A synthetic fluid, manufactured to specific characteristics. Normally will not support combustion if heated to a specific temperature.

PILOT LINE
A passage in a fluid power system that is used to transport a fluid at a pressure lower than the normal operating pressure to facilitate controlled shifting of spool valves.

PILOT-OPERATED CHECK VALVE
A special check valve that may be opened against a checked load by applying pilot pressure from a secondary source to open the check to free reverse flow.

PILOT PRESSURE
The pressure in the pilot circuit.

PILOT VALVE
A valve applied to operate another valve or control. The controlling stage of a 2-stage valve.

PIPE
A passage in a fluid power system that is constructed of metal and conforms dimensionally to standards established by the ANSI. May be acquired by size and schedule, where increase in wall thickness does not increase the outside diameter. See TUBE.

PISTON, CYLINDER
A cylinder in which the movable element has a greater cross-sectional area than the piston rod.

PISTON RING
A metal ring that is used to seal high pressure fluid inside a passage to prevent (limit) leakage across the passage. Normally found in cylinders.

PLUNGER, CYLINDER
A cylinder in which the movable element has the same cross-sectional area as the piston rod (Ram).

POPPET
That part of certain valves which blocks flow when it closes against a seat.

PORT
An internal or external terminus of a passage in a component.

POSITIVE DISPLACEMENT
A characteristic of a pump or motor when a constant volume is delivered for each revolution or stroke.

POTENTIOMETER
An electrical device that changes its internal resistance when moved to a specified point. Most commonly found in electronic control panels. Used to change the voltage in a control system for required control changes (position, speed, pressure, etc.).

POUNDS PER SQ. INCH, GAUGE (PSIG) & ABSOLUTE (PSIA)
The measure of pressure, corrected for atmospheric pressure, that is 'zero' psig = 14.65 psia. 'Zero' psia = absolute zero vacuum.

POWER SUPPLY
Term used to describe a fluid power source. A hydraulic power unit.

PRE-FILL VALVE
A valve that is arranged so its inlet port is connected to a reservoir or tank and so that fluid will flow from the inlet of the valve into a cylinder or ram when opened. When closed, the valve must close off the ram or cylinder from the reservoir or tank to permit application of high pressure from another source on the cylinder side of the valve. Most commonly used to fill large rams on presses to take up non-operating stroke.

PRECHARGE PRESSURE
The pressure of compressed gas in an accumulator prior to the admission of liquid.

PRESSURE COMPENSATOR
A hydro-mechanical device fitted to a pump or other flow producing/controlling device that reduces flow when pressure rises and increases flow as pressure decreases, to preset limits.

PRESSURE DIFFERENTIAL
The difference in pressure between any two points in a system or a component.

PRESSURE DIFFERENTIAL SWITCH
A digital device that opens or closes a switch when the internal pressure differential changes state. Most commonly used to sense clogging of filter elements.

PRESSURE DROP
See Pressure, Differential.

PRESSURE GAUGE
A visual indicator of pressure that is set for 'zero' psi at atmospheric pressure and includes a dial which will continue to indicate the level of pressure above atmospheric pressure. See VACUUM GAUGE and COMPOUND GAUGE.

PRESSURE LINE
A passage that carries fluid from the source of flow to various operating elements of a fluid power system. Rated for operating pressure at the maximum expected pressure of the system.

PRESSURE OVERRIDE
The measure of pressure increase over the nominal setting of a device when additional fluid flow is passed over the device after it initially opens.

PRESSURE REDUCING VALVE
A pressure control valve whose primary function is to limit outlet pressure.

PRESSURE SWITCH
A digital device that opens or closes a switch when the internal pressure changes state.

PRESSURE TRANSDUCER
An analog device that produces a change in voltage or current when the internal pressure changes state. Normally a fast response device for use in servo control systems. See PRESSURE TRANSMITTER.

PRESSURE TRANSMITTER
An analog device that produces a change in voltage or current when the internal pressure changes state. Normally a slow acting device for use in display systems where update time is not crucial. See PRESSURE REDUCER.

PROOF PRESSURE
The level of pressure at which a component, pipe, tube, hose or other fluid passage will not yield during application of internal pressure. Normally 1.5 times working pressure. See WORKING PRESSURE and BURST PRESSURE.

PROPORTIONAL FLOW
In a filter, the condition where part of the flow passes through the filter element in proportion to pressure drop.

PUMP, AIR-OIL
A mechanical device containing two sets of isolated pistons and control valving that are used to intensify fluid pressure by use of a multiplication effect across the two sets of pistons. The air piston being larger than the fluid piston.

PUMP, FIXED DISPLACEMENT
A mechanical device that creates a flow of fluid when its shaft is rotated in the proper direction and when its inlet is connected to a chamber filled with fluid (a reservoir or tank). The outlet port may be connected to a passage leading to a fluid power system or exhausted into another chamber that is at a higher pressure. The higher pressure chamber must be equipped with a pressure limiting device. The output flow rate is fixed by the pump displacement per revolution.

PUMP, VACUUM
A mechanical device that creates a pressure that is lower than atmospheric at its inlet when the shaft is rotated. The outlet port is normally connected to a higher pressure chamber or atmosphere.

PUMP, VARIABLE DISPLACEMENT
A mechanical device that creates a flow of fluid when its shaft is rotated in the proper direction and when its inlet is connected to a chamber filled with fluid (a reservoir or tank). The outlet port may be connected to a passage leading to a fluid power system or exhausted into another chamber that is at a higher pressure. The higher pressure chamber must be equipped with a pressure limiting device. The output flow rate is fixed by the pump displacement per revolution but variable by the operator in a manual or servo controlled system, depending on the design.

q

QUICK DISCONNECT
A mechanical device that may be engaged or dis-engaged to attach two fluid passages. Typically, dis-engagement is possible by manual means.

r

RECIPROCATION
Back-and-forth straight line motion or oscillation.

RAM
A cylinder that has an extend port only. Usually accompanied by auxiliary cylinders that are mechanically linked to the ram to facilitate retraction action.

RAMP
The rate of change of a specific output, such as the ramp of a pressure compensator.

RAMP MODULE
An electronic device that controls the rate of rise of a servo or proportional valve by using capacitors to limit the rate of voltage or current change to the servo or proportional valve.

RATED FLOW
The maximum flow that a manufacturer assigns to a specific component as the maximum desirable flow at which the device will function properly. Also the flow that a designer assigns to a system as the nominal maximum flow. See WORKING PRESSURE.

RATED PRESSURE
The maximum pressure that a manufacturer assigns to a specific component as the maximum desirable pressure at which the device will function properly. See WORKING PRESSURE.

REDUCING VALVE
A valve that decreases the downstream pressure (at the valve outlet) in order to control the flow and therefore the outlet pressure to some preset level. Normally accomplished by balancing the outlet pressure against a precision spring.

REGENERATIVE CIRCUIT
A piping arrangement for a differential type cylinder in which discharge fluid from the rod end combines with pump delivery to be directed into the head end.

REGULATOR
A term used to describe a valve or device that limits the pressure in a passage.

RELIEF VALVE
A valve that limits the pressure at its inlet port by exhausting flow present at its inlet port to another chamber of lower pressure potential through its outlet port.

REPLENISH
To add fluid to maintain a full hydraulic system.

RESERVOIR
A chamber used to store fluid.

RESPONSE TIME
The elapsed time that occurs after the beginning of a function until its completion. For example, the time elapsed between application of electrical power to a solenoid and its full excursion or stroke.

RESTRICTION
A reduced cross-sectional area in a line of passage producing a pressure drop.

RESTRICTOR
See ORIFICE.

RETURN LINE
A passage that is used to route fluid to a reservoir or tank after use in some function. Normally limited to low pressures of 0-150 psig, but may be higher in special applications if so designed.

REVERSING VALVE
A four-way directional valve used to reverse a double-action cylinder or reversible motor.

ROTARY VARIABLE DIFFERENTIAL TRANSFORMER (RVDT)
An electro-mechanical rotary device that produces an analog signal in proportion to the difference in distance between a magnet and a separate fixed coil.

ROTARY VARIABLE TRANSFORMER (RVT)
An electro-mechanical rotary device that produces an analog signal in proportion to the difference in velocity between a magnet and a separate fixed coil.

ROTARY ACTUATOR
A hydro-mechanical device that converts fluid flow into incremental rotary motion as compared to a fluid motor which produces infinite numbers of turns. See FLUID MOTOR.

ROTARY JOINT
A connector or fitting that is equipped with seals or o-rings that allow it to rotate while passing one or more fluid paths through sealed internal passages.

S

SAE 4 BOLT PORT, CODE 61
A system for flange and surface mounting configurations that are used to attach pipes, tubes or hoses to a component or manifold. Normally rated at 3000 psig.
See SAE 4 BOLT PORT, CODE 62.

SAE 4 BOLT PORT, CODE 62
A system of flange and surface mounting configurations that are used to attach pipes, tubes or hoses to a component or manifold. Nominally rated at 6000 psig, although larger sizes are only rated for 500 psig.
See SAE 4 BOLT PORT, CODE 61.

SAE PORT
A threaded hole and stud system that may be used to attach fittings to a component or manifold. Sealed with an o-ring or gasket.

SAFETY FACTOR
The ratio of burst pressure to rated pressure under specific static pressure and temperature conditions.
See BURST PRESSURE.

SCRAPER RING
A metal or synthetic ring that is fitted to the shaft of a cylinder to remove particles from the shaft so to prevent them from entering the cylinder seal chamber.

SEAL
See O-RING and GASKET.

SENSITIVITY
The minimum input signal required to produce a specified output signal.

SEQUENCE VALVE
A valve that is normally closed or normally open and changes to the opposite state when pilot pressure is applied to its spring chamber at a preset pressure level. Normally used to initiate a secondary set of operations in a system, based on application of the pilot signal.

SERVO CONTROL
A term used to describe the type of electronic system used for finite, analog control of a function.
See SOLENOID CONTROL.

SERVO VALVE
A valve that uses a torque motor type coil to control a small stream of fluid. Direction of the fluid stream is used to position a large spool. Therefore a low level power signal may provide precise spool position. Normally, the spool has mechanical feedback of spool position to the torque motor, creating a closed loop spool position system.

SHUTTLE VALVE
A valve that has three ports and a common ball or spool check valve. When flow is applied at either of the two inlet ports, the third or output port receives flow from the higher pressure inlet port.

SILENCER
See MUFFLER.

SILT
Fine particles of debris. Normally found in chambers with little or no circulation, such as at the bottom of a reservoir or tank. See SLUDGE.

SLIP
Internal leakage of hydraulic fluid.

SLUDGE
Partially hardened silt. See SILT.

SOLENOID
A coil of metallic wire, usually copper, wound around a bobbin. Used to magnetize the bobbin and produce linear motion of a companion spool when electricity is applied.

SPOOL
A term loosely applied to almost any moving cylindrically shaped part of a hydraulic component which moves to direct flow through the component.

STABILITY
Ability of a system to maintain control when subject to severe outside disturbances.

STATIC BEHAVIOR
Describes how a control system, or an individual unit, carries on under fixed conditions (As contrasted to dynamic behavior which refers to behavior under changing conditions).

STEP CHANGE
The change from one value to another in a single step.

STATIC HEAD
A measurement of pressure that is present when no fluid flow exists in a passage. The static head is normally expressed in feet of water.

STRAINER
A series of wire or fabric meshes which are bonded together by caps or perforated cylinders and are fitted into hydraulic system passages to strain particles from fluid passed through the passage.

SUBPLATE
A metal base to which a specific valve may be attached using a specified bolt kit.

SUCTION LINE
A passage that leads from a reservoir or tank to the inlet port of a pump.

SUPERCHARGE
To replenish a hydraulic system above atmospheric pressure.

SURGE
An increase in pressure that occurs for a specified short period of time over the normal expected working pressure.

SWASH PLATE
A stationary canted plate in an axial type-piston pump which causes the pistons to reciprocate as the cylinder barrel rotates.

SWITCH
A digital device which closes or opens a discrete set of contacts at a pre-set condition.

SWIVEL JOINT
A connector or fitting that is equipped with seals or o-rings that allow it to partially rotate while passing a fluid path through a sealed internal passage.

SYNCHRO
A rotary electromagnetic device generally used as an AC feedback signal generator which indicates position. It can also be used as a reference signal generator.

SYNTHETIC FLUID
A hydraulic oil (fluid) that is made from a synthetic base. A fluid, manufactured to specified characteristics. Normally will not support combustion if heated to a specific temperature.

SYSTEM PRESSURE
See OPERATING PRESSURE.

TACHOMETER
A digital or analog device that produces a pulse train of electrical signals that is proportional to its rotational speed.

TANK
See RESERVOIR.

TEMPERATURE SWITCH
A digital device that opens or closes a switch when the internal temperature changes state to a preset temperature limit.

THERMOCOUPLE
A precision resistive element that changes resistance in proportion to the temperature of the element. May be used therefore, with proper DC electrical voltage to indicate temperature on a voltmeter style indicator.

THREE WAY
A term used to describe a valve that has three ports, normally a pressure (inlet) port, a normally closed (n.c.) port and a normally open (n.o.) port. Used to block or open a common flow passage.

THREE WAY VALVE, PROPORTIONAL CONTROL & DIRECT OPERATED
A valve having a three way functional capability that may be proportionately actuated by a solenoid to control the operating spool in infinite resolution. Movement of the spool from extreme end to extreme rod completely reverses the flow paths of the ports. See THREE WAY.

THREE WAY VALVE, MANUALLY & DIRECT OPERATED
A valve having a three way functional capability that may be manually activated to directly control the operating spool. Movement of the spool from extreme end to extreme end reverses the flow paths of the ports. See THREE WAY.

THREE WAY VALVE, SOLENOID & PILOT OPERATED
A valve having a three way functional capability that may be solenoid activated to directly control the operating spool which then controls a secondary, larger spool. Movement of the secondary spool from extreme end to extreme end reverses the flow paths of ports. See THREE WAY.

THREE WAY VALVE, SOLENOID & DIRECT OPERATED
A valve having a three way functional capability that may be solenoid activated to directly control the operating spool. Movement of the spool from extreme end to extreme end reverses the flow paths of the ports. See THREE WAY.

THROTTLE
To permit passing of a restricted flow. May control flow rate or create a deliberate pressure drop.

TIE ROD
A metal rod that is used to prevent two or more components from separating. Normally used to restrain the end plates of cylinders against the cylinder tube.

TORQUE
The measure of force applied to a lever arm. Normally expressed in lb.-ft. (pound-feet) or lb.-in. (pound-inch).

TORQUE CONVERTER
A rotary fluid coupling that is capable of multiplying torque.

TORQUE MOTOR
A coil of wire and bobbin assembly used in a servo valve that causes the internal mechanism of the servo valve to be offset when current passes through the coil.

TRANSDUCER
An analog device which produces a change in signal level during state changes. Normally used for high speed control systems.

TRANSFER FUNCTION
A mathematical expression of the relationship between the outgoing and incoming signals of a process or control element.

TRANSMITTER
An analog device which produces a change in signal level during state changes. Normally used for indication systems.

TUBE
A term used to describe a passage for fluid in a hydraulic system. Normally specified by outside diameter, wall thickness, material type and material strength.

TURBINE
A rotary device that is actuated by the impact of a moving fluid against blades or vanes.

TURBULENT FLOW
A condition of flow in a passage that is typified by rapid movement of fluid in a passage, where the fluid is churning and bouncing off the passage walls. See LAMINAR FLOW.

TWO WAY
A term used to describe a valve that has two ports, normally a pressure (inlet) port and an outlet port. Used to open or close a flow passage. May be configured as normally closed (n.c.) or normally open (n.o.).

TWO WAY VALVE, SOLENOID & PILOT OPERATED
A valve having a two way functional capability that may be solenoid activated to directly control the operating spool which then controls a secondary, larger spool. Movement of the secondary spool from extreme end to extreme end opens or closes the flow paths of the ports. See TWO WAY.

TWO WAY VALVE, SOLENOID & DIRECT OPERATED
A valve having a two way functional capability that may be solenoid activated to directly control the operating spool. Movement of the spool from extreme end to extreme end opens or closes the flow paths of the ports. See TWO WAY.

TWO WAY VALVE, MANUALLY & DIRECT OPERATED
A valve having two way functional capability that may be manually activated to directly control the operating spool. Movement of the spool form extreme end to extreme end opens or closes the flow paths of the ports. See TWO WAY.

UNDERLAP
The condition of a spool and body in a servo valve or other spool valve wherein the spool is displaced a specified amount (the underlap) to expose two adjacent cavities to each other.

UNLOAD
To release flow (usually directly to the reservoir), to prevent pressure being imposed on the system or portion of the system.

UNLOADING VALVE
A valve that is normally closed and opens from a separate fluid source on rising pressure that is balanced against a precision spring. Re-set point is normally fixed.

UPSTREAM
The passage ahead of a device, normally at the inlet of direction of flow.

VACUUM
Pressure less than atmospheric pressure. It is usually expressed in inches of mercury (Hg) as referred to the existing atmospheric pressure.

VACUUM GAUGE
A visual indicator of pressure that is set for 'zero' psi at atmospheric pressure and includes a dial which will continue to indicate the level of pressure below atmospheric pressure.

VALVE
A mechanical device that is used in a fluid power system, which is used to provide some change of state of the fluid.

VAPOR PRESSURE
The measure of pressure at which a specific fluid will change to a gas.

VARIABLE
A factor or condition which can be measured, altered or controlled, i.e., temperature, pressure, flow, liquid level, humidity, weight, chemical composition, color, etc.

VELOCITY
The speed of fluid flow through a hydraulic line. Expressed in feet per second (fps), inches per second (ips), or meters per second (mps). Also, the speed or a rotating component measured in revolutions per minute (rpm).

VENT VALVE
A valve that may be manually opened to allow air or fluid or a combination of both to be exhausted into a lower pressure chamber or to the atmosphere.

VISCOSITY
The measure of resistance to flow of a fluid against an established standard. See SUS and SSU.

VISCOSITY INDEX
A measure of the viscosity-temperature characteristics of a fluid as referred to that of two arbitrary reference fluids (ASTM Designation D2270-64).

VOLUME
The size of a space or chamber in cubic units. Loosely applied to the output of a pump in gallons per minute.

WAFER VALVE
A two way valve that may be opened or closed to block the flow of fluid in a passage. Normally manually operated, but may be automated, especially for larger sizes. Normally designed so that when open, the opening of the passage is only restricted by the thickness of the wafer. There will be some pressure loss. See GATE VALVE, GLOBE VALVE and NEEDLE VALVE.

WATER GLYCOL FLUID
A hydraulic fluid that is comprised of a mix of distilled or other pure water and glycol to form a fluid that has enough lubricity to function as a fluid power fluid, but is relatively fire-resistant, i.e. will not support combustion.

WIPER RING
A rubber or other synthetic seal that is fitted around a moving shaft to form a low pressure seal. Normally used to prevent fluid from entering the sealed volume.

WORK
The transfer of power from one state to another. The movement of weight over a specified distance.

Abbreviations

AAR .... Association of American Railroads
abs .... absolute
acfm .... actual cubic feet per minute
AISI .... American Iron and Steel Institute
alt .... altitude
amb .... ambient
amp .... ampere
AN .... Army Navy Aeronautical Standards
AND .... Air Force-Navy-Aeronautical Design Standards
ANSI .... American National Standards Institute
API .... American Petroleum Institue
ARP .... Aeronautical Recommended Practice
ASA .... American Standards Association
ASAE .... American Society of Agriculture Engineers
ASHRAE .... American Society of Heating, Refrigerating and Air Conditioning
ASME .... American Society of Mechanical Engineers
ASTM .... American Society for Testing Materials
ASTME .... American Society of Tool and Manufacturing Engineers
atm .... atmosphere
bhp .... brake horsepower
BTU .... British thermal unit
C .... circumference
C or cent. .... centigrade
cSt .... centistokes
cal .... calorie
cc .... cubic centimeter
cm .... centimeter
cfm .... cubic feet per minute
cu .... cubic
cu ft .... cubic feet
cu in .... cubic inch
cyl .... cylinder
db .... decibel
deg .... degrees
D or dia .... diameter
F or fahr .... Fahrenheit
fpm .... feet per minute
fps .... feet per second
FPS .... Fluid Power Society
ft .... foot
flgd .... flanged
FM .... approved by Factory Mutual Laboratories
gal .... gallons
gpm .... gallons per minute
gm .... gram
GN_2 .... gaseous nitrate
hp or HP .... horsepower
"hg .... inches mercury
hr .... hour
Hz .... Hertz (cycles per second)
ICC .... Interstate Commerce Commission
in .... inch
ID .... inside diameter
Imp gal .... imperial gallon
IPS .... iron pipe size
ISO .... International Organization for Standardization
IEC .... International Electrotechnical Committee
JIC .... Joint Industry Conference
kip .... thousand pounds
kg .... kilograms
km .... kilometer
kva .... thousand volt amperes
kw .... kilowatt
kwhr .... kilowatt hour
kHz .... thousand Hertz (cycles per second)
lbs .... pounds
liq .... liquid
log .... logarithm (common)
LOX .... liquid oxygen
lb per cu ft .... pounds per cubic foot
MS .... military standard
MIL .... military
M .... Meter
mm .... milimeter
µm .... micron (micro-meter)
max .... maximum
mech .... mechanical
min .... minute
min .... minimum
MSS .... Manufacturers Standardization Society of the Valve and Fittings Industry
NAS .... National Aerospace Standard
NASA .... National Aeronautics and Space Administration
NEMA .... National Electrical Manufacturers Association
NFPA .... National Fluid Power Association
NMTBA .... National Machine Tool Builders Association
NEC .... National Electrical Code
NPT .... National Standard Pipe - Tapered
NPTF .... National Standard Pipe, Tapered - Fuel (dryseal)
NPSC .... National Standard Pipe, Straight-Couplings
NPSH .... National Standard Pipe, Straight-Hose Couplings and Nipples
NPSI .... National Standard Pipe, Straight-Internal (dryseal)
NPSL .... National Standard Pipe, Straight-Lock Nuts
NPSM .... National Standard Pipe, Straight-Mechanical
NPTR .... National Standard Pipe, Tapered-Railroad
od .... outside diameter
OS & Y .... outside screw & yoke
oz .... ounces
psi .... pounds per square inch
psia .... pounds per square inch absolute
psig .... pounds per square inch, gauge
π .... (Pi) 3.1416
QPL .... Military Qualified Products List
r or rad .... radius
rpm .... revolutions per minute
red .... reducing
RMS .... Root Mean - Square
SAE .... Society of Automotive Engineers
scfm .... standard cubic feet per minute
sec .... second
SSU .... Saybolt Universal seconds (or SUS)
SSF .... Saybolt Furol seconds (or SFS)
shp .... shaft horsepower
sol .... solenoid
sp gr .... specific gravity
sq ft .... square foot
std .... standard
sch .... schedule
temp .... temperature
UL .... Underwriters Laboratory
USASI .... USA Standards Institute
va .... volt amperes
vel .... velocity
VI .... viscosity index

Prefixes

Prefix	U.S. Term
deka	Ten
hecto	Hundred
kilo	Thousand
mega	Million
giga	Billion
tera	Trillion
deci	Tenth
centi	Hundredth
milli	Thousandth
micro	Millionth
nana	Billionth
pico	Trillionth
femto	Quadrillionth
atto	Quintillionth

To Convert	Into	Multiply By
a		
Abcoulomb	Statcoulombs	2.998×10^{16}
Acre	Sq. chain (Gunters)	10.0
Acre	Rods	160.0
Acre	Sq. links (Gunters)	1.0×10^{5}
Acre	Hectare or Sq. hectometer	0.4047
Acres	Sq. feet	43,560.0
Acres	Sq. meters	4,047.0
Acres	Sq. miles	1.562×10^{-3}
Acres	Sq. yards	4,840.0
Acre-feet	Cu. feet	43,560.0
Acre-feet	Gallons	3.259×10^{5}
Amperes/sq. cm.	Amps./sq. in.	6.452
Amperes/sq. cm.	Amps./sq. meter	10^{4}
Amperes/sq. in.	Amps./sq. cm.	0.1550
Amperes/sq. in.	Amps./sq. meter	1,550.0
Amperes/sq. meter	Amps./sq. cm.	10^{-4}
Amperes/sq. meter	Amps./sq. in.	6.452×10^{-4}
Ampere-hours	Coulombs	3,600.0
Ampere-hours	Faradays	0.03731
Ampere-turns	Gilberts	1.257
Ampere-turns/cm.	Amp-turns/in.	2.540
Ampere-turns/cm.	Amp-turns/meter	100.0
Ampere-turns/cm.	Gilberts/cm.	1.257
Ampere-turns/in.	Amp-turns/cm.	0.3937
Ampere-turns/in.	Amp-turns/meter	39.37
Ampere-turns/in.	Gilberts/cm.	0.4950
Ampere-turns/meter	Amp/turns/cm.	0.01
Ampere-turns/meter	Amp-turns/in.	0.0254
Ampere-turns/meter	Gilberts/cm.	0.01257
Angstrom unit	Inch	3937.0×10^{-9}
Angstrom unit	Meter	1.0×10^{-10}
Angstrom unit	Micron or (Mu)	1.0×10^{-4}
Are	Acre (U.S.)	0.02471
Ares	Sq. yards	119.60
Ares	Acres	0.02471
Ares	Sq. meters	100.0
Astronomical unit	Kilometers	1.495×10^{8}
Atmospheres	Bar	1.013
Atmospheres	Ton/sq. inch	0.007348
Atmospheres	Cms. of mercury	76.0
Atmospheres	Mm. of mercury	760.0
Atmospheres	Ft. of water (at 4° C)	33.90
Atmospheres	In. of mercury (at 0° C)	29.92
Atmospheres	Kgs./sq. cm.	1.0333
Atmospheres	Kgs./sq. meter	10,332.0
Atmospheres	Pounds/sq. in.	14.70
Atmospheres	Tons/sq. ft.	1.058
b		
Barrels (U.S., dry)	Cu. inches	7056.0
Barrels (U.S., dry)	Quarts (dry)	105.0
Barrels (U.S., liquid)	Gallons	31.5
Barrels (Oil)	Gallons (oil)	42.0
Bars	Atmospheres	0.9869
Bars	At. (tech.)	1.0197
Bars	Dynes/sq. cm.	10^{4}
Bars	Kgs./sq. meter	1.020×10^{4}
Bars	Kilopascal	100.0
Bars	Newtons/Sq. meter	10^{5}
Bars	Pounds/sq. ft.	2,089.0
Bars	Pounds/sq. in.	14.504
Bars	Psi	14.504
Baryl	Dyne/sq. cm.	1.000
Bolt (U.S. cloth)	Meters	36.576
Btu	Liter-Atmosphere	10.409
Btu	Ergs	1.0550×10^{10}
Btu	Foot-lbs.	778.3
Btu	Gram-calories	252.0
Btu	Horsepower-hrs.	3.931×10^{-4}
Btu	Joules	1,054.8
Btu	Kilogram-calories	0.2520
Btu	Kilogram-meters	107.5
Btu	Kilowatt-hrs.	2.928×10^{-4}
Btu/hr.	Foot-pounds/sec.	0.2162
Btu/hr.	Gram-cal./sec.	0.0700
Btu/hr.	Horsepower-hrs.	3.929×10^{-4}
Btu/hr.	Watts	0.2931
Btu/lb. °F	Kg. - calories/kg. °C	1.0
Btu/min.	Foot-lbs./sec.	12.96
Btu/min	Horsepower	0.02356
Btu/min.	Kilowatts	0.01757
Btu/min.	Watts	17.57
Btu/sq. ft./min.	Watts/sq. in.	0.1221
Bucket (Br. dry)	Cubic cm.	1.818×10^{4}
Bushels	Cu. ft.	1.2445
Bushels	Cu. in.	2,150.4
Bushels	Cu. meters	0.03524
Bushels	Liters	35.24
Bushels	Pecks	4.0
Bushels	Pints (dry)	64.0
Bushels	Quarts (dry)	32.0
c		
Calories, gram (mean)	Btu (mean)	3.9685×10^{-3}
Candle/sq. cm.	Lamberts	3.142
Candle/sq. inch	Lamberts	0.4870
Centares (centiares)	Sq. meters	1.0
Centigrade	Fahrenheit	(C° x 9/5) +32°
Centigrade	Kelvin	C° + 273°
Centigrams	Grams	0.01
Centiliter	Ounce fluid (U.S.)	0.3382
Centiliter	Cubic inch	0.6103
Centiliter	Drams	2.705
Centiliters	Liters	0.01
Centimeters	Feet	3.281×10^{-2}
Centimeters	Inches	0.3937
Centimeters	Kilometers	10^{-5}
Centimeters	Meters	0.01
Centimeters	Miles	6.214×10^{-6}
Centimeters	Millimeters	10.0
Centimeters	Mils	393.7
Centimeters	Yards	1.024×10^{-2}
Centimeter-dynes	Cm.-grams	1.020×10^{-3}
Centimeter-dynes	Meter/kgs.	1.020×10^{-8}
Centimeter-dynes	Pound-feet	7.376×10^{-8}
Centimeter-grams	Cm.-dynes	980.7
Centimeter-grams	Meter-kgs.	10^{-5}
Centimeter-grams	Pound-feet	7.233×10^{-5}
Centimeters of mercury	Atmospheres	0.01316
Centimeters of mercury	Feet of water	0.4461
Centimeters of mercury	Kgs./sq. meter	136.0
Centimeters of mercury	Pounds/sq. ft.	27.85
Centimeters of mercury	Pounds/sq. in.	0.1934
Centimeters/sec.	Feet/min.	1.1969
Centimeters/sec.	Feet/sec.	0.03281
Centimeters/sec.	Kilometers/hr.	0.036
Centimeters/sec.	Knots	0.1943
Centimeters/sec.	Meters/min.	0.6
Centimeters/sec	Miles/hr	0.02237
Centimeters/sec.	Miles/min.	3.728×10^{-4}
Centimeters/sec./sec.	Feet/sec./sec.	0.03281
Centimeters/sec./sec.	Kms./hr./sec.	0.036
Centimeters/sec./sec.	Meters/sec./sec	0.01
Centimeters/sec./sec.	Miles/hr./sec.	0.02237
Centipoise	Gram/cm. sec.	0.01
Centipoise	Pound mass/ft. sec.	0.000672
Centistokes	Sq. feet/sec.	1.076×10^{-6}
Chain	Inches	792.00
Chain	Meters	20.12
Chains (surveyors' or Gunter's)	Yards	22.00
Circular mils	Sq. cms.	5.067×10^{-6}
Circular mils	Sq. mils	0.7854
Cirumference	Radians	6.283
Circular mils	Sq. inches	7.854×10^{-7}
Cords	Cord feet	8.0
Cord feet	Cu. feet	16.0
Coulomb	Statcoulombs	2.998×10^{9}

To Convert	Into	Multiply By
Coulombs	Faradays	1.036×10^{-5}
Coulombs/sq. cm.	Coulombs/sq. in.	64.52
Coulombs/sq. cm.	Coulombs/sq. meter	10^{4}
Coulombs/sq. in.	Coulombs/sq. cm	0.1550
Coulombs/sq. in.	Coulombs/sq. meter	1,550.0
Coulombs/sq. meter	Coulombs/sq. cm.	10^{-4}
Coulombs/sq. meter	Coulombs/sq. in.	6.452×10^{-4}
Cubic centimeters	Cu. feet	3.531×10^{-5}
Cubic centimeters	Cu. inches	0.06102
Cubic centimeters	Cu. meters	10^{-6}
Cubic centimeters	Cu. yards	1.308×10^{-6}
Cubic centimeters	Gallons (U.S. liq.)	2.642×10^{-4}
Cubic centimeters	Liters	0.001
Cubic centimeters	Pints (U.S. liq.)	2.113×10^{-3}
Cubic centimeters	Quarts (U.S. liq.)	1.057×10^{-3}
Cubic feet	Bushels (dry)	0.8036
Cubic feet	Cu. cms.	28,320.0
Cubic feet	Cu. inches	1728.0
Cubic feet	Cu. meters	0.02832
Cubic feet	Cu. yards	0.03704
Cubic feet	Gallons (U.S. liq.)	7.48052
Cubic feet	Liters	28.32
Cubic feet	Pints (U.S. liq.)	59.84
Cubic feet	Quarts (U.S. liq.)	29.92
Cubic feet/min.	Cu. cms./sec.	472.0
Cubic feet/min.	Gallons/sec.	0.1247
Cubic feet/min.	Liters/sec.	0.4720
Cubic feet/min.	Pounds of water/min.	62.43
Cubic feet/sec.	Million gal./day	0.646317
Cubic feet/sec.	Gallons/min.	448.831
Cubic inches	Cu. cms.	16.39
Cubic inches	Cu. feet	5.787×10^{-4}
Cubic inches	Cu. meters	1.639×10^{-5}
Cubic inches	Cu. yards	2.143×10^{-5}
Cubic inches	Gallons	4.329×10^{-3}
Cubic inches	Liters	0.01639
Cubic inches	Mil.-feet	1.061×10^{5}
Cubic inches	Pints (U.S. liq.)	0.03463
Cubic inches	Quarts (U.S. liq.)	0.01732
Cubic meters	Bushels (dry)	28.38
Cubic meters	Cu. cms.	10^{6}
Cubic meters	Cu. feet	35.31
Cubic meters	Cu. inches	61,023.0
Cubic meters	Cu. yards	1.308
Cubic meters	Gallons (U.S. liq.)	264.2
Cubic meters	Liters	1,000.0
Cubic meters	Pints (U.S. liq.)	2,113.0
Cubic meters	Quarts (U.S. liq.)	1,057.0
Cubic yards	Cu. cms.	7.646×10^{5}
Cubic yards	Cu. feet	27.0
Cubic yards	Cu. inches	46,656.0
Cubic yards	Cu. meters	0.7646
Cubic yards	Gallons (U.S. liq.)	202.0
Cubic yards	Liters	764.6
Cubic yards	Pints (U.S. liq.)	1,615.9
Cubic yards	Quarts (U.S. liq.)	807.9
Cubic yards/min.	Cubic ft./sec.	0.45
Cubic yards/min.	Gallons/sec.	3.367
Cubic yards/min.	Liters/sec.	12.74

d

To Convert	Into	Multiply By
Dalton	Gram	1.650×10^{-24}
Days	Seconds	86,400.0
Decigrams	Grams	0.1
Deciliters	Liters	0.1
Decimeters	Meters	0.1
Degrees (angle)	Minutes	60.0
Degrees (angle)	Quadrants	0.01111
Degrees (angle)	Radians	0.01745
Degrees (angle)	Seconds	3,600.0
Degrees/sec.	Radians/sec.	0.01745
Degrees/sec.	Revolutions/min.	0.1667
Degrees/sec.	Revolutions/sec.	2.778×10^{-3}
Dekagrams	Grams	10.0
Dekaliters	Liters	10.0
Dekameters	Meters	10.0
Drams (apothecaries' or troy)	Ounces (avoidupois)	0.1371429
Drams (apothecaries' or troy)	Ounces (troy)	0.125
Drams (U.S., fluid or apothecaries')	Cubic cm.	3.6967
Drams	Grams	1.7718
Drams	Grains	27.3437
Drams	Ounces	0.0625
Dyne/cm.	Erg/sq. millimeter	0.01
Dyne/sq. cm.	Atmospheres	9.869×10^{-7}
Dyne/sq. cm.	Inch of mercury at 0°C	2.953×10^{-5}
Dyne/sq. cm.	Inch of water at 4°C	4.015×10^{-4}
Dynes	Grams	1.020×10^{-3}
Dynes	Joules/cm.	10^{-7}
Dynes	Joules/meter (newtons)	10^{-5}
Dynes	Kilograms	1.020×10^{-6}
Dynes	Poundals	7.233×10^{-5}
Dynes	Pounds	2.248×10^{-6}
Dynes/sq. cm.	Bars	10^{-6}

e

To Convert	Into	Multiply By
Ell	Cm.	114.30
Ell	Inches	45.0
Em, Pica	Inch	0.167
Em, Pica	Cm.	0.4233
Erg/sec.	Dyne-cm./sec.	1.000
Ergs	Btu	9.480×10^{-11}
Ergs	Dyne-centimeters	1.0
Ergs	Foot-pounds	7.367×10^{-8}
Ergs	Gram-calories	0.2389×10^{-7}
Ergs	Grams-cms.	1.020×10^{-3}
Ergs	Horsepower-hrs.	3.7250×10^{-14}
Ergs	Joules	10^{-7}
Ergs	Kg.-calories	2.389×10^{-11}
Ergs	Kg.-meters	1.020×10^{-8}
Ergs	Kilowatt-hrs.	0.2778×10^{-13}
Ergs	Watt-hours	0.2778×10^{-10}
Ergs/sec.	Btu/min.	$5{,}688.0 \times 10^{-9}$
Ergs/sec.	Ft.-lbs./min.	4.427×10^{-6}
Ergs/sec.	Ft.-lbs./sec.	7.3756×10^{-8}
Ergs/sec.	Horsepower	1.341×10^{-10}
Ergs/sec.	Kg.-calories/min.	1.433×10^{-9}
Ergs/sec.	Kilowatts	10^{-10}

f

To Convert	Into	Multiply By
Fahrenheit	Centigrade	5/9 (F° - 32°)
Fahrenheit	Rankine	F° + 460°
Farads	Microfarads	10^{6}
Faraday/sec.	Ampere (absolute)	9.6500×10^{4}
Faradays	Ampere-hours	26.80
Faradays	Coulombs	9.649×10^{4}
Fathom	Meter	1.828804
Fathoms	Feet	6.0
Feet	Centimeters	30.48
Feet	Inches	12.0
Feet	Kilometers	3.048×10^{-4}
Feet	Meters	0.3048
Feet	Miles (naut.)	1.645×10^{-4}
Feet	Miles (stat.)	1.894×10^{-4}
Feet	Millimeters	304.8
Feet	Mils	1.2×10^{4}
Feet	Yards	1/3
Feet of water	Atmospheres	0.02950
Feet of water	In. of mercury	0.8826
Feet of water	Kgs./sq. cm.	0.03048
Feet of water	Kgs./sq. meter	304.8
Feet of water	Pounds/sq. ft.	62.43
Feet of water	Pounds/sq. in.	0.4335
Feet/min.	Cms./sec.	0.5080
Feet/min.	Feet/sec.	0.01667
Feet min.	Kms./hr.	0.01829
Feet/min.	Meters/min.	0.3048
Feet/min.	Miles/hr.	0.01136

To Convert	Into	Multiply By
Feet/sec.	Cms./sec.	30.48
Feet/sec.	Kms./hr.	1.097
Feet/sec.	Knots	0.5921
Feet/sec.	Meters/min.	18.29
Feet/sec.	Miles/hr.	0.6818
Feet/sec.	Miles/min.	0.01136
Feet/sec./sec.	Cms./sec./sec.	30.48
Feet/sec./sec.	Kms./hr./sec.	1.097
Feet/sec./sec.	Meters/sec./sec.	0.3048
Feet/sec./sec.	Miles/hr./sec.	0.6818
Feet/100 feet	Per cent grade	1.0
Foot-candle	Lumen/sq. meter	10.764
Foot-pounds	Btu	1.286×10^{-3}
Foot-pounds	Ergs	1.356×10^{7}
Foot-pounds	Gram-calories	0.3238
Foot-pounds	Hp-hrs	5.050×10^{-7}
Foot-pounds	Joules	1.356
Foot-pounds	Kg.-calories	3.24×10^{-4}
Foot-pounds	Kg.-meters	0.1383
Foot-pounds	Kilowatt-hrs.	3.766×10^{-7}
Foot-pounds	Newton-meters	1.356
Foot-pounds/min.	Btu/min.	1.286×10^{-3}
Foot-pounds/min.	Foot-pounds/sec.	0.01667
Foot-pounds/min.	Horsepower	3.030×10^{-5}
Foot-pounds/min.	Kg.-calories/min.	3.24×10^{-4}
Foot-pounds/min.	Kilowatts	2.260×10^{-5}
Foot-pounds/sec.	Btu/hr.	4.6263
Foot-pounds/sec.	Btu/min.	0.07717
Foot-pounds/sec.	Horsepower	1.818×10^{-3}
Foot-pounds/sec.	Kg.-calories/min.	0.01945
Foot-pounds/sec.	Kilowatts	1.356×10^{-3}
Foot-pounds/sec.	Newton-meters/Sec.	1.356
Furlongs	Miles (U.S.)	0.125
Furlongs	Rods	40.0
Furlongs	Feet	660.0

g

To Convert	Into	Multiply By
Gallons	Cu. cms.	3,785.0
Gallons	Cu. feet	0.1337
Gallons	Cu. inches	231.0
Gallons	Cu. meters	3.785×10^{-3}
Gallons	Cu. yards	4.951×10^{-3}
Gallons	Liters	3.785
Gallons	Pints (liq.)	8.0
Gallons	Quarts (liq.)	4.0
Gallons (liq. Br. Imp.)	Gallons (U.S. liq.)	1.20095
Gallons (U.S.)	Gallons (Imp.)	0.83267
Gallons of water	Pound of water	8.3453
Gallons/min.	Cu. ft./sec.	2.228×10^{-3}
Gallons/min.	Liters/sec.	0.06308
Gallons/min.	Cu. ft./hr.	8.0208
Gallons/min.	Pounds of water/hr.	500.0
Gausses	Lines/sq. in.	6.452
Gausses	Webers/sq. cm.	10^{-8}
Gausses	Webers/sq. in.	6.452×10^{-8}
Gausses	Webers/sq. meter	10^{-4}
Gilberts	Ampere-turns	0.7958
Gilberts/cm.	Amp-turns/cm.	0.7958
Gilberts/cm.	Amp-turns/in.	2.021
Gilberts/cm.	Amp-turns/meter	79.58
Gills (British)	Cubic cm.	142.07
Gills	Liters	0.1183
Gills	Pints (liq.)	0.25
Grade	Radian	0.01571
Grains	Drams (avoirdupois)	0.03657143
Grains (troy)	Grains (avdp.)	1.0
Grains (troy)	Grams	0.06480
Grains (troy)	Ounces (avdp.)	2.0833×10^{-3}
Grains (troy)	Pennyweight (troy)	0.04167
Grains/U.S. gal.	Parts/million	17.118
Grains/U.S. gal.	Pounds/million gal.	142.86
Grains/Imp. gal.	Parts/million	14.286
Grams	Dynes	980.7
Grams	Grains	15.43
Grams	Joules/cm.	9.807×10^{-5}
Grams	Joules/meter (newtons)	9.807×10^{-3}
Grams	Kilograms	0.001
Grams	Milligrams	1,000.0
Grams	Ounces (avdp.)	0.03527
Grams	Ounces (troy)	0.03215
Grams	Poundals	0.07093
Grams	Pounds	2.205×10^{-3}
Grams/cm.	Pounds/inch	5.600×10^{-3}
Grams/cu. cm.	Pounds/cu. ft.	62.43
Grams/cu. cm.	Pounds/cu. in.	0.03613
Grams/cu. cm.	Pounds/mil.-foot	3.405×10^{-7}
Grams/liter	Grains/gal.	58.417
Grams/liter	Pounds/1,000 gal.	8.345
Grams/liter	Pounds/cu. ft.	0.062427
Grams/liter	Parts/million	1,000.0
Grams/sq. cm.	Pounds/sq. ft.	2.0481
Gram-calories	Btu	3.9683×10^{-3}
Gram-calories	Ergs	4.1868×10^{7}
Gram-calories	Foot-pounds	3.0880
Gram-calories	Horsepower-hrs.	1.5596×10^{-6}
Gram-calories	Kilowatt-hrs	1.1630×10^{-6}
Gram-calories	Watt-hrs.	1.1630×10^{-3}
Gram-calories/sec.	Btu/hr.	14.286
Gram-centimeters	Btu	9.297×10^{-8}
Gram-centimeters	Ergs	980.7
Gram-centimeters	Joules	9.807×10^{-5}
Gram-centimeters	Kg.-cal.	2.343×10^{-8}
Gram-centimeters	Kg.-meters	10^{-5}

h

To Convert	Into	Multiply By
Hand	Cm.	10.16
Hectares	Acres	2.471
Hectares	Sq. feet	1.076×10^{5}
Hectograms	Grams	100.0
Hectoliters	Liters	100.0
Hectometers	Meters	100.0
Hectowatts	Watts	100.0
Henries	Millihenries	1,000.0
Hogsheads	(British)Cubic ft.	10.114
Hogsheads (U.S.)	Cubic ft.	8.42184
Hogsheads (U.S.)	Gallons (U.S.)	63
Horsepower	Btu/min.	42.44
Horsepower	Foot-lbs./min.	33,000.0
Horsepower	Foot-lbs./sec.	550.0
Horsepower (metric) (542.5 ft. lb./sec.)	Horsepower (550 ft. lb./sec.)	0.9863
Horsepower (550 ft. lb./sec.)	Horsepower (metric) (542.5 ft. lb./sec.)	1.014
Horsepower	Kg.-calories/min.	10.68
Horsepower	Kilowatts	0.7457
Horsepower	Watts	745.7
Horsepower (boiler)	Btu/hr.	33,479
Horsepower (boiler)	Kilowatts	9.803
Horsepower-hrs.	Btu	2,547.0
Horsepower-hrs.	Ergs	2.6845×10^{13}
Horsepower-hrs.	Foot-lbs.	1.98×10^{6}
Horsepower-hrs.	Gram-calories	641,190.0
Horsepower-hrs.	Joules	2.684×10^{6}
Horsepower-hrs.	Kg.-calories	641.1
Horsepower-hrs.	Kg.-meters	2.737×10^{5}
Horsepower-hrs.	Kilowatt-hrs.	0.7457
Hours	Days	4.167×10^{-2}
Hours	Weeks	5.952×10^{-3}
Hundredweights (long)	Pounds	112.0
Hundredweights (long)	Tons (long)	0.05
Hundredweights (short)	Ounces (avoirdupoids)	1600.0
Hundredweights (short)	Pounds	100.0
Hundredweights (short)	Tons (metric)	0.0453592
Hundredweights (short)	Tons (long)	0.0446429

i

To Convert	Into	Multiply By
Inches	Centimeters	2.540
Inches	Meters	2.540×10^{-2}
Inches	Miles	1.578×10^{-5}
Inches	Millimeters	25.40
Inches	Mils	1,000.0

To Convert	Into	Multiply By
Inches	Yards	2.778×10^{-2}
Inches of mercury	Atmospheres	0.03342
Inches of mercury	Feet of water	1.133
Inches of mercury	Kgs./sq. cm.	0.03453
Inches of mercury	Kgs./sq. meter	345.3
Inches of mercury	Pounds/sq. ft.	70.73
Inches of mercury	Pounds/sq. in.	0.4912
Inches of water (at 4°C)	Atmospheres	2.458×10^{-3}
Inches of water (at 4°C)	Inches of mercury	0.07355
Inches of water (at 4°C)	Kgs./sq. cm.	2.540×10^{-3}
Inches of water (at 4°C)	Ounces/sq. in.	0.5781
Inches of water (at 4°C)	Pounds/sq. ft.	5.204
Inches of water (at 4°C)	Pounds/sq. in.	0.03613
International Ampere	Ampere (absolute)	0.9998
International Volt	Volts (absolute)	1.0003
International Volt	Joules	9.654×10^{4}

j

To Convert	Into	Multiply By
Joules	Btu	9.480×10^{-4}
Joules	Ergs	10^{7}
Joules	Foot-pounds	0.7376
Joules	Kg.-calories	2.389×10^{-4}
Joules	Kg.-meters	0.1020
Joules	Watt-hrs	2.778×10^{-4}
Joules/cm.	Grams	1.020×10^{4}
Joules/cm.	Dynes	10^{7}
Joules/cm.	Joules/meter (newtons)	100.0
Joules/cm.	Poundals	723.3
Joules/cm.	Pounds	22.48

k

To Convert	Into	Multiply By
Kilograms	Dynes	980,665.0
Kilograms	Grams	1,000.0
Kilograms	Joules/cm.	0.09807
Kilograms	Joules/meter (newtons)	9.807
Kilograms	Poundals	70.93
Kilograms	Pounds	2.205
Kilograms	Tons (long)	9.842×10^{-4}
Kilograms	Tons (short)	1.102×10^{-3}
Kilograms/cu. meter	Grams/cu. cm.	0.001
Kilograms/cu. meter	Pounds/cu. ft.	0.06243
Kilograms/cu. meter	Pounds/cu. in.	3.613×10^{-5}
Kilograms/cu. meter	Pounds/mil.-foot	3.405×10^{-10}
Kilograms/meter	Pounds/ft.	0.6720
Kilograms/sq. cm.	Bar	0.981
Kilograms/sq. cm.	Dynes	980,665.0
Kilograms/sq. cm.	Atmospheres	0.9678
Kilograms/sq. cm.	Feet of water	32.81
Kilograms/sq. cm.	Inches of mercury	28.96
Kilograms/sq. cm.	Pounds/sq. ft.	2,048.0
Kilograms/sq. cm.	Pounds/sq. in.	14.223
Kilograms/sq. meter	Atmospheres	9.678×10^{-5}
Kilograms/sq. meter	Bars	98.07×10^{-6}
Kilograms/sq. meter	Feet of water	3.281×10^{-3}
Kilograms/sq. meter	Inches of mercury	2.896×10^{-3}
Kilograms/sq. meter	Pounds/sq. ft.	0.2048
Kilograms/sq. meter	Pounds/sq. in.	1.422×10^{-3}
Kilograms/sq. mm.	Kgs./sq. meter	10^{6}
Kilogram-calories	Btu	3.968
Kilogram-calories	Foot-pounds	3,088.0
Kilogram-calories	Hp.-hrs.	1.560×10^{-3}
Kilogram-calories	Joules	4,186.0
Kilogram-calories	Kg.-meters	426.9
Kilogram-calories	Kilojoules	4.186
Kilogram-calories	Kilowatt-hrs.	1.163×10^{-3}
Kilogram meters	Btu	9.294×10^{-3}
Kilogram meters	Ergs	9.804×10^{7}
Kilogram meters	Foot-pounds	7.233
Kilogram meters	Joules	9.804
Kilogram meters	Kg.-calories	2.342×10^{-3}
Kilogram meters	Kilowatt-hrs.	2.723×10^{-6}
Kilolines	Maxwells	1,000.0
Kiloliters	Liters	1,000.0
Kilometers	Centimeters	10^{5}
Kilometers	Feet	3,281.0
Kilometers	Inches	3.937×10^{4}
Kilometers	Meters	1,000.0
Kilometers	Miles	0.6214
Kilometers	Millimeters	10^{6}
Kilometers	Yards	1,094.0
Kilometers/hr.	Cms./sec.	27.78
Kilometers/hr.	Feet/min.	54.68
Kilometers/hr.	Feet/sec.	0.9113
Kilometers/hr.	Knots	0.5396
Kilometers/hr.	Meters/min.	16.67
Kilometers/hr.	Miles/hr.	0.6214
Kilometers/hr./sec.	Cms./sec./sec.	27.78
Kilometers/hr./sec.	Ft./sec./sec.	0.9113
Kilometers/hr./sec.	Meters/sec./sec.	0.2778
Kilometers/hr./sec.	Miles/hr./sec.	0.6214
Kilowatts	Btu/min.	56.92
Kilowatts	Foot-lbs./min.	4.426×10^{4}
Kilowatts	Foot-lbs./sec.	737.6
Kilowatts	Horsepower	1.341
Kilowatts	Kg.-calories/min.	14.34
Kilowatts	Watts	1,000.0
Kilowatt-hrs.	Btu	3,413.0
Kilowatt-hrs.	Ergs	3.600×10^{13}
Kilowatt-hrs.	Foot-lbs.	2.655×10^{6}
Kilowatt-hrs.	Gram-calories	859,850.0
Kilowatt-hrs.	Horsepower-hrs.	1.341
Kilowatt-hrs.	Joules	3.6×10^{6}
Kilowatt-hrs.	Kg.-calories	860.5
Kilowatt-hrs.	Kg.-meters	3.671×10^{5}
Kilowatt-hrs	Pounds of water evaporated from and at 212°F	3.53
Kilowatt-hrs.	Pounds of water raised from 62° to 212°F	22.75
Knots	Feet/hr.	6,080.0
Knots	Kilometers/hr.	1.8532
Knots	Nautical miles/hr.	1.0
Knots	Statute miles/hr.	1.151
Knots	Yards/hr.	2,027.0
Knots	Feet/sec.	1.689

l

To Convert	Into	Multiply By
League	Miles (approx.)	3.0
Light year	Miles	5.9×10^{12}
Light year	Kilometers	9.46091×10^{12}
Lines/sq. cm.	Gausses	1.0
Lines/sq. in.	Gausses	0.1550
Lines/sq. in.	Webers/sq. cm.	1.550×10^{-9}
Lines/sq. in.	Webers/sq. in.	10^{-8}
Lines/sq. in.	Webers/sq. meter	1.550×10^{-5}
Links (engineer's)	Inches	12.0
Links (surveyor's)	Inches	7.92
Liters	Bushels (U.S. dry)	0.02838
Liters	Cu. cm.	1,000.0
Liters	Cu. decimeters	1.0
Liters	Cu. feet	0.03531
Liters	Cu. inches	61.02
Liters	Cu. meters	0.001
Liters	Cu. yards	1.308×10^{-3}
Liters	Gallons (U.S. liq.)	0.2642
Liters	Pints (U.S. liq.)	2.113
Liters	Quarts (U.S. liq.)	1.057
Liters/min.	Cu. ft./sec.	5.886×10^{-4}
Liters/min.	Gals./sec.	4.403×10^{-3}
Lumens/sq. ft.	Foot-candles	1.0
Lumen	Spherical candle power	0.07958
Lumen	Watt	0.001496
Lumen/sq. ft.	Lumen/sq. meter	10.76
Lux	Foot-candles	0.0929

m

To Convert	Into	Multiply By
Maxwells	Kilolines	0.001
Maxwells	Webers	10^{-8}

To Convert	Into	Multiply By
Megalines	Maxwells	10^6
Megohms	Microhms	10^{12}
Megohms	Ohms	10^6
Meters	Centimeters	100.0
Meters	Feet	3.281
Meters	Inches	39.37
Meters	Kilometers	0.001
Meters	Miles (naut.)	5.396×10^4
Meters	Miles (stat.)	6.214×10^{-4}
Meters	Millimeters	1,000.0
Meters	Yards	1.094
Meters	Varas	1.179
Meters/min.	Cms./sec.	1.667
Meters/min.	Feet/min.	3.281
Meters/min.	Feet/sec.	0.05468
Meters/min.	Kms./hr.	0.06
Meters/min.	Knots	0.03238
Meters/min.	Miles/hr.	0.03728
Meters/sec.	Feet/min.	196.8
Meters/sec.	Feet/sec.	3.281
Meters/sec.	Kilometers/hr.	3.6
Meters/sec.	Kilometers/min.	0.06
Meters/sec.	Miles/hr.	2.237
Meters/sec.	Miles/min.	0.03728
Meters/sec./sec.	Cms./sec./sec.	100.0
Meters/sec./sec.	Ft./sec./sec.	3.281
Meters/sec./sec.	Kms./hr./sec.	3.6
Meters/sec./sec.	Miles/hr./sec.	2.237
Meter-kilograms	Cm-dynes	9.807×10^7
Meter-kilograms	Cm-grams	10^5
Meter-kilograms	Pound-feet	7.233
Microfarad	Farads	10^{-6}
Micrograms	Grams	10^{-6}
Microhms	Megohms	10^{-12}
Microhms	Ohms	10^{-6}
Microliters	Liters	10^{-6}
Microns	Inches	39×10^{-6}
Microns	Meters	1×10^{-6}
Miles (naut.)	Feet	6,080.27
Miles (naut.)	Kilometers	1.853
Miles (naut.)	Meters	1,853.0
Miles (naut.)	Miles (statute)	1.1516
Miles (naut.)	Yards	2,027.0
Miles (statute)	Centimeters	1.609×10^5
Miles (statute)	Feet	5,280.0
Miles (statute)	Inches	6.336×10^4
Miles (statute)	Kilometers	1.609
Miles (statute)	Meters	1,609.0
Miles (statute)	Miles (naut.)	0.8684
Miles (statute)	Yards	1,760.0
Miles/hr.	Cms./sec.	44.70
Miles/hr.	Feet/min.	88.0
Miles/hr.	Feet/sec.	1.467
Miles/hr.	Kms./hr.	1.609
Miles/hr.	Kms./min.	0.02682
Miles/hr.	Knots	0.8684
Miles/hr.	Meters/min.	26.82
Miles/hr.	Miles/min.	0.1667
Miles/hr./sec.	Cms./sec./sec.	44.70
Miles/hr./sec.	Feet/sec./sec.	1.467
Miles/hr./sec.	Kms./hr./sec.	1.609
Miles/hr./sec.	Meters/sec./sec.	0.4470
Miles/min.	Cms./sec.	2,682.0
Miles/min.	Feet/sec.	88.0
Miles/min.	Kms./min.	1.609
Miles/min.	Knots/min.	0.8684
Miles/min.	Miles/hr.	60.0
Mil.-feet	Cu. inches	9.425×10^{-6}
Milliers	Kilograms	1,000.0
Millimicrons	Meters	1×10^{-9}
Milligrams	Grains	0.01543236
Milligrams	Grams	0.001
Milligrams/liter	Parts/million	1.0
Millihenries	Henries	0.001
Milliliters	Liters	0.001
Millimeters	Centimeters	0.1
Millimeters	Feet	3.281×10^{-3}
Millimeters	Inches	0.03937
Millimeters	Kilometers	10^{-6}

To Convert	Into	Multiply By
Millimeters	Meters	0.001
Millimeters	Miles	6.214×10^{-7}
Millimeters	Mils	39.37
Millimeters	Yards	1.094×10^{-3}
Millimeters of mercury	Psi	0.0194
Million gals./day	Cu. ft./sec.	1.54723
Mils	Centimeters	2.540×10^{-3}
Mils	Feet	8.333×10^{-5}
Mils	Inches	0.001
Mils	Kilometers	2.540×10^{-8}
Mils	Yards	2.778×10^{-5}
Miner's inches	Cu. ft./min.	1.5
Minims (British)	Cubic cm.	0.059192
Minims (U.S., fluid)	Cubic cm.	0.061612
Minutes (angles)	Degrees	0.01667
Minutes (angles)	Quadrants	1.852×10^{-4}
Minutes (angles)	Radians	2.909×10^{-4}
Minutes (angles)	Seconds	60.0
Myriagrams	Kilograms	10.0
Myriameters	Kilometers	10.0
Myriawatts	Kilowatts	10.0

n

To Convert	Into	Multiply By
Nepers	Decibels	8.686
Newton	Dynes	1 x 105.0
Newton	Kilograms	0.1020
Newton	Pounds	8.85
Newton/sq. meter	Pascal	1.0
Newton-meter	Foot-pounds	0.7375
Newton-meter	Joule	1.0
Newton-meter/sec.	Foot-pounds/sec.	0.7375
Newton-meter/sec.	Watts	1.0

o

To Convert	Into	Multiply By
OHM (International)	OHM (absolute)	1.0005
Ohms	Megohms	10^{-6}
Ohms	Microhms	10^6
Ounces	Drams	16.0
Ounces	Grains	437.5
Ounces	Grams	28.349527
Ounces	Pounds	0.0625
Ounces	Ounces (troy)	0.9115
Ounces	Tons (long)	2.790×10^{-5}
Ounces	Tons (metric)	2.835×10^{-5}
Ounces (fluid)	Cu. inches	1.805
Ounces (fluid)	Liters	0.02957
Ounces (troy)	Grains	480.0
Ounces (troy)	Grams	31.103481
Ounces (troy)	Ounces (avdp.)	1.09714
Ounces (troy)	Pennyweights (troy)	20.0
Ounces (troy)	Pounds (troy)	0.08333
Ounce/sq. inch	Dynes/sq. cm.	4309.0
Ounces/sq.in.	Pounds/sq. in.	0.0625

p

To Convert	Into	Multiply By
Parsec	Miles	19×10^{12}
Parsec	Kilometers	3.084×10^{13}
Parts/million	Grains/U.S. gal.	0.0584
Parts/million	Grains/Imp.gal.	0.07016
Parts/million	Pounds/million gal.	8.345
Pecks (British)	Cubic inches	554.6
Pecks (British)	Liters	9.091901
Pecks (U.S.)	Bushels	0.25
Pecks (U.S.)	Cubic inches	537.605
Pecks (U.S.)	Liters	8.809582
Pecks (U.S.)	Quarts (dry)	8
Pennyweights (troy)	Grains	24.0
Pennyweights (troy)	Ounces (troy)	0.05
Pennyweights (troy)	Grams	1.55517
Pennyweights (troy)	Pounds (troy)	4.1667×10^{-3}
Pints (dry)	Cu. inches	33.60

To Convert	Into	Multiply By
Pints (liq.)	Cu. cms.	473.2
Pints (liq.)	Cu. feet	0.01671
Pints (liq.)	Cu. inches	28.87
Pints (liq.)	Cu. meters	4.732×10^{-4}
Pints (liq.)	Cu. yards	6.189×10^{-4}
Pints (liq.)	Gallons	0.125
Pints (liq.)	Liters	0.4732
Pints (liq.)	Quarts (liq.)	0.5
Planck's Quantum	Erg-second	6.624×10^{-27}
Poise	Gram/cm. sec.	1.00
Pounds (avoirdupois)	Ounces (troy)	14.5833
Poundals	Dynes	13,826.0
Poundals	Grams	14.10
Poundals	Joules/cm.	1.383×10^{-3}
Poundals	Joules/meter (newtons)	0.1383
Poundals	Kilograms	0.01410
Poundals	Pounds	0.03108
Pounds	Drams	256.0
Pounds	Dynes	44.4823×10^{4}
Pounds	Grains	7,000.0
Pounds	Grams	453.5924
Pounds	Joules/cm.	0.04448
Pounds	Joules/meter (newtons)	4.448
Pounds	Kilograms	0.4536
Pounds	Newtons (N)	4.44
Pounds	Ounces	16.0
Pounds	Ounces (troy)	14.5833
Pounds	Poundals	32.17
Pounds	Pounds (troy)	1.21528
Pounds	Tons (short)	0.0005
Pounds (troy)	Grains	5,760.0
Pounds (troy)	Grams	373.24177
Pounds (troy)	Ounces (avdp.)	13.1657
Pounds (troy)	Ounces (troy)	12.0
Pounds (troy)	Pennyweights (troy)	240.0
Pounds (troy)	Pounds (avdp.)	0.822857
Pounds (troy)	Tons (long)	3.6735×10^{-4}
Pounds (troy)	Tons (metric)	3.7324×10^{-4}
Pounds (troy)	Tons (short)	4.1143×10^{-4}
Pounds of water	Cu. feet	0.01602
Pounds of water	Cu. inches	27.68
Pounds of water	Gallons	0.1198
Pounds of water/min.	Cu. ft./sec.	2.670×10^{-4}
Pounds/inch	Newton-meters	0.113
Pound-feet	Cm.-dynes	1.356×10^{7}
Pound-feet	Cm.-grams	13,825.0
Pound-feet	Meter-kgs.	0.1383
Pounds/foot	Newton-meters	1.356
Pounds/cu. ft.	Grams/cu. cm.	0.01602
Pounds/cu. ft.	Kgs/cu. meter	16.02
Pounds/cu. ft.	Pounds/cu. in.	5.787×10^{-4}
Pounds/cu. ft.	Pounds/mil-foot	5.456×10^{-9}
Pounds/cu. in.	Gms./cu. cm.	27.68
Pounds/cu. in.	Kgs./cu. meter	2.768×10^{4}
Pounds/cu. in.	Pounds/cu. ft.	1,728.0
Pounds/cu. in.	Pounds/mil-foot	9.425×10^{-6}
Pounds/ft.	Kgs./meter	1.488
Pounds/in.	Gms./cm.	178.6
Pounds/mil.-foot	Gms./cu. cm.	2.306×10^{6}
Pounds/sq. ft.	Atmopsheres	4.725×10^{-4}
Pounds/sq. ft.	Feet of water	0.01602
Pounds/sq. ft.	Inches of mercury	0.01414
Pounds/sq. ft.	Kgs./sq. meter	4.882
Pounds/sq. ft.	Pounds/sq. in.	6.944×10^{-3}
Pounds/sq. in.	Atmospheres	0.06804
Pounds/sq. in.	Bar	0.0690
Pounds/sq. in.	Feet of water	2.307
Pounds/sq. in.	Inches of mercury	2.036
Pounds/sq. in.	Inches of water	27.7
Pounds/sq. in.	Kgs./sq. meter	703.1
Pounds/sq. in.	Kilopascal	6.895
Pounds/sq. in.	Pounds/sq. ft.	144.0
Pounds/hr.	Kilograms/hr.	0.454
Pounds/sec.	Kilograms/hr.	1,633.0
Pounds-sec./sq. ft.	Pound mass/ft. sec.	32.2

q

To Convert	Into	Multiply By
Quadrants (angle)	Degrees	90.0
Quadrants (angle)	minutes	5,400.0
Quadrants (angle)	Radians	1.571
Quadrants (angle)	Seconds	3.24×10^{5}
Quarts (dry)	Cu. inches	67.20
Quarts (liq.)	Cu. cms.	946.4
Quarts (liq.)	Cu. feet	0.03342
Quarts (liq.)	Cu. inches	57.75
Quarts (liq.)	Cu. meters	9.464×10^{-4}
Quarts (liq.)	Cu. yards	1.238×10^{-3}
Quarts (liq.)	Gallons	0.25
Quarts (liq.)	Liters	0.9463

r

To Convert	Into	Multiply By
Radians	Degrees	57.30
Radians	Minutes	3,438.0
Radians	Quadrants	0.6366
Radians	Seconds	2.063×10^{5}
Radians/sec.	Degrees/sec.	57.30
Radians/sec.	Revolutions/min.	9.549
Radians/sec.	Revolutions/sec.	0.1592
Radians/sec./sec.	Revs./min./min.	573.0
Radians/sec./sec.	Revs./min./sec.	9.549
Radians/sec./sec.	Revs./sec./sec.	0.1592
Revolutions	Degrees	360.0
Revolutions	Quadrants	4.0
Revolutions	Radians	6.283
Revolutions/min.	Degrees/sec.	6.0
Revolutions/min.	Radians/sec.	0.1047
Revolutions/min.	Revs./sec.	0.01667
Revolutions/min./min.	Radians/sec./sec.	1.745×10^{-3}
Revolutions/min./min.	Revs./min./sec.	0.01667
Revolutions/min./min.	Revs./sec./sec.	2.778×10^{-4}
Revolutions/sec.	Degrees/sec.	360.0
Revolutions/sec.	Radians/sec.	6.283
Revolutions/sec.	Revs./min.	60.0
Revolutions/sec./sec.	Radians/sec./sec.	6.283
Revolutions/sec./sec.	Revs./min./min.	3,600.0
Revolutions/sec./sec.	Revs./min./sec.	60.0
Rod	Chain (Gunters)	0.25
Rod	Meters	5.029
Rods (Surv. meas.)	Yards	5.5
Rods	Feet	16.5

s

To Convert	Into	Multiply By
Scruples	Grains	20.0
Seconds (angle)	Degrees	2.778×10^{-4}
Seconds (angle)	Minutes	0.01667
Seconds (angle)	Quadrants	3.087×10^{-6}
Seconds (angle)	Radians	4.848×10^{-6}
Slug	Kilograms	14.59
Slug	Pounds	32.17
Sphere	Steradians	12.57
Square centimeters	Circular mils.	1.973×10^{5}
Square centimeters	Sq. feet	1.076×10^{-3}
Square centimeters	Sq. inches	0.1550
Square centimeters	Sq. meters	0.0001
Square centimeters	Sq. miles	3.861×10^{-11}
Square centimeters	Sq. millimeters	100.0
Square centimeters	Sq. yards	1.196×10^{-4}
Square feet	Acres	2.296×10^{-5}
Square feet	Circular mils.	1.833×10^{8}
Square feet	Sq. cms.	929.0
Square feet	Sq. inches	144.0
Square feet	Sq. meters	0.09290
Square feet	Sq. miles	3.587×10^{-8}
Square feet	Sq. millimeters	9.290×10^{4}
Square feet	Sq. yards	0.1111
Square feet/sec.	Centistokes	92,903.0
Square inches	Circular mils.	1.273×10^{6}
Square inches	Sq. cms.	6.452

To Convert	Into	Multiply By
Square inches	Sq. feet	6.944×10^{-3}
Square inches	Sq. millimeters	645.2
Square inches	Sq. mils.	10^{6}
Square inches	Sq. yard	7.716×10^{-4}
Square kilometers	Acres	247.1
Square kilometers	Sq. cms.	10^{10}
Square kilometers	Sq. ft.	10.76×10^{4}
Square kilometers	Sq. inches	1.550×10^{9}
Square kilometers	Sq. meters	10^{6}
Square kilometers	Sq. miles	0.3861
Square kilometers	Sq. yards	1.196×10^{4}
Square meters	Acres	2.471×10^{-4}
Square meters	Sq. cms.	10^{4}
Square meters	Sq. feet	10.76
Square meters	Sq. inches	1,550.0
Square meters	Sq. miles	3.861×10^{-7}
Square meters	Sq. millimeters	10^{6}
Square meters	Sq. yards	1.196
Square miles	Acres	640.0
Square miles	Sq. feet	27.88×10^{6}
Square miles	Sq. kms.	2.590
Square miles	Sq. meters	2.590×10^{6}
Square miles	Sq. yards	3.098×10^{6}
Square millimeters	Circular mils.	1,973.0
Square millimeters	Sq. cms.	0.01
Square millimeters	Sq. feet	1.076×10^{-5}
Square millimeters	Sq. inches	1.550×10^{-3}
Square mils	Circular mils.	1.273
Square mils	Sq. cms.	6.452×10^{-6}
Square mils	Sq. inches	10^{-6}
Square yards	Acres	2.066×10^{-4}
Square yards	Sq. cms.	8,361.0
Square yards	Sq. feet	9.0
Square yards	Sq. inches	1,296.0
Square yards	Sq. meters	0.8361
Square yards	Sq. miles	3.228×10^{-7}
Square yards	Sq. millimeters	8.361×10^{5}

t

To Convert	Into	Multiply By
Temperature (°C) +273	Absolute Temp. (°C)	1.0
Temperature (°C) +17.78	Temperature (°F)	1.8
Temperature (°F) +460	Absolute Temp. (°F)	1.0
Temperature (°F) -32	Temperature (°C)	5/9
Tons (long)	Kilograms	1,016.0
Tons (long)	Pounds	2,240.0
Tons (long)	Tons (short)	1.120
Tons (metric)	Kilograms	1,000.0
Tons (metric)	Pounds	2,205.0
Tons (short)	Kilograms	907.1848
Tons (short)	Ounces	32,000.0
Tons (short)	Ounces (troy)	29,166.66
Tons (short)	Pounds	2,000.0
Tons (short)	Pounds (troy)	2,430.56
Tons (short)	Tons (long)	0.89287
Tons (short)	Tons (metric)	0.9078
Tons (short)/sq. ft.	Kgs./sq. meter	9,765.0
Tons (short)/sq. ft.	Pounds/sq. in.	2,000.0
Tons of water/24 hrs.	Pounds of water/hr.	83.333
Tons of water/24 hrs.	Gallons/min.	0.16643
Tons of water/24 hrs.	Cu. ft./hr.	1.3349

V

To Convert	Into	Multiply By
Volt/inch	Volt/cm.	0.39370
Volt (absolute)	Statvolts	0.003336
Watts	Btu/hr.	3.4129
Watts	Btu/min.	0.05688
Watts	Ergs/sec.	107.0
Watts	Foot-lbs./min.	44.27
Watts	Foot lbs./sec.	0.7378
Watts	Horsepower	1.341×10^{-3}
Watts	Horsepower (metric)	1.360×10^{-3}
Watts	Kg.-calories/min.	0.01433
Watts	Kilowatts	0.001
Watts (Abs.)	Btu (mean)/min.	0.056884
Watts (Abs.)	Joules/sec.	1.0
Watt-hours	Btu	3.413
Watt-hours	Ergs	3.60×10^{10}
Watt-hours	Foot-pounds	2,656.0
Watt-hours	Gram-calories	859.85
Watt-hours	Horsepower-hrs.	1.341×10^{-3}
Watt-hours	Kilograms-calories	0.8605
Watt-hours	Kilograms-meters	367.2
Watt-hours	Kilowatt-hrs.	0.001
Watt (International)	Watt (absolute)	1.0002
Webers	Maxwells	10^{8}
Webers	Kilolines	10^{5}
Webers/sq. in.	Gausses	1.550×10^{7}
Webers/sq. in.	Lines/sq. in.	10^{6}
Webers/sq. in.	Webers/sq. meter	1,550.0
Webers/sq. meter	Gausses	10^{4}
Webers/sq. meter	Lines/sq. in.	6.452×10^{4}
Webers/sq. meter	Webers/sq. cm.	10^{-4}
Webers/sq. meter	Webers/sq. in.	6.452×10^{-4}

y

To Convert	Into	Multiply By
Yards	Centimeters	91.44
Yards	Kilometers	9.144×10^{-4}
Yards	Meters	0.9144
Yards	Miles (naut.)	4.934×10^{-4}
Yards	Miles (stat.)	5.682×10^{-4}
Yards	Millimeters	914.4

Reference Specifications

I. ISO Standards (International Standards Organization)

Spec. #	Date	Title
1219-1	1991	Fluid Power Systems — Graphic Symbols and Circuit Diagrams
5598-1985	1985	Fluid Power Terms
5596	1982	Accumulator Pressure/Volume Ranges, Characteristics and Identification
4412-1	1991	Airborne Pump Noise Levels
3019-1	1975	Pump/Motor Flange Shaft Dimensions, Inch Series
9632	1992	Pump Sensitivity to Contamination
4021	1992	Extracting Fluid Samples
3938	1986	Reporting Contamination Analysis Data
4402	1991	Particle Counter Calibration ACFTD
4406	1987	ACFTD Calibration Solid Contamination Level Code
4407	1991	Fluid Contamination Counting Method Using a Microscope
2942	1985	Hydraulic Filter Element Integrity
4572	1981	Hydraulic Filter Fine Element Multipass Test
7745	1989	Fire-Resistant Guidelines
4413	1979	Hydraulic Power Equipment
7368	1989	Slip-In Cartridge Valves
4397	1993	O.D. of Tubes and I.D. of Hoses
4399	1977	Nominal Pressure for Connectors and Related Components
6605	1988	Hydraulic Hose Assembly Test Requirements
4401	1980	Hydraulic Valve Interface
4400	1985	3-Pin Electrical Plug Connector
529	1993	Short Machine Tap and Hand Taps

II. ANSI Standards (American National Standards Institute)

Spec. #	Date	Title
A13.1-1981	1985	Scheme For Identification of Piping Systems
B18.2.1	1992	Square and Hex Bolts and Screws
B18.2.2	1987	Square and Hex Nuts, Inch Series
B18.3	1986	Socket and Shoulder Set Screws, Inch Series
B18.21.1	1994	Lock Washers, Inch Series
B31.1	1977	Hydraulic Line Seamless Steel Pipe
B93.4M-1981	1988	Hydraulic Line Welded Tubing
B93.7M-1986	1986	Hydraulic Valve Interfaces
B93.9-1969	1988	Fluid Power Valve Connection Symbols
B93.9.4	1987	3-Pin Electrical Plug Connector
B93.11M-1981	1988	Hydraulic Line Seamless Tubing
B93.22-1987	1993	Hydraulic Filter Element Integrity
B93.1-1964	1970	Code for Fluid Power Cylinders
B93.2-1986	1986	Hydraulic Filter/Separator Terms
B93.18M1973	1987	Hydraulic Reservoir Requirements
B93-41M-1976	1987	Hydraulic Power Unit Practice
B93.19M-1972	1993	Extracting Fluid Samples
B93.30M-1980	1993	Reporting Contamination Analysis Data
B93.50-1979	1988	Synthetic Lubricants
B93.55M-1981	1988	Hydraulic Valve Electrical Connector
B93.5R3-1989	1989	Fire-Resistant Fluids
B93.59M-1982	1988	O.D. of Tubes and I.D. Hoses
B93.60M-1982	1988	Nominal Pressure for Connectors and Related Components
B93.100M-1987	1987	Hydraulic Hose Assembly Test Requirements
B93.5M-1979	1979	Practice for the Use of Fire Resistant Fluids In Industrial Fluid Power Systems
B93.71M-1986	1986	Airborne Pump Noise Levels
B93.76M-1988	1993	Hydraulic Fluid Power Housing for Piston and Rod Seal
T2.13.5-1991	1991	High Water Content Fluids
T2.24.1-1991	1991	NFPA/JIC Hydraulic Systems
T3.9.2 R2-1990	1990	Pump/Motor Flange Shaft Dimensions, Inch Series
T3.9.17 R1-19901	1990	Pump Test
T3.9.20-1992	1992	Pressure Compensated Pump Test
T3.10.4 R1-1991	1991	Hydraulic Filter/Separator Graphic Symbols
T3.10.5.1-1976	1989	Hydraulic Filter/Separator Housing Pressure Rating
T3.10.8.8 R1-1990	1990	Hydraulic Filter Fine Element Multipass Test
Y14.1M-1992	1992	Metric Drawing Sheet Size and Format
Y14.2M-1992	1992	Line Conventions and Lettering
Y14.3-1975	1987	Multi and Sectional View Drawings
Y14.4M-1989	1989	Pictorial Drawing
Y14.5M-1982	1988	Dimensioning and Tolerancing

Reference Specifications

II. ANSI Standards (American National Standards Institute, continued)

Spec. #	Date	Title
Y14.6-1981	1978	Screw Thread Representation
Y14.7.1-19741	1988	Gear Drawing Standards Part One for Spur, Helical, Double Helical and Rack
Y14.7.2-1978	1989	Gear and Spline Drawing Standards Part Two, Bevel and Hyphoid Gears
Y14.8M-1989	1989	Castings and Forgings
Y14.13M-1981	1981	Mechanical Spring Representation
Y14-18M-1986	1986	Optical Parts
Y14.34M-1989	1989	Parts Lists, Data Lists and Other Indexes
Y14.26M-1989	1989	Digital Representation for Communication of Product Definition Data
Y14.34M-1989	1989	Parts Lists, Data Lists and Index Lists
Y14.24M-1989	1989	Types and Applications of Engineering Drawings
Z535.1-1991	1991	American National Standard for Safety Color Code

III. ASTM Standards (American Society for Testing and Materials)

Spec. #	Date	Title
A36	1993	Specification for Structural Steel
A53-90b	1993	Specification for Pipe Steel, Black and Hot-Dipped, Zinc-Coated Welded and Seamless
A105/A105M-92	1991	Specifications for Forgings, Carbon Steel, for Piping Components
A193A	1993	Standard Specifications for Alloy Steel and Stainless Steel Bolting Material for High Temperature Service
A307-92a	1993	Specification for Carbon Steel Bolts and Studs, 60,000 PSI Tensile
A354-92a	1993	Specification for Quenched and Tempered Alloy Steel Bolts, Studs and Other Externally Threaded Fasteners
A449-93	1993	Specifications for Quenched and Tempered Steel Bolts and Studs
D665	1992	Test Method for Rust Prevention Characteristics of Inhibited Mineral Oil in the Presence of Water
A822-90	1990	Specifications for Seamless, Cold Drawn Carbon Steel Tubing for Hydraulic System Service
D951-88	1988	Standard Test Method for Water Resistance of Shipping Containers by Spray Method
D1251-79	1987	Standard Test Method for Water Vapor Permeability of Packages by Cycle Method
D1974-92	1994	Standard Practice for Methods of Closing, Sealing and Reinforcing Fiberboard Shipping Containers
D3951-90	1990	Standard Practice for Commercial Packaging
D4169-92a	1993	Standard Practice for Performance Testings of Shipping Containers and Systems
D4675-87	1993	Standard Guide for Selection and use of Flat Strapping Materials
F912-91	1991	Specification for Alloy Steel Socket Set Screws
D4174	1982	Standard Practice for Cleaning, Flushing and Purification of Petroleum Fluid Hydraulic Systems

IV. JIC Standards (Joint Industrial Council)

Spec. #	Date	Title
H-1-1973	1973	Hydraulic Standard for Industrial Equipment and General Purpose Machine Tools

V. SAE Standards (Society of Automotive Engineers)

Spec. #	Date	Title
J343	1993	Tests and Procedures for SAE 100R Series Hydraulic Hose and Hose Assemblies
J429	1983	Mechanical and Material Requirements, Externally Threaded Fasteners
J496	1972	Spring Type Straight Pins
J1402	1985	Automotive Air Brakes Hose and Hose Assemblies
J1165	1986	Reporting Cleanliness Levels of Hydraulic Fluids
J1127	1986	Battery Cable
J516	1993	Hydraulic Hose Fittings
J517	1993	Hydraulic Hose
J1926	1988	Specification for Straight Thread O-Ring Boss Port
J514	1993	Hydraulic Tube Fittings
J1453	1993	Fittings, O-Ring Face Seal
J518	1991	Hydraulic Flange Tube, Pipe and Hose Connections, Four-Bolt Split Flange Type
J223	1980	Maintenance Instructions, Container and Filler
J356	1991	Welded Flash Controlled Low Carbon Steel Tubing Normalized for Bending, Double Flaring and Beading
J524	1991	Seamless Low Carbon Steel Tubing Annealed for Bending and Flaring
J525	1991	Welded and Cold Drawn Low Carbon Steel Tubing Annealed for Sending and Flaring
J526	1991	Welded Low Carbon Steel Tubing
J527	1991	Brazed Double Wall Low Carbon Tubing
J528	1991	Seamless Copper Tube

V.	SAE Standards (Society of Automotive Engineers continued)		
	Spec. #	**Date**	**Title**
	J598	1988	Sealed Lighting Units for Construction, Industrial
	J1273	1991	Selection, Installation and maintenance of Hose and Hose Assemblies
	100R1A	1991	Specification, Hydraulic Hose, 1-Wire Braid
	100R1AT	1991	Specification, Hydraulic Hose, 1-Wire Braid, Thin Cover
	100R2A	1991	Specification, Hydraulic Hose, 2-Wire Braid
	100R2AT	1991	Specification, Hydraulic Hose, 2-Wire Braid, Thin Cover
	100R4	1991	Specification, Hydraulic Suction Hose, Wire Reinforced
	100R12	1991	Specification, Hydraulic Suction Hose, 4-Wire Braid
VI.	**NEMA Standards (National Electric Manufacturers Association)**		
	Spec. #	**Date**	**Title**
	MG-1-1993	1993	Motors and Generators
	250-1991	1991	Enclosures for Electrical Equipment (1,000 Volts Maximum)
	ICS-6-1988	1988	Enclosures for Industrial Controls and Systems
VII.	**NFPA Standards (National Fire Protection Association)**		
	Spec. #	**Date**	**Title**
	70	1993	National Electric Code (NEC)
VIII.	**ISA Standards (Instrument Society of America)**		
	Spec. #	**Date**	**Title**
	S5.1-1984	1986	Instrumentation Symbols and Identification
	S20-1981	1981	Specification Forms for Process Measurements and Control Instruments, Primary Elements and Control Valves
IX.	**SSPC SP Specification (Steel Structures Painting Council)**		
	Spec. #	**Date**	**Title**
	SP 1	1982	Solvent Cleaning
	SP 2	1989	Hand Tool Cleaning
	SP 3	1989	Power Tool Cleaning
	SP 5	1991	Commercial Blast Cleaning
	PS 13.01	1991	Epoxy-Polyamide Painting Systems
	PAINT 2	1991	Red Lead, Iron Oxide, Raw Linseed Oil and Alkyd Primer
	PAINT 22	1991	Epoxy-Polyamide Paints (Primers, Intermediate and Topcoat)
	PA2	1991	Measurements of Dry Paint Thickness With Magnetic Gauges
X.	**IEC Standards (Internationale Electrotechnique Commission)**		
	Spec. #	**Date**	**Title**
	529	1989	Degrees of Protection Provided by Enclosures (IP Code)
	947-4-1	1990	Low Voltage Switch Gear and Control Gear Fluid 4: Contractors and Motor Starters
XI.	**NFPA Standards (National Fluid Power Association)**		
	Spec. #	**Date**	**Title**
	T2.6.1B1-1991	1991	Pressure Rating
	T2.6.1MS1-1976	1981	Hydraulic Filter/Separator Housing Pressure Rating
	T2.6.1M58-1977	1990	Hydraulic Valve Pressure Rating
	T2.6.1N56-1988	1988	Pump/Motor Pressure Rating
	T3.4.7-1975	1980	Accumulator Pressure Rating
	T2.74M-1988	1983	Airborne Pump Noise Levels
	T3.9.2R2-1989	1989	Pump/Motor Flange Shaft Dimensions, Inch Series
	T3.9.13-1982	1982	Pump/Motor Terms
	T3.9.17R1-1989	1989	Pump Test
	T3.5.26-1977	1990	Hydraulic Valve Pressure Rating
	T3.10.8.8R1-1990	1990	Hydraulic Filter Fine Element Multipass Test
	T2.13.2R3-1989	1989	Fire-Resistant Fluids Trade Names
	T2.13.3-1979	1979	Hydraulic Fluid Index
XII.	**MS (Military Specification)**		
	Spec. #	**Date**	**Title**
	E-22118B	1974	Enamel, Electrical-Insulating
	H-13528B	1993	Hydrochloric Acid, Inhibited, Rust Removing
	MLI-45208	1981	Inspection Systems Requirements

Reference Specifications

XIII. NAS (National Aerospace Standards)

Spec. #	Date	Title
NAS1638	1964	Cleanliness Requirements of Parts Used In Hydraulic System

XIV. MIL-STD (Military Standards)

Spec. #	Date	Title
105	1989	Sampling Procedures and Tables for Inspection by Attributes
129	1993	Marking for Shipment and Storage
147	1994	Palletized and Containerized Unit Loads, Pallets, Skids, Runners or Pallets Type Base
1168	1984	Commercial Packaging of Suppliers and Equipment

XV. AWS (American Welding Society)

Spec. #	Date	Title
A2.4-93	1993	Standard Symbol for Welding Brazing and Nondestructive Examination
A3.0-89	1989	Welding Terms and Definitions
B2.1-84	1984	Standard Welding Procedures and Performance
D1.1-94	1994	Structural Welding Code-Steel

XVI. OSHA Industrial Federal (Occupational Safety and Health Administration)

Spec. #	Date	Title
29CFR1910.95	1993	Occupational Safety Exposure

XVII. FED-STD (Federal Standards)

Spec. #	Date	Title
313	1988	Material Safety Data, Transportation Data and Disposal Data for Hazardous Materials Furnished to Government Activities

XVIII. Specifications Federal

Spec. #	Date	Title
QQ-S-637	1980	Steel Bar, Carbon, Cold Finished, Standard Quality, Free Machining
QQ-S-593	1963	Steel, Sheet and Strip, Low Carbon

XIX. Code of Federal Regulations

Spec. #	Date	Title
29	1994	Code of Federal Regulations